Lecture Notes in Computer Science 12847

More information about this subseries at http://www.springer.com/series/7407

Thierry Lecroq · Svetlana Puzynina (Eds.)

Combinatorics on Words

13th International Conference, WORDS 2021
Rouen, France, September 13–17, 2021
Proceedings

 Springer

Editors
Thierry Lecroq 🆔
University of Rouen Normandy
Mont-St-Aignan, France

Svetlana Puzynina
Saint Petersburg State University
Saint Petersburg, Russia

ISSN 0302-9743 ISSN 1611-3349 (electronic)
Lecture Notes in Computer Science
ISBN 978-3-030-85087-6 ISBN 978-3-030-85088-3 (eBook)
https://doi.org/10.1007/978-3-030-85088-3

LNCS Sublibrary: SL1 – Theoretical Computer Science and General Issues

This Springer imprint is published by the registered company Springer Nature Switzerland AG
The registered company address is: Gewerbestrasse 11, 6330 Cham, Switzerland

Preface

This volume contains the papers presented at WORDS 2021, the 13th International Conference on Words, held online during September 13–17, 2021, at the University of Rouen Normandy, France. The meeting was supported financially by the Normastic FR CNRS 3638 Federation, the LITIS EA4108 Laboratory, the GREYC UMR CNRS 6072, the UFR Sciences and Techniques, the University of Rouen Normandy and the city of Rouen.

WORDS is a biannual conference devoted entirely to combinatorics on words. WORDS is the main conference series devoted to the mathematical theory of words, i.e., finite or infinite sequences of letters. In particular, the combinatorial, algebraic, and algorithmic aspects of words are emphasized. Motivations may also come from other domains such as theoretical computer science, bioinformatics, digital geometry, symbolics dynamics, numeration systems, text processing, number theory, automata theory, etc. Previous WORDS conferences took place in Rouen (France) in 1997 and 1999, Palermo (Italy) in 2001, Turku (Finland) in 2003 and 2013, Montréal (Canada) in 2005 and 2017, Marseille (France) in 2007, Salerno (Italy) in 2009, Prague (Czech Republic) in 2011, Kiel (Germany) in 2015, and Loughborough (UK) in 2019.

Using different e-mail lists, the WORDS 2021 call for papers was distributed around the world, resulting in 18 submissions. The EasyChair system was used to facilitate management of submissions and refereeing, with three referees from the 18-member Program Committee assigned to each paper. A total of 14 papers (77%) were accepted, subject to revision, for presentation at the conference.

Six invited talks were given:

- Nathalie Aubrun (CNRS, Université Paris-Saclay, France): "1D substitutions as tilings and applications";
- Golnaz Badkobeh (Goldsmiths, University of London, UK): "Avoidability of Patterns";
- Julien Leroy (Université de Liège, Belgium): "Rauzy graphs, S-adicity and dendricity";
- Zuzana Masáková (Czech Technical University in Prague, Czech Republic): "Infinite words connected to numeration: β-integers and Erdös spectrum";
- Jeffrey Shallit (University of Waterloo, Canada): "Synchronized Sequences";
- Luca Zamboni (Université Claude Bernard Lyon 1, France): "Continuants with equal values, a combinatorial approach".

These proceedings contain all 14 presented papers, together with two extended versions of the invited talks. We thank the authors for their valuable combinatorial contributions and the referees for their thorough, constructive, and enlightening comments on the manuscripts.

July 2021   Thierry Lecroq
                                                                    Svetlana Puzynina

Organization

Program Committee Co-chairs

Thierry Lecroq   University of Rouen Normandy, France
Svetlana Puzynina   Saint Petersburg State University, Russia

Program Committee

Hideo Bannai   Tokyo Medical and Dental University, Japan
Elena Barcucci   Università degli Studi di Firenze, Italy
Marie-Pierre Béal   Université Gustave Eiffel, France
Gabriele Fici   Università di Palermo, Italy
Anna Frid   Aix-Marseille Université, France
Sylvie Hamel   University of Montreal, Canada
Manfred Kufleitner   University of Stuttgart, Germany
Thierry Lecroq   University of Rouen Normandy, France
Florin Manea   University of Göttingen, Germany
Edita Pelantova   Czech Technical University in Prague, Czech Republic
Svetlana Puzynina   Saint Petersburg State University, Russia
Narad Rampersad   University of Winnipeg, Canada
Michael Rao   CNRS and LIP, ENS Lyon, France
Daniel Reidenbach   Loughborough University, UK
Gwenaël Richomme   Université Paul-Valéry Montpellier 3, France
Michel Rigo   Université de Liège, Belgium
Aleksi Saarela   University of Turku, Finland
Arseny Shur   Ural Federal University, Russia

Steering Committee

Valérie Berthé   CNRS and IRIF, Université de Paris, France
Srečko Brlek   Université du Québec à Montréal, Canada
Julien Cassaigne   Institut de Mathématiques de Luminy and CNRS, France
Maxime Crochemore   King's College London, UK, and Université Gustave Eiffel, France
Aldo de Luca   University of Naples, Italy
Anna Frid   Aix-Marseille Université, France
Juhani Karhumäki (Chair)   University of Turku, Finland
Jean Néraud   Université de Rouen Normandie, France
Dirk Nowotka   Kiel University, Germany

Edita Pelantová	Czech Technical University in Prague, Czech Republic
Dominique Perrin	Université Gustave Eiffel, France
Daniel Reidenbach	Loughborough University, UK
Antonio Restivo	University of Palermo, Italy
Christophe Reutenauer	Université du Québec à Montréal, Canada
Jeffrey Shallit	University of Waterloo, Canada
Mikhail Volkov	Ural Federal University, Russia

Organizing Committee

Saïd Abdeddaim	LITIS, Université de Rouen Normandie, France
Nicolas Bedon (Co-chair)	LITIS, Université de Rouen Normandie, France
Caroline Bérard	LITIS, Université de Rouen Normandie, France
Julien Clément	GREYC, Université de Caen Normandie, France
Thierry de la Rue	LMRS, Université de Rouen Normandie, France
Richard Groult	LITIS, Université de Rouen Normandie, France
Giovanna Guaiana	LITIS, Université de Rouen Normandie, France
Elise Janvresse	LAMFA, Université de Picardie Jules Verne, France
Thierry Lecroq	LITIS, Université de Rouen Normandie, France
Arnaud Lefebvre	LITIS, Université de Rouen Normandie, France
Laurent Mouchard	LITIS, Université de Rouen Normandie, France
Elise Prieur-Gaston (Co-chair)	LITIS, Université de Rouen Normandie, France
Carla Selmi	LITIS, Université de Rouen Normandie, France
Véronique Terrier	GREYC, Université de Caen Normandie, France
Brigitte Vallée	GREYC, Université de Caen Normandie, France
Nicolas Vergne	LMRS, Université de Rouen Normandie, France

Additional Reviewers

Andres, Eric	Lejeune, Marie
Charlier, Emilie	Łopaciuk, Szymon
Chen, Yu-Fang	Lutfalla, Victor
Day, Joel	Mercas, Robert
Fernique, Thomas	Nakashima, Yuto
I, Tomohiro	Parshina, Olga
Inenaga, Shunsuke	Radoszewski, Jakub
Jeandel, Emmanuel	Whiteland, Markus

Sponsors

Contents

Synchronized Sequences

Jeffrey Shallit[✉][iD]

School of Computer Science, University of Waterloo, Waterloo, ON N2L 3G1, Canada
shallit@uwaterloo.ca

Abstract. The notion of synchronized sequence, introduced by Carpi and Maggi in 2002, has turned out to be a very useful tool for investigating the properties of words. Moreover, if sequence is synchronized, then one can use a theorem-prover such as Walnut to "automatically" prove many results about it, with little human intervention. In this paper I will prove some of the basic properties of synchronization, and give a number of applications to combinatorics on words.

Keywords: Synchronized sequence · Automata · Automatic sequence · Regular sequence · Combinatorics on words · Theorem-prover

1 Introduction

Let us recall the notions of automatic and regular sequences. In what follows, k is an integer ≥ 2, and $\Sigma_k = \{0, 1, \ldots, k-1\}$.

A sequence $(a(n))_{n \geq 0}$ taking its values in a finite alphabet Δ is said to be k-*automatic* if there is a deterministic finite automaton with output (DFAO) that, on input n represented in base k, reaches a state with output $a(n)$. Classical examples of automatic sequences include the Thue-Morse sequence and the Rudin-Shapiro sequence [3].

The notion of automatic sequence can be generalized to representations other than base k—for example, Fibonacci representation, Tribonacci representation, and Pell representation.

A sequence $(a(n))_{n \geq 0}$ taking its values in $\mathbb{Q}$ is said to be k-*regular* if there exist an integer d, vectors $v, w \in \mathbb{Q}^d$, and a matrix-valued morphism $\zeta : \Sigma_k^* \to \mathbb{Q}^{d \times d}$ such that $a(n) = v\zeta(x)w^T$, where x is the base-k representation of n. The triple (v, ζ, w) is called a *linear representation* for $(a(n))_{n \geq 0}$. In this paper we will only be interested in sequences taking their values in $\mathbb{N}$ or a finite set. Every k-automatic sequence is also k-regular.

Classical examples of k-regular sequences include $\nu_2(n)$, the exponent of the highest power of 2 dividing n, and $s_2(n)$, the sum of the bits of the base-2 representation of n [4,5]. Once again, this notion can easily be generalized to representations other than base k.

Carpi and Maggi [10] introduced a third class of sequences lying strictly in between these two: the synchronized sequences. We say a sequence $(a(n))_{n \geq 0}$

© Springer Nature Switzerland AG 2021
T. Lecroq and S. Puzynina (Eds.): WORDS 2021, LNCS 12847, pp. 1–19, 2021.
https://doi.org/10.1007/978-3-030-85088-3_1

is *k-synchronized* if there exists a deterministic finite automaton (DFA) that recognizes the language of base-k representations of n and $a(n)$, in parallel. Here the shorter of the two representation is padded with leading zeros, if necessary.

As an example, the sequence $(a(n))_{n \geq 0}$ defined by $a(n) = 2n + 1$ is 2-synchronized. (Actually, it is k-synchronized for all $k \geq 2$, but we need a different automaton for each k.) To see this, examine the DFA in Fig. 1. For example, it accepts the input $[0, 1][1, 1][1, 0][0, 0][0, 1]$. Here the first components spell out 01100, which is a base-2 representation of 12, while the second components spell out 11001, the base-2 representation of 25.

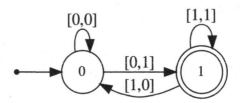

Fig. 1. Synchronized automaton computing the sequence $a(n) = 2n + 1$.

My goal in this paper is to prove some of the basic properties of synchronized sequences and demonstrate their utility in combinatorics on words.

2   Connection with Logic

There is a close connection between synchronized sequences and fundamental results of Bruyère et al. [7]. The basic idea is that if we can write a first-order formula for an integer $s = f(n)$ in terms of k-automatic sequences, logical operations, universal and existential quantifiers, and comparisons and addition on integers, then the pairs (n, s) are k-synchronized. In fact, we can "compile" the formula directly into a synchronized DFA. For example, the formula $s = 2n + 1$ corresponds to the DFA in Fig. 1. This immediately implies many useful properties of f, such as bounds on its growth rate (as we will see in Sect. 8).

Furthermore, if a sequence is synchronized, then many of its properties can be proved "automatically" using a theorem-prover, such as Walnut [16], with almost no work.

3   Closure Properties of Synchronized Sequences

Before we start discussing the properties of synchronized sequences, let us generalize the concept a bit. Let $k, \ell \geq 2$ be two natural numbers. We say a sequence $(a(n))_{n \geq 0}$ is (k, ℓ)-*synchronized* if there exists a DFA recognizing, in parallel, the base-k representation of n and the base-ℓ representation of $a(n)$. We write $(n)_{k,\ell}$ for this representation, consisting of a word over the alphabet $\Sigma_k \times \Sigma_\ell$.

Theorem 1. *Suppose* $(a(n))_{n\geq 0}$ *and* $(b(n))_{n\geq 0}$ *are* (k,ℓ)*-synchronized sequences. Then so are the sequences*

(a) $(a(n)+b(n))_{n\geq 0}$;
(b) $(a(n) \dotdiv b(n))_{n\geq 0}$, *where* $x \dotdiv y$ *is the "monus" function, defined by* $\max(0, x-y)$;
(c) $(|a(n)-b(n)|)_{n\geq 0}$;
(d) $(\lfloor \alpha a(n) \rfloor)_{n\geq 0}$, *where* α *is a non-negative rational number;*
(e) $(\max(a(n),b(n)))_{n\geq 0}$;
(f) $(\min(a(n),b(n)))_{n\geq 0}$;
(g) *running maximum, defined by* $c(n) = \max_{0\leq i\leq n} a(n)$;
(h) *running minimum, defined by* $d(n) = \min_{0\leq i\leq n} a(n)$.

Proof. Let A be a (k,ℓ)-synchronized DFA computing $a(n)$ and B be a (k,ℓ)-synchronized DFA computing $b(n)$. Here we treat A as a boolean function that returns TRUE if A accepts $(n)_k$ and $(x)_\ell$ in parallel and similarly for B. To prove these results, it suffices to provide first-order formulas asserting $s = f(n)$ for each transformation $f(n)$. We do this as follows:

(a) $\forall x,y\ (A(n,x) \wedge B(n,y)) \implies s = x+y$.
(b) $\forall x,y\ (A(n,x) \wedge B(n,y)) \implies ((x \geq y \implies x = s+y) \wedge (x < y \implies s = 0))$.
(c) $\forall x,y\ (A(n,x) \wedge B(n,y)) \implies ((x \geq y \implies x = s+y) \wedge (x < y \implies y = s+x))$.
(d) Write $\alpha = p/q$. Then the formula is $\forall x\ A(n,x) \implies (qs \geq px \wedge qs < px+q)$. As stated, this is not a first-order formula, but it becomes one if q and p are replaced by their particular values. Here we understand $2x$ to mean "$x+x$", $3x$ to mean "$x+x+x$", etc.
(e) $\forall x,y\ (A(n,x) \wedge B(n,y)) \implies ((x \geq y \implies s = x) \wedge (x < y \implies s = y))$.
(f) $\forall x,y\ (A(n,x) \wedge B(n,y)) \implies ((x \geq y \implies s = y) \wedge (x < y \implies s = x))$.
(g) $(\exists i\ (i \leq n) \wedge A(i,s)) \wedge (\forall j,t\ (j \leq n \wedge A(j,t)) \implies s \geq t)$.
(h) $(\exists i\ (i \leq n) \wedge A(i,s)) \wedge (\forall j,t\ (j \leq n \wedge A(j,t)) \implies s \leq t)$.

∎

Remark 1. Not all of these properties hold for k-regular sequences. For example, the k-regular sequences are not closed under absolute value, min, or max.

Assume that $(a(n))_{n\geq 0}$ is unbounded. The first discrete inverse is defined by $g(n) = \min\{i : a(i) \geq n\}$. If further we have $\lim_{n\to\infty} a(n) = \infty$, the second discrete inverse is defined to be $h(n) = \max\{i : a(i) \leq n\}$.

Theorem 2. *Suppose* $(a(n))_{n\geq 0}$ *is* k*-synchronized. Then so are the first and second discrete inverses.*

Proof. Again, it suffices to provide first-order formulas. For $g(n)$ it is

$$\exists i\ A(i,s) \wedge s \geq n \wedge (\forall j\ (A(j,t) \wedge t \geq n) \implies j \geq i),$$

and for $h(n)$ it is

$$\exists i \; A(i,s) \; \wedge \; s \le n \; \wedge \; (\forall j \; (A(j,t) \; \wedge \; t \le n) \implies j \le i).$$

∎

Our next theorem concerns composition of synchronized sequences.

Theorem 3. *Suppose $(a(n))_{n \ge 0}$ is (k, ℓ)-synchronized and $(b(n))_{n \ge 0}$ is (ℓ, m)-synchronized. Then $(b(a(n)))_{n \ge 0}$ is (k, m)-synchronized.*

Proof. Let A be an automaton recognizing pairs $(n, a(n))$ represented in bases k and ℓ, respectively, and similarly B for $(t, b(t))$ in bases ℓ and m. Consider the first-order formula $\exists t \; A(n,t) \; \wedge \; B(t,s)$; it is true iff $s = b(a(n))$. ∎

Remark 2. A priori, given an automaton A with inputs drawn from $(\Sigma_k \times \Sigma_\ell)^*$, we do not necessarily know that A represents a sequence (that is, a function f from $\mathbb{N}$ to $\mathbb{N}$) and not just a relation. However, we can check this property of being a sequence with a first-order formula asserting that for each n we have $A(n, s)$ for one and only one value of s, as follows:

$$(\forall n \; \exists s \; A(n,s)) \; \wedge \; (\forall n, s, t \; (A(n,s) \; \wedge \; A(n,t)) \implies s = t).$$

At first glance, it might seem that a synchronized representation for a sequence $f(n)$ might not be very useful because it doesn't give an explicit way to compute f on any specific n. But it does! We can, for example, intersect the synchronized automaton with an automaton of $O(\log_k n)$ states that recognizes all words that spell out $0^*(n)_k$ in the first component, and anything in the second component. The resulting automaton accepts exactly one word with no leading $[0,0]$'s, which we can easily find in $O(\log_k n)$ time using breadth-first search. Thus we have proved

Proposition 1. *The value of a synchronized sequence f at n can be computed in linear time in the number of bits of n.*

4 Relationship Between Automatic, Synchronized, and Regular Sequences

Theorem 4. *Suppose $(a(n))_{n \ge 0}$ is a (k, ℓ)-synchronized sequence taking values in $\mathbb{N}$, and $a(n) = O(1)$. Then $(a(n))_{n \ge 0}$ is k-automatic.*

Proof. If $a(n) = O(1)$ then it takes only finitely many values. By intersecting the synchronized DFA with a DFA that spells out each of these values c expressed in base-ℓ in the second component, and anything in the first component, we get a DFA M_c recognizing those n in base k corresponding to $a(n) = c$. Now, using the familiar cross-product construction, we can simply combine all these different DFA's M_c into one DFAO computing $(a(n))_{n \ge 0}$.

Theorem 5. *Suppose* $\mathbf{a} = (a(n))_{n \geq 0}$ *is a* (k, ℓ)-*synchronized sequence taking values in* $\mathbb{N}$, *and let* S *be the range of the sequence* $\mathbf{a}$. *Then the characteristic sequence* $(\chi_S(n))_{n \geq 0}$, *defined to be* 1 *if* $n \in S$ *and* 0 *otherwise, is* ℓ-*automatic.*

Proof. Let A be a (k, ℓ) synchronized DFA computing $a(n)$. Consider the first-order formula

$$\exists n \, A(n, s);$$

then the set of values of s making this formula evaluate to **true** is ℓ-automatic.

Theorem 6. *Let* $(a(n))_{n \geq 0}$ *be a* (k, ℓ)-*synchronized sequence. Then it is* k-*regular.*

Proof. Suppose A is a DFA recognizing the pairs $(n, a(n))$ with n represented in base k and $a(n)$ in base ℓ. Consider the first-order formula

$$\exists s \, i < s \, \wedge \, A(n, s);$$

the set of pairs (i, n) for which this formula evaluates to **true** forms an (ℓ, k)-automatic sequence and we can constructively find the appropriate DFA B. Furthermore, for a given n, the number of i making this formula true is $a(n)$. Therefore, using a theorem of [11], we can easily compute a linear representation for $a(n)$ directly from B, and hence $(a(n))_{n \geq 0}$ is k-regular. (Actually, the proof in that paper assumes $\ell = k$, but the generalization given here is clear.)

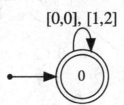

Fig. 2. $(2, 3)$-synchronized DFA for the Cantor sequence.

Example 1. Consider the DFA in Fig. 2. This one-state DFA A, which is $(2, 3)$-synchronized, computes the Cantor sequence $(c(n))_{n \geq 0} = (0, 2, 6, 8, 18, 20, 24, 26, \ldots)$ of numbers having no 1's in their base-3 representation; it is sequence A005823 in the *On-Line Encyclopedia of Integer Sequences* (OEIS) [23]. The characteristic sequence of the range of $(c(n))_{n \geq 0}$ is easily seen to be 3-automatic.

However, using the formula $\exists s \, i < s \, \wedge \, A(n, s)$ given above, and computing the corresponding linear representation, gives

$$v = \begin{bmatrix} 1 & 0 \end{bmatrix}; \quad \zeta(0) = \begin{bmatrix} 1 & 0 \\ 0 & 3 \end{bmatrix}; \quad \zeta(1) = \begin{bmatrix} 1 & 2 \\ 0 & 3 \end{bmatrix}; \quad w = \begin{bmatrix} 0 \\ 1 \end{bmatrix},$$

so the sequence $(c(n))_{n \geq 0}$ itself is 2-regular.

5   Propp's Sequence

Propp [17] studied an increasing sequence $\mathbf{s} = (s(n))_{n\geq 0}$ having the property that $s(s(n)) = 3n$ for all n. In fact, there is only one such sequence, and the first few values are given in Table 1 below [2]. It is sequence A003605 in the OEIS.

Table 1. First few values of Propp's sequence.

n	0	1	2	3	4	5	6	7	8	9	10	11	12	13	14	15	16	17	18	19	20
$s(n)$	0	2	3	6	7	8	9	12	15	18	19	20	21	22	23	24	25	26	27	30	33

Let us look at three ways of describing it.

First, as an automatic sequence. Consider the characteristic sequence $\chi_{\mathbf{s}}(n)$. This sequence is 3-automatic, and is generated by the DFAO given in Fig. 3. The meaning of a/b in the state is that a is the state name and b is the output.

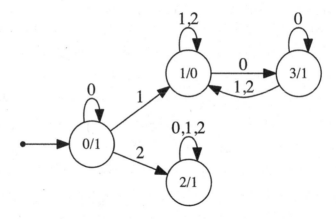

Fig. 3. DFAO generating $\chi_{\mathbf{s}}(n)$.

Second, as a 3-regular sequence. Here its linear representation is given as follows:

$$v = [0\ 2\ 3]; \quad \zeta(0) = \begin{bmatrix} 3 & 6 & 6 \\ 0 & 1 & 0 \\ 0 & 0 & 1 \end{bmatrix}; \quad \zeta(1) = \begin{bmatrix} 0 & 0 & 0 \\ 1 & 2 & 1 \\ 0 & 1 & 2 \end{bmatrix}; \quad \zeta(2) = \begin{bmatrix} 0 & -6 & -12 \\ 0 & 3 & 6 \\ 1 & 2 & 1 \end{bmatrix}; \quad w = \begin{bmatrix} 1 \\ 0 \\ 0 \end{bmatrix}.$$

Third, the most useful representation is as a 3-synchronized sequence $\mathrm{prop}(n, x)$, embodied by the automaton in Fig. 4. With the aid of this DFA (stored as `prop.txt` in the Automata Library of Walnut) we can prove that s does indeed satisfy the identity $s(s(n)) = 3n$.

```
eval proppcheck "?msd_3 An,x,y ($prop(n,x) & $prop(x,y)) => y=3*n":
```

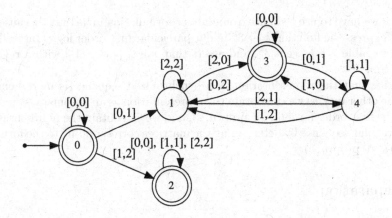

Fig. 4. Synchronized DFA for Propp's function, base 3.

This may be a good time to briefly explain the syntax of `Walnut`.

- `eval` is the instruction to evaluate the formula that follows and decide `TRUE` or `FALSE`.
- `?msd_3` specifies that the formula should be evaluated using base-3 representations of the objects.
- `&` is logical AND ($\land$) and `=>` is logical implication ($\Longrightarrow$). Similarly, `Walnut` uses `|` for logical OR ($\lor$), `<=>` for $\Longleftrightarrow$, and `~` for logical NOT ($\neg$).
- `A` is `Walnut`'s way of writing the universal quantifier $\forall$. Similarly, the existential quantifier $\exists$ is written `E`.
- Values of sequences are written `@0`, `@1`, etc.
- DFA's can be defined in terms of regular expressions using the `reg` command.
- Once defined, a DFA can be referenced by placing a `$` before the name.

We can also check that our DFAO for the range of s was correct:

`eval proppcheck2 "?msd_3 An PRO[n]=@1 <=> Em $prop(m,n)":`

Finally, we can verify yet another formula for s, namely

$$s(n) = \begin{cases} 0, & \text{if } n = 0; \\ n + 3^k, & \text{if } 3^k \le n < 2 \cdot 3^k \text{ for } k \ge 0; \\ 3(n - 3^k), & \text{if } 2 \cdot 3^k \le n < 3^{k+1} \text{ for } k \ge 0. \end{cases}$$

with the following `Walnut` code:

```
reg power3 msd_3 "0*10*":
def pow3n "?msd_3 $power3(x) & x<=n & ~Et $power3(t) & x<t & t<=n":
    # x is the largest power of 3 that is <= n
eval proppcheck3 "?msd_3 An,x (($pow3n(n,x) & n<2*x) =>
    $prop(n,n+x)) | (($pow3n(n,x) & n>=2*x) => $prop(n,(3*n)-3*x))":
```

Here we have to use a slightly roundabout formulation in Walnut, because we cannot express the function $k \to 3^k$ in our particular first-order logic. Instead we let the variable x represent 3^k, and assert that x is a power of 3 with a regular expression.

Notice that the representation as a synchronized sequence is more general than the other two, as we can obtain the representation as an automatic sequence using the first-order formula $\exists n \; \mathrm{prop}(n, s)$, and we can obtain the representation as a k-regular sequence by determining the matrices corresponding to the formula $\exists s \; i < s \land \mathrm{prop}(n, s)$.

6   Separation

Let **x** be a sequence over a finite alphabet. We say that a factor w of **x** is *recurrent* if w appears infinitely often in **x**. Obviously for each n there is at least one recurrent factor of **x** of length n. If every finite factor of **x** is recurrent, then we say **x** is recurrent. If further for each factor w there is a constant $c = c(w)$ such that two consecutive occurrences of w are separated by at most $c(w)$ symbols, then we say **x** is *uniformly recurrent*.

Let w be a recurrent factor and let $i_j(w)$ be the starting position of the j'th occurrence of w in **x**. Then $i_{j+1}(w) - i_j(w)$ is the distance between two consecutive (possibly overlapping) occurrences of w in **x**.

We can consider the four quantities

$$S_1(n) = \min_{|w|=n} \; \min_j (i_{j+1}(w) - i_j(w)),$$

$$S_2(n) = \min_{|w|=n} \; \max_j (i_{j+1}(w) - i_j(w)),$$

$$S_3(n) = \max_{|w|=n} \; \min_j (i_{j+1}(w) - i_j(w)),$$

$$S_4(n) = \max_{|w|=n} \; \max_j (i_{j+1}(w) - i_j(w)).$$

The function $S_1(n)$ was called the *repetitivity index* by Carpi and D'Alonzo [9]. Note that if **x** is not uniformly recurrent, then the functions S_2 and S_4 may not be well-defined for all n.

Theorem 7. *Suppose* **x** *is k-automatic. Then S_1 and S_3 are k-synchronized, and if* **x** *is uniformly recurrent, so are S_2 and S_4.*

Proof. For each of these four interpretations, $1 \le i \le 4$, we can easily write a first-order formula giving the value $t = S_i(n)$ for each corresponding n. ∎

Here is the Walnut code for the four formulas above in the case of the Thue-Morse sequence, represented in Walnut with the capital letter T. Here tmfactoreq is used to assert that the two factors $\mathbf{t}[i..i+n-1]$ and $\mathbf{t}[j..j+n-1]$ are equal, and tmconsec asserts that $\mathbf{t}[j..j+n-1]$ and $\mathbf{t}[k..k+n-1]$ are two consecutive occurrences of $\mathbf{t}[i..i+n-1]$.

```
def tmfactoreq "At (t<n) => T[i+t]=T[j+t]":
def tmconsec "(k>j) & $tmfactoreq(i,j,n) & $tmfactoreq(i,k,n) &
    Al (j<l&l<k) => ~$tmfactoreq(i,l,n)":
def mindist "(Aj,k $tmconsec(i,j,k,n) => s+j<=k) &
    (Ej,k $tmconsec(i,j,k,n) & s+j=k)":
def maxdist "(Aj,k $tmconsec(i,j,k,n) => s+j>=k) &
    (Ej,k $tmconsec(i,j,k,n) & s+j=k)":
def s1 "(Ei $mindist(i,n,t)) & (Ai,s $mindist(i,n,s) => s>=t)":
def s2 "(Ei $maxdist(i,n,t)) & (Ai,s $maxdist(i,n,s) => s>=t)":
def s3 "(Ei $mindist(i,n,t)) & (Ai,s $mindist(i,n,s) => s<=t)":
def s4 "(Ei $maxdist(i,n,t)) & (Ai,s $maxdist(i,n,s) => s<=t)":
```

The first few values for the Thue-Morse sequence are as follows:

n	1	2	3	4	5	6	7	8	9	10	11	12	13	14	15	16	17
$S_1(n)$	1	2	3	4	6	6	8	8		12	12	12	12	16	16	16	16	24
$S_2(n)$	3	4	8	8		16	16	16	16	32	32	32	32	32	32	32	32	64
$S_3(n)$	1	4	4	8		8		16	16	16	16	32	32	32	32	32	32	32	32
$S_4(n)$	3	8	9	18	18	36	36	36	36	72	72	72	72	72	72	72	72

We now illustrate one huge advantage to a synchronized representation for a sequence: we can verify "guessed" formulas with ease. For example, studying the values of $S_4(n)$ for the Thue-Morse sequence given above suggests that

$$S_4(n) = 9 \cdot 2^k \text{ provided } 2^k + 2 \leq n < 2^{k+1} + 2 \text{ and } n \geq 3;$$
$$S_4(n) \leq 9n - 18 \text{ for } n \geq 3.$$

We can now verify both of these with Walnut, as follows:

```
reg power2 msd_2 "0*10*":
eval s4check1 "An,x (n>=3 & $power2(x) & x+2<=n & n<2*x+2) =>
    $s4(n,9*x)":
eval s4check2 "An,s (n>=3 & $s4(n,s)) => s+18<=9*n":
```

and Walnut returns TRUE for both. Here we are using the same trick as we used for Propp's sequence, with x here representing a power of 2 (instead of a power of 3, as was the case for Propp's sequence).

7 Other Synchronized Sequences

There are many other aspects of automatic sequences **x** that are synchronized. For example

- the position of the first occurrence of two identical symbols exactly n symbols apart [20];

- the starting position of the first run of length $\geq n$;
- length of shortest prefix of **x** containing all factors of length n (the *appearance function*) [3];
- the longest distance between two consecutive length-n palindromes appearing in **x**;
- the order $|y|$ of the largest square yy centered at position n of **x** [11];
- the length of the longest palindromic suffix of $\mathbf{x}[0..n-1]$ [6];
- the length of the shortest prefix of **x** containing two (possibly overlapping) occurrences of a length-n word [8];
- the number of distinct factors of length n (called *subword complexity* or *factor complexity*) [13];
- the number of distinct primitive factors of length n [13].

See the cited papers for more details about them. First-order formulas are relatively easy to construct for all.

8   Growth Rate of Synchronized Sequences

A k-regular sequence $(a(n))_{n \geq 0}$ can grow as fast as any polynomial in n, but no faster. However, for synchronized sequences, their growth rate is much more constrained.

Let $k, \ell \geq 2$ be integers, and define $\beta = (\log \ell)/(\log k)$.

Theorem 8. *Let $(a(n))_{n \geq 0}$ be a (k, ℓ)-synchronized sequence. Then*

(a) $a(n) = O(n^{\beta})$;
(b) If $a(n) = o(n^{\beta})$, then $a(n) = O(1)$;
(c) If there exists an increasing subsequence $0 < n_1 < n_2 < \cdots$ such that $\lim_{i \to \infty} a(n_i)/n_i^{\beta} = 0$, then there exists a constant C such that $a(n) = C$ for infinitely many n.

Proof.

(a) Suppose $f \neq O(n^{\beta})$. Then there exists an increasing subsequence $(n_i)_{i \geq 0}$ such that $f(n_i)/n_i^{\beta} \to \infty$. Suppose the DFA accepting $\{(n, f(n))_{k,\ell} : n \geq 0\}$ has t states; then t is the pumping lemma constant. Choose i such that $n_i \geq k^t$ and $f(n_i)/n_i^{\beta} > \ell^{t+1}$, and in the pumping lemma let $z = (n_i, f(n_i))_{k,\ell}$. Then $|z| > t$, and furthermore we have

$$|f(n_i)_{\ell}| > \log_{\ell} f(n_i) > \log_{\ell}(n_i^{\beta} \ell^{t+1}) = (\log_{\ell} n_i^{\beta}) + t + 1 = (\beta \log_{\ell} n_i) + t + 1$$
$$= (\log_k n_i) + t + 1 \geq |(n_i)_k| + t.$$

Hence the first component of z starts with at least t 0's, while the second component starts with a nonzero digit. When we pump (that is, write $z = uvw$ with $|uv| \leq t$ and $|v| \geq 1$ and consider uv^2w) we only add to the number of leading 0's in the first component, but the second component's base-ℓ value increases in size (since it starts with a nonzero digit). This implies that f is not a function, a contradiction.

(b) We prove the contrapositive. Since $L = \{(n, f(n))_{k,\ell} : n \geq 0\}$ is regular, so is the reversed language $L^R = \{(n, f(n))^R_{k,\ell} : n \geq 0\}$. Let M be a DFA recognizing L^R, and let t be the number of states of M (which is the pumping lemma constant). Assume that $f \neq O(1)$. Then there must be an $n_0 > 0$ for which $f(n_0) > \ell^t$. Let $z = (n_0, f(n_0))^R_{k,\ell}$. Then $|z| > t$. Using the pumping lemma, write $z = uvw$ with $|uv| \leq t$ and $|v| \geq 1$, and consider the sequence of words $z_i = uv^{i+1}w \in L^R$ for $i \geq 0$. Then $z_i = (a_i, b_i)^R_{k,\ell}$ for some integers a_i, b_i and hence $f(a_i) = b_i$. Let r, s be integers such that $k^r \leq n_0 < k^{r+1}$ and $\ell^s \leq f(n_0) < \ell^{s+1}$. Then $a_i < k^{r+1+i|v|}$ and $b_i > \ell^{s+i|v|}$. Thus $\log_\ell b_i - \log_k a_i > s - r - 1$, and so $f(a_i) = b_i > \ell^{s-r-1} a_i^\beta$. Thus $f(n) > cn^\beta$ for infinitely many n, with $c = \ell^{s-r-1}$, and hence $f(n) \neq o(n^\beta)$.

(c) Suppose the synchronized DFA for f has t states. Since $\lim_{i \to \infty} f(n_i)/n_i^\beta = 0$, there must exist some $n_i > k^t$ for which $f(n_i)/n_i^\beta < \ell^{-(t+1)}$. Then

$$|f(n_i)_\ell| \leq (\log_\ell f(n_i)) + 1 \leq (\log_\ell n_i^\beta \ell^{-(t+1)}) + 1 = (\beta \log_\ell n_i) - t$$
$$= (\log_k n_i) - t \leq |(n_i)_k| - t.$$

Let z be the base k representation of the pair $(n_i, f(n_i))$. The second component of z then starts with at least t 0's. Applying the pumping lemma to z then implies there are infinitely many n for which $f(n) = f(n_i)$. Take $C = f(n_i)$ to obtain the result. ∎

Corollary 1. *Let $f(n)$ be a (k, ℓ)-synchronized sequence that is increasing. Then either $f = O(1)$ or $f = \Theta(n^\beta)$, where $\beta = (\log \ell)/(\log k)$.*

Let's now apply some of these ideas to the infinite fixed point

$$\mathbf{vn} = aabaabbaabaabbbb\cdots$$

of the morphism $a \to aab$, $b \to b$. This sequence was previously studied by Allouche et al. [1], where the authors prove it is not 2-automatic. Their proof was somewhat involved, but with the bounds of Theorem 8, we can prove a stronger result rather easily.

Theorem 9. *The sequence $\mathbf{vn}$ is not k-automatic for any base $k \geq 2$.*

Proof. Suppose it is k-automatic. Then, as in Sect. 7, the starting position of the first occurrence of a run of n b's would be k-synchronized and hence is $O(n)$. However, the first occurrence of b^n appears at position $2^{n+1} - n - 1$, a contradiction. ∎

As another consequence of these ideas, we can now prove a theorem about the asymptotic critical exponent of an automatic sequence. Recall that a finite word w of length n has period p if $w[i] = w[i+p]$ for $1 \leq i \leq n-p$. The smallest positive period is called *the* period, and is denoted by $\mathrm{per}(w)$. The *exponent* of a word w, denoted $\exp(w)$, is defined to be $|w|/\mathrm{per}(w)$.

The asymptotic critical exponent $\mathrm{ace}(\mathbf{x})$ of an infinite word is defined to be

$$\sup\{\alpha \;:\; \text{there exist arbitrarily long factors } w \text{ of } \mathbf{x} \text{ with } \exp(\mathbf{w}) \geq \alpha\}.$$

Theorem 10. *If $\mathbf{x}$ is a k-automatic sequence, then $\mathrm{ace}(\mathbf{x}) > 1$.*

Proof. We saw in Theorem 7 that if $\mathbf{x}$ is k-automatic, then the function $S_1(n)$ is k-synchronized. By Theorem 8 (a), we have $S_1(n) = O(n)$. Hence there exists a constant c such that for all sufficiently large n, there are two consecutive occurrences of some length-n factor z at distance d from each other, with $d \leq cn$. So $n \geq d/c$. Let $w = w_n$ be a word of length at most $d + n$ with two distinct occurrences of z, one as prefix and one as suffix. Then

$$\exp(w) = |w|/\mathrm{per}(w) \geq (d+n)/d \geq (d+d/c)/d = 1 + 1/c > 1.$$

∎

Here is yet another application of synchronization. Recall that a *maximal run* is a sequence of consecutive symbols $\mathbf{x}[i..i+n-1]$ all equal to a, where $\mathbf{x}[i+n] \neq a$ and (if $i > 0$) $\mathbf{x}[i-1] \neq a$.

Theorem 11. *Let $\mathbf{x}$ be a k-automatic sequence. There exists a constant C such that for all $i \geq 0$, there are in $\mathbf{x}$ at most C different maximal run lengths ℓ lying in the interval $[k^i, k^{i+1})$.*

Proof. Let $f(i)$ be the number of different maximal run lengths lying in the interval $[k^i, k^{i+1})$.

For each different run length $r \in [k^i, k^{i+1})$ there exists a factor of $\mathbf{x}$ of the form $ab^r c$ in $\mathbf{x}$ with $a \neq b \neq c$. These different factors can only overlap, at most, at the endpoints a, c, so even with the most efficient "packing" of these factors together, the last such maximal run to appear in $\mathbf{x}$ must appear at a starting position $p \geq (f(i) - 1)k^i$. But since maximal run lengths are synchronized, by Theorem 8 there exists a constant c such that the first occurrence of any run of length t must occur in $\mathbf{x}$ at a starting position $p \leq ct$. Hence $(f(i) - 1)k^i \leq p \leq ct \leq ck^{i+1}$. It follows that $f(i) \leq 1 + ck$. Taking $C = 1 + ck$, the result follows.

Corollary 2. *Let $\mathbf{x}$ be a k-automatic sequence, and let $\ell_1 < \ell_2 < \cdots$ be the lengths of all maximal runs appearing in $\mathbf{x}$. Then the base-k representations of $\{\ell_1, \ell_2, \ldots\}$ can be written as a finite union of sets of the form uv^*w.*

Proof. From Theorem 11 we know that the number of different maximal run lengths in the interval $[k^n, k^{n+1})$ is $O(1)$. Hence, letting L be the language of base-k representations of all these different maximal run lengths, the language L contains at most a constant number of different words of each length: it is "slender". Furthermore, since there is a first-order formula for specifying that n is the length of a maximal run, we know that L is a regular language. By a theorem of [21, 24], a slender regular language is the union of a finite number of regular languages of the form uv^*w.

9   Fibonacci Synchronization

We can talk about Fibonacci synchronization of a sequence $(a(n))_{n \geq 0}$ in analogy with base-k synchronization. Here both n and $a(n)$ are expressed in Fibonacci representation [15,26]. Defining $F_0 = 0$, $F_1 = 1$, and $F_n = F_{n-1} + F_{n-2}$ for $n \geq 2$, the Fibonacci representation of a natural number n is defined to be a binary string $w = a_1 a_2 \cdots a_t$ with $a_1 \neq 0$ and $a_i a_{i+1} = 0$ for $1 \leq i < t$ such that $n = \sum_{1 \leq i \leq t} a_i F_{t+2-i}$.

Theorem 12. *Let $\varphi = (1 + \sqrt{5})/2$, the golden ratio. The following functions are Fibonacci-synchronized:*

(a) $n \to \lfloor \varphi n \rfloor$;
(b) $n \to \lfloor \varphi^2 n \rfloor$;
(c) $n \to \lfloor n/\varphi \rfloor$.

Proof. We start from the identities

$$[(n)_F 0]_F = \lfloor (n+1)\varphi \rfloor - 1$$
$$[(n)_F 00]_F = \lfloor (n+1)\varphi^2 \rfloor - 2$$

for $n \geq 0$, whose proof can be found, for example, in [18]. First we define a DFA shift that accepts two inputs in parallel if the second is the left shift of the first, with a 0 in the last position. Then we can construct synchronized DFA's for $\lfloor \varphi n \rfloor$ and $\lfloor \varphi^2 n \rfloor$ as follows:

```
reg shift {0,1} {0,1} "([0,0]|[0,1][1,1]*[1,0])*":
def phin "?msd_fib (s=0 & n=0) | Ex $shift(n-1,x) & s=x+1":
def phi2n "?msd_fib (s=0 & n=0) | Ex,y $shift(n-1,x) & $shift(x,y)
        & s=y+2":
```

with the automata depicted in Fig. 5. Next, using the fact that $n/\varphi = n\varphi - n$, we get a synchronized automaton for (c) as follows:

```
def noverphi "?msd_fib Et $phin(n,t) & s+n=t":
```

giving us the automaton in Fig. 6.   ∎

Corollary 3. *Let $\gamma \in \mathbb{Q}(\sqrt{5}) = \mathbb{Q}(\varphi)$ be positive. Then the sequence $(\lfloor \gamma n \rfloor)_{n \geq 0}$ is Fibonacci-synchronized.*

Proof. Write $\gamma = (a + b\varphi)/c$ for integers a, b, c with c positive.

First, note that for all real x and integers $c \geq 1$ we have $\lfloor x/c \rfloor = \lfloor \frac{\lfloor x \rfloor}{c} \rfloor$. Let a, c be integers with a, c positive. Then, writing $f(n) = \lfloor \varphi n \rfloor$, we have

$$\lfloor \gamma n \rfloor = \left\lfloor \left(\frac{a + b\varphi}{c} \right) n \right\rfloor = \left\lfloor \frac{an + \varphi bn}{c} \right\rfloor = \left\lfloor \frac{\lfloor an + \varphi bn \rfloor}{c} \right\rfloor$$
$$= \left\lfloor \frac{an + \lfloor \varphi bn \rfloor}{c} \right\rfloor = \left\lfloor \frac{an + f(bn)}{c} \right\rfloor.$$

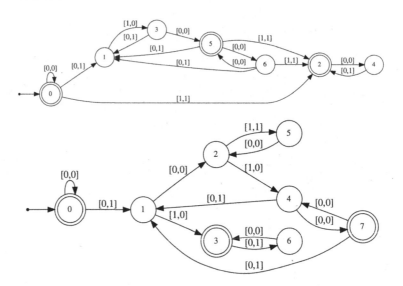

Fig. 5. Synchronized Fibonacci automata for $\lfloor\varphi n\rfloor$ (top) and $\lfloor\varphi^2 n\rfloor$ (bottom).

Since $f(n)$ is synchronized, so is $f(bn)$. And hence so is $an + f(bn)$. And hence so is $\lfloor(an + f(bn))/c\rfloor$.

If b is negative then we use the fact that $\lfloor-x\rfloor = -1 - \lfloor x\rfloor$ if $x \notin \mathbb{Z}$. ∎

Now we can use all this to prove some recent conjectures of Don Reble about letters in 3-term arithmetic progressions in the infinite Fibonacci word $\mathbf{f} = (f_n)_{n\geq 0} = 01001001\cdots$, which is sequence A003849 in the OEIS. Letting

$$\text{AP}_0 := \{n \ : \ \exists i \ \mathbf{f}[i] = \mathbf{f}[i + n] = \mathbf{f}[i + 2n] = 0\} = \{0, 2, 3, 4, 5, 6, 8, 9, 10, 11, 12, \ldots\}$$
$$\text{AP}_1 := \{n \ : \ \exists i \ \mathbf{f}[i] = \mathbf{f}[i + n] = \mathbf{f}[i + 2n] = 1\} = \{0, 3, 5, 8, 10, 13, 16, 18, 21, 24, \ldots\}.$$

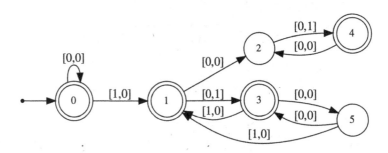

Fig. 6. Synchronized Fibonacci automaton for $\lfloor n/\varphi\rfloor$.

we can define the sets

$$R_1 := AP_0 \setminus AP_1 = \{2, 4, 6, 9, 11, 12, 15, 17, 19, 22, 23, 25, \ldots\}$$
$$R_2 := \mathbb{N} \setminus (AP_0 \cup AP_1) = \{1, 7, 14, 20, 27, 35, 41, 48, 54, \ldots\}$$
$$R_3 := AP_0 \cap AP_1 = \{0, 3, 5, 8, 10, 13, 16, 18, 21, 24, 26, \ldots\}.$$

Defining $r = 2$, $s = (\sqrt{5} - 1)/2$, $t = (1 + \sqrt{5})/2$, Reble conjectured that for $x \geq 1$ we have

$$x + 1 \in R_1 \iff x \in \underline{A189377} = \{n + \lfloor sn/r \rfloor + \lfloor tn/r \rfloor : n \geq 1\}$$
$$x + 1 \in R_2 \iff x \in \underline{A189378} = \{n + \lfloor rn/s \rfloor + \lfloor tn/s \rfloor : n \geq 1\}$$
$$x + 1 \in R_3 \iff x \in \underline{A189379} = \{n + \lfloor rn/t \rfloor + \lfloor sn/t \rfloor : n \geq 1\}.$$

So we can prove all the Reble conjectures as follows. Here F is `Walnut`'s way of representing the Fibonacci word **f**.

```
def ap0 "?msd_fib Ei F[i]=@0 & F[i]=F[i+n] & F[i]=F[i+2*n]":
def ap1 "?msd_fib Ei F[i]=@1 & F[i]=F[i+n] & F[i]=F[i+2*n]":

def r1 "?msd_fib $ap0(n) & ~$ap1(n)":
def r2 "?msd_fib ~$ap0(n) & ~$ap1(n)":
def r3 "?msd_fib $ap0(n) & $ap1(n)":

def fibsr "?msd_fib Er $phin(n,r) & s=(r-n)/2":
def fibtr "?msd_fib Er $phin(n,r) & s=r/2":
def fibrs "?msd_fib $phin(2*n,s)":
def fibts "?msd_fib Er $phin(n,r) & s=r+n":
def fibrt "?msd_fib Er $phin(2*n,r) & s=r-2*n":
def fibst "?msd_fib Er $phin(n,r) & s=2*n-(r+1)":

def a189377 "?msd_fib En,x,y $fibsr(n,x) & $fibtr(n,y) & z=n+x+y":
def a189378 "?msd_fib En,x,y $fibrs(n,x) & $fibts(n,y) & z=n+x+y":
def a189379 "?msd_fib En,x,y $fibrt(n,x) & $fibst(n,y) & z=n+x+y":

eval rebleconj1 "?msd_fib Ax (x>=1) => ($r1(x+1) <=> $a189377(x))":
eval rebleconj2 "?msd_fib Ax (x>=1) => ($r2(x+1) <=> $a189378(x))":
eval rebleconj3 "?msd_fib Ax (x>=1) => ($r3(x+1) <=> $a189379(x))":
```

and `Walnut` returns `TRUE` for the last three.

The positions of the 1's in $\underline{A003849}$ form sequence $\underline{A003622}$, namely (as is well known) $\lfloor n\varphi^2 \rfloor - 1$, where $\varphi = (1 + \sqrt{5})/2$. We can easily prove this with `Walnut`, as follows.

```
eval pos1 "?msd_fib An (F[n]=@1) <=> (Er $phi2n(r,n+1))":
```

Now let's look at the possible distances between all occurrences of the 1's. This is the set

$$D_1 = \{n : \exists i \ \mathbf{f}[i] = \mathbf{f}[i+n] = 1\} = \{0, 2, 3, 5, 6, 7, 8, 10, 11, 13, 14, 15, 16, 18, \ldots\},$$

which we can define in `Walnut` as follows:

```
def dist1 "?msd_fib Ei (F[i]=@1 & F[i+n]=@1)":
```

Notice that D_1 is not cofinite, which we can check as follows:

```
eval d1notcofinite "?msd_fib Am En (n>m & ~$dist1(n))":
```

The set D_1 is (up to the inclusion of 0) sequence A307295 in the OEIS. This latter sequence is defined in the OEIS to be A001950$(n/2 + 1)$ if n is even, and $a(n) = $ A001950$((n+1)/2)+1$ if n is odd, where A001950 is the sequence $\lfloor \varphi^2 n \rfloor$. We can prove the equality of these two sequences as follows:

```
def even "?msd_fib Em n=2*m":
def odd "?msd_fib Em n=2*m+1":
def a307295 "?msd_fib (Em $even(m) & $phi2n(m/2 + 1,n)) |
    (Em,r $odd(m) & $phi2n((m+1)/2,r) & n=r+1)":
eval checkdist1 "?msd_fib An (n>=1) => ($dist1(n) <=> $a307295(n))":
```

The complementary sequence of D_1 (that is, $\mathbb{N} \setminus D_1$) is

$$\overline{D_1} = \{1, 4, 9, 12, 17, 22, 25, \ldots\},$$

which is sequence A276885. There the formula $2\lfloor (n - 1)\varphi \rfloor + n$ is given for A276885, which we can prove as follows:

```
def altc "?msd_fib Em,r $phin(m-1,r) & n=2*r+m":
eval test276885 "?msd_fib An $dist1(n) <=> ~$altc(n)":
```

See, for example, [12].

R. J. Mathar conjectured (see A276885) that

$$\overline{D_1} = \{1\} \cup \text{A089910},$$

where

$$\text{A089910} = \{n : \mathbf{f}[n - 1] = \mathbf{f}[n - 2]\}.$$

We can prove this as follows:

```
eval mathar "?msd_fib An (n>=2) => ($altc(n) <=> F[n-1]=F[n-2])":
```

Finally, let's prove the classical characterization of the sequences $A_n = \lfloor \varphi n \rfloor$ and $B_n = \lfloor \varphi n^2 \rfloor$ in terms of the "mex" or "minimal excluded number" function, due to Wythoff [25]. For a set $S \subsetneq \mathbb{N}$, we define $\text{mex}(S) = \min\{n : n \notin S\}$. Then

$$A_n = \text{mex}\{A_i, B_i : 0 \le i < n\}$$
$$B_n = A_n + n.$$

We can check this as follows:

```
def incl "?msd_fib Ei i<n & ($phin(i,s) | $phi2n(i,s))":
    # s appears in {A_i, B_i : 0 <= i < n }
def mex "?msd_fib (~$incl(n,s)) & At (t<s) => $incl(n,t)":
    # s equals mex {A_i, B_i : 0 <= i < n }
eval mexchk1 "?msd_fib An,s $mex(n,s) <=> $phin(n,s)":
eval mexchk2 "?msd_fib An,s $phi2n(n,s) <=> (Et $phin(n,t) & s=t+n)":
```

10   Unsynchronized Sequences

Although a synchronized representation for a sequence is often the most useful one to have, not all k-regular sequences have one, even if their growth rate might permit it.

For example, we could consider the number of unbordered length-n factors of the characteristic sequence of the powers of 2. This is a 2-regular sequence, but in [13], the authors proved that it is not 2-synchronized.

Here is another example. Consider the function $f(n) = n^2$. The bounds on growth rate in Theorem 8 do not rule out the possibility that f could be (k, k^2)-synchronized for some k. Assume it is. Then $f(n+1)$ would be (k, k^2)-synchronized by Theorem 3, and $g(n) = f(n+1) - f(n)$ would be (k, k^2)-synchronized by Theorem 1 (b). But $g(n) = 2n+1$, which cannot be (k, k^2)-synchronized because its growth rate is $\Theta(n)$, which violates Corollary 1.

11   Hilbert's Space-Filling Curve

As lagniappe, we offer one more example.

In 1891 David Hilbert famously described the construction of a continuous curve that fills the unit square [14]. Instead of filling the unit square, we consider a sequence $(x_n, y_n)_{n \geq 0}$ that visits every non-negative pair of integers exactly once, starting from the origin $(x_0, y_0) = (0, 0)$.

It turns out that the coordinates (x_n, y_n) of the Hilbert curve are synchronized, but only if we represent n, x_n, and y_n in the right way. The right way is to represent n in base 4, but represent x_n and y_n in base 2! In other words, the triple (n, x_n, y_n) is $(4, 2, 2)$-synchronized, by a 10-state automaton HS; see [19] for details.

From the synchronized automaton, given $(n)_4 = a_1 a_2 \cdots a_t$, the base-4 representation of n, we can easily determine (x_n, y_n) as explained in the proof of Proposition 1. The reverse is also true: given the base-2 representations of (x, y), we can easily determine the n for which $(x_n, y_n) = (x, y)$, using the same idea.

With the aid of the synchronized representation for HS, we can easily produce a bitmap image of each generation of the Hilbert curve, as previously done in [22, Fig. 6].

To do so, we "expand" the curve, inserting rows and column that are blank, except for when they connect two consecutive points of the curve. The following Walnut code produces a DFA $hp describing a bitmap image of the Hilbert curve.

```
def even "Em n=2*m":
def odd "Em n=2*m+1":
def hp "($even(x) & $even(y)) | ($even(x) & $odd(y) &
(En (HS[?msd_4 n][x/2][(y-1)/2]=@1 &
HS[?msd_4 n+1][x/2][(y+1)/2]=@1) |(HS[?msd_4 n][x/2][(y+1)/2]=@1
& HS[?msd_4 n+1][x/2][(y-1)/2]=@1)) | ($odd(x) & $even(y) &
(En (HS[?msd_4 n][(x-1)/2][y/2]=@1 &
HS[?msd_4 n+1][(x+1)/2][y/2]=@1) |(HS[?msd_4 n][(x+1)/2][y/2]=@1
& HS[?msd_4 n+1][(x-1)/2][y/2]=@1))":
```

For example, for generation 7 we get the image in Fig. 7.

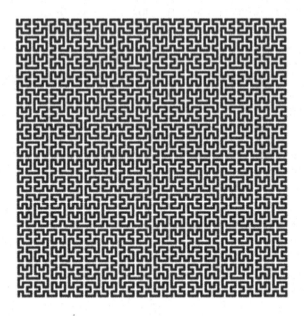

Fig. 7. Generation 7 of the Hilbert curve.

Acknowledgments. Thanks to Jean-Paul Allouche and Narad Rampersad for their helpful comments.

References

1. Allouche, J.-P., Bétréma, J., Shallit, J.: Sur des points fixes de morphismes du monoïde libre. RAIRO Inform. Théor. App. **23**, 235–249 (1989)
2. Allouche, J.-P., Rampersad, N., Shallit, J.: On integer sequences whose first iterates are linear. Aequationes Math. **69**, 114–127 (2005)
3. Allouche, J.-P., Shallit, J.: Automatic Sequences: Theory, Applications, Generalizations. Cambridge University Press, Cambridge (2003)
4. Allouche, J.-P., Shallit, J.O.: The ring of k-regular sequences. Theoret. Comput. Sci. **98**, 163–197 (1992)
5. Allouche, J.-P., Shallit, J.O.: The ring of k-regular sequences. II. Theoret. Comput. Sci. **307**, 3–29 (2003)
6. Blondin Massé, A., Brlek, S., Garon, A., Labbé, S.: Combinatorial properties of f-palindromes in the Thue-Morse sequence. Pure Math. Appl. **19**(2–3), 39–52 (2008). http://puma.dimai.unifi.it/19_2_3/4.pdf
7. Bruyère, V., Hansel, G., Michaux, C., Villemaire, R.: Logic and p-recognizable sets of integers. Bull. Belgian Math. Soc. **1**, 191–238 (1994). corrigendum, Bull. Belgian Math. Soc. **1**, 577 (1994)

8. Bugeaud, Y., Kim, D.H.: A new complexity function, repetitions in Sturmian words, and irrationality exponents of Sturmian numbers. Trans. Amer. Math. Soc. **371**, 3281–3308 (2019)
9. Carpi, A., D'Alonzo, V.: On the repetitivity index of infinite words. Int. J. Algebra Comput. **19**, 145–158 (2009)
10. Carpi, A., Maggi, C.: On synchronized sequences and their separators. RAIRO Inform. Théor. App. **35**, 513–524 (2001)
11. Charlier, E., Rampersad, N., Shallit, J.: Enumeration and decidable properties of automatic sequences. Int. J. Found. Comp. Sci. **23**, 1035–1066 (2012)
12. Dekking, F.M.: The Frobenius problem for homomorphic embeddings of languages into the integers. Theoret. Comput. Sci. **732**, 73–79 (2018)
13. Goč, D., Schaeffer, L., Shallit, J.: Subword complexity and k-synchronization. In: Béal, M.-P., Carton, O. (eds.) DLT 2013. LNCS, vol. 7907, pp. 252–263. Springer, Heidelberg (2013). https://doi.org/10.1007/978-3-642-38771-5_23
14. Hilbert, D.: Über die stetige Abbildung einer Linie auf ein Flächenstück. Math. Annalen **38**, 459–460 (1891)
15. Lekkerkerker, C.G.: Voorstelling van natuurlijke getallen door een som van getallen van Fibonacci. Simon Stevin **29**, 190–195 (1952)
16. Mousavi, H.: Automatic theorem proving in Walnut (2016). Preprint available at http://arxiv.org/abs/1603.06017
17. Propp, J.: Problem proposal 474. Crux Math. **5**, 229 (1979). Solution by G. Patruno, Crux Math. 6, 198 (1980)
18. Reble, D.: Zeckendorf vs. Wythoff representations: comments on A007895 (2008). Manuscript available at https://oeis.org/A007895/a007895.pdf
19. Shallit, J.: Hilbert's spacefilling curve described by automatic, regular, and synchronized sequences (2021). Preprint, https://arxiv.org/abs/2106.01062
20. Shallit, J.: The Logical Approach to Automatic Sequences: Exploring Combinatorics on Words with Walnut. Cambridge University Press (2022, to appear)
21. Shallit, J.O.: Numeration systems, linear recurrences, and regular sets. Inform. Comput. **113**, 331–347 (1994)
22. Shallit, J.O., Stolfi, J.: Two methods for generating fractals. Comput. Graph. **13**, 185–191 (1989)
23. Sloane, N.J.A., et al.: The on-line encyclopedia of integer sequences (2021). https://oeis.org
24. Szilard, A., Yu, S., Zhang, K., Shallit, J.: Characterizing regular languages with polynomial densities. In: Havel, I.M., Koubek, V. (eds.) MFCS 1992. LNCS, vol. 629, pp. 494–503. Springer, Heidelberg (1992). https://doi.org/10.1007/3-540-55808-X_48
25. Wythoff, W.A.: A modification of the game of Nim. Nieuw Archief voor Wiskunde **7**, 199–202 (1907)
26. Zeckendorf, E.: Représentation des nombres naturels par une somme de nombres de Fibonacci ou de nombres de Lucas. Bull. Soc. Roy. Liège **41**, 179–182 (1972)

Continuants with Equal Values, a Combinatorial Approach

Gerhard Ramharter[1] and Luca Q. Zamboni[2(✉)]

[1] Department of Convex and Discrete Geometry, Technische Universität Wien,
Wiedner Hauptstrasse 8-10, 1040 Vienna, Austria
gerhard.ramharter@tuwien.ac.at
[2] Institut Camille Jordan, Université Claude Bernard Lyon 1,
43 boulevard du 11 novembre 1918, 69622 Villeurbanne Cedex, France
zamboni@math.univ-lyon1.fr

Abstract. A regular continuant is the denominator K of a terminating regular continued fraction, interpreted as a function of the partial quotients. We regard K as a function defined on the set of all finite words on the alphabet $1 < 2 < 3 < \cdots$ with values in the positive integers. Given a word $w = w_1 \cdots w_n$ with $w_i \in \mathbb{N}$ we define its multiplicity $\mu(w)$ as the number of times the value $K(w)$ is assumed in the Abelian class $\mathcal{X}(w)$ consisting of all permutations of the word w. We prove that there is an infinity of different lacunary alphabets of the form $\{b_1 < \cdots < b_t < l+1 < l+2 < \cdots < s\}$ with $b_j, t, l, s \in \mathbb{N}$ and s sufficiently large such that μ takes arbitrarily large values for words on these alphabets. The method of proof relies in part on a combinatorial characterisation of the word w_{max} in the class $\mathcal{X}(w)$ where K assumes its maximum.

Keywords: Values of continuants · Regular continued fractions · Combinatorial word problems

1 Introduction

Given a sequence $w = (w_1, \ldots, w_n)$, of positive w_i, let $K(w)$ be the continuant of w, i.e., the denominator of the finite regular continued fraction

$$[w] = \cfrac{1}{w_1 + \cfrac{1}{w_2 + \cfrac{1}{\ddots + \cfrac{1}{w_n}}}}$$

We shall regard w as a *word* of length n over the alphabet $\{1 < 2 < 3 < \ldots\}$ and write $w = w_1 \cdots w_n$. Since $K(w) = K(\bar{w})$, where $\bar{w} = w_n \cdots w_1$ denotes the reversal of w, we shall henceforth identify each word w with its reverse $\bar{w}$.

© Springer Nature Switzerland AG 2021
T. Lecroq and S. Puzynina (Eds.): WORDS 2021, LNCS 12847, pp. 20–26, 2021.
https://doi.org/10.1007/978-3-030-85088-3_2

Let $\mathcal{X}(w)$ denote the Abelian class of w consisting of all permutations of w. The following problem has attracted much attention and led to a number of applications (see e.g. [1,4,5,7,8]): Let $A = \{a_1 < \cdots < a_s\}$ be a finite ordered alphabet with $a_j \in \mathbb{N}$. Given a word $w = w_1 w_2 \cdots w_n$ with $w_i \in A$, find the arrangements $w_{\max}, w_{\min} \in \mathcal{X}(w)$ maximizing resp. minimizing the function $K(\cdot)$ on $\mathcal{X}(w)$. The first author [3] gave an explicit description of both extremal arrangements $w_{\max}$ and $w_{\min}$ and showed that in each case the arrangement is unique (up to reversal) and independent of the actual values of the positive integers a_i. He also investigated the analogous problem for the semi-regular continuant K' defined as the denominator of the semi-regular continued fraction

$$[w]^\bullet = \cfrac{1}{w_1 - \cfrac{1}{w_2 - \cfrac{1}{\ddots - \cfrac{1}{w_n}}}}$$

with entries $w_i \in \{2, 3, \ldots\}$. He gave a fully combinatorial description of the minimizing arrangement $w'_{\min}$ for $K'(\cdot)$ on $\mathcal{X}(w)$ and showed that the arrangement is unique (up to reversal) and independent of the actual values of the positive integers a_i. However, the determination of the maximizing arrangement $w'_{\max}$ for the semi-regular continuant turned out to be more difficult. He showed that in the special case of a 2-digit alphabet $\{(2 \le)\, a_1 < a_2\}$, the maximizing arrangement $w'_{\max}$ is a Sturmian word and is independent of the values of the a_i. Recently the second author together with M. Edson and A. De Luca [8] developed an algorithm for constructing $w'_{\max}$ over any ternary alphabet $\{(2 \le)\, a_1 < a_2 < a_3\}$, and showed that the maximizing arrangement is independent of the choice of the digits. In contrast, they exhibited examples of words $w = w_1 \cdots w_n$ over a 4-digit alphabet $A = \{(2 \le)\, a_1 < a_2 < a_3 < a_4\}$ for which the maximizing arrangement for $K'(\cdot)$ is not unique and depends on the actual values of the positive integers a_1 through a_4. In the course of these investigations the following problem came up: given an alphabet A of positive integers, we say that a word w on A has multiplicity $\mu = \mu(w)$ if the value $K(w)$ occurs precisely μ times in the multi-set $\{K(x) : x \in \mathcal{X}(w)\}$. The multiplicity $\mu'(w)$ is defined analogously for the semi-regular continuant $K'(w)$. Thus each Abelian class $\mathcal{X}(w)$ is split into subclasses of equally valued words. Question: is it true that μ can take arbitrarily large values for infinitely many alphabets and is there a combinatorial proof of this? Our aim here is to give a positive answer to this question in the case of regular continued fractions:

Theorem 1. *Fix positive integers $1 \le t \le l < s$, $b_1 < \ldots < b_t \le l$ and let $A = \{b_1 < \cdots < b_t < l+1 < \cdots < s\}$. Then for all s sufficiently large, there exists an infinite sequence of words w_k over the alphabet A with multiplicities $\mu(w_k) \to \infty$ as $k \to \infty$.*

It should be noted that for fixed s one obtains the largest possible alphabet $A' = \{1 < 2 < \cdots < s\}$ by choosing $b_1 = t = l = 1$ ($< s$). Our proof makes use of the combinatorial structure of $w_{\max}$ found by the first author in [3].

2   Preliminaries

We introduce some notation. Let $w = w_1 \cdots w_n$ be a word of length $n \geq 2$ with $w_j \in \mathbb{N}$ ($j = 1, \ldots, n$). The regular continuant of w has a matrix representation

$$K(w_1) = w_1 \quad \text{and} \quad K(w) = \det \begin{pmatrix} w_1 & -1 & 0 & \cdots & & 0 \\ 1 & w_2 & -1 & \ddots & & \vdots \\ 0 & 1 & \ddots & \ddots & & 0 \\ \vdots & & \ddots & \ddots & w_{n-1} & -1 \\ 0 & \cdots & & 0 & 1 & w_n \end{pmatrix}, \quad n \geq 2$$

It can also be defined recursively by $K(\{\,\}) = 1$ ($\{\,\} = $ empty word), $K(w_1) = w_1$ and $K(w_1 \cdots w_j) = w_j K(w_1 \cdots w_{j-1}) + K(w_1 \cdots w_{j-2})$ for $j \geq 2$. For each $1 \leq k \leq m \leq n$ we set $w_{k,m} := w_k \cdots w_m$ and $W := W_{1,n}$, $W_{k,m} := K(w_{k,m})$. The following fundamental formula goes back to the late 19th century and can be found in Perron [2], p. 11, (4): ($W =$) $W_{1,n} = W_{1,j} W_{j+1,n} + W_{1,j-1} W_{j+2,n}$ ($j \in \{1, \ldots, n-1\}$). From this we infer the simple but useful inequality

$$W_{1,n} < 2 W_{1,j} W_{j+1,n}. \tag{1}$$

Let $A = \{a_1 < \cdots < a_s\} \subset \mathbb{N}$. We consider a word $w = w_1 \cdots w_n := a_1^{p_1} \cdots a_s^{p_s}$ of length n with Parikh vector $\mathbf{p} = (p_1, \ldots, p_s)$ with $p_1 + \cdots + p_s = n$ where

$$a^r = \underbrace{aa \cdots a}_{r\text{-times}}$$

denotes a sequence of r equal elements a. Let $\mathcal{X} = \mathcal{X}(A, \mathbf{p})$ denote the set of all permutations of w where we identify each word v with its reverse $\bar{v}$. Let $N(A, \mathbf{p})$ denote the cardinality of $\mathcal{X}$. Then, $N(A, \mathbf{p}) \geq \frac{n!}{2p_1! \ldots p_s!}$. We put $W_{\max} = W_{\max}(A, \mathbf{p}) := \max\{K(v) : v \in \mathcal{X}\}$. It was shown in [3] (see (3), p. 190) that $W_{\max}$ is uniquely attained (up to reversal) by the arrangement

$$a_s L_{s-1} a_{s-2} L_{s-3} \cdots a_1^{p_1} \cdots a_{s-3} L_{s-2} a_{s-1} L_s \tag{2}$$

where $L_i = a_i^{p_i - 1}$. Let $P = P(A, \mathbf{p}) = \#\{K(v) : v \in \mathcal{X}\}$.

3   Proof of Theorem 1

Our first goal is to describe how to specify the last digit s (≥ 2) in an alphabet $A : \{b_1 < \cdots < b_t < l + 1 < \cdots < s\}$. We consider 'equipartitioned' words

$$w = w_1 \cdots w_n := b_1^m \cdots b_t^m (l+1)^m \cdots s^m.$$

corresponding to the Parikh vector $\mathbf{p} = (m, m, \ldots, m)$ in which each digit of A occurs precisely m-times in w. We will give a lower bound for s (see (7) below). To this end, we introduce the quantities $Q_{r,m-1} := K(r^{m-1})$ $(r \in 1, 2, \ldots)$. They are the elements of the r-th generalized Fibonacci sequence which is determined by the recursion $Q_{r,0} := 1$, $Q_{r,1} := K(r) = r$, $Q_{r,j+1} := rQ_{r,j} + Q_{r,j-1}$ $(j = 1, 2, \ldots)$.

Claim: $Q_{r,j-1} < (r+1)^j$ for each fixed $r \geq 1$ and all $j \geq 1$.

To prove the claim, we proceed by induction on j: This is obviously true for $j = 1$ and $j = 2$. Then by the induction hypothesis

$$\begin{aligned} Q_{r,j-1} = rQ_{r,j-2} + Q_{r,j-3} &< r(r+1)^{j-1} + (r+1)^{j-2} \\ &= (r+1)^{j-2}(r(r+1)+1) < (r+1)^{j-2}(r+1)^2 \\ &= (r+1)^j. \end{aligned}$$

In order to obtain an upper bound for the number $P(A, \mathbf{p})$, it suffices to consider words over the largest allowed s-digit alphabet $A' : \{1 < \cdots < s\}$, $b_1 = t = l = 1$ $(< s)$, with Parikh vector $\mathbf{p}' = (\underbrace{m, m, \ldots, m}_{s\text{-times}})$. Clearly

$$P(A, \mathbf{p}) < W_{\max}(A, \mathbf{p}) \leq W_{\max}(A', \mathbf{p}'),$$

and by (2)

$$w_{\max}(A', \mathbf{p}') = s \cdot (s-1)^{m-1} \cdot (s-2) \cdots 1 \cdot 1^{m-1} \cdots (s-2)^{m-1} \cdot (s-1) \cdot s^{m-1}. \tag{3}$$

By iteration of (1) applied to the decomposition in (3) we obtain the inequalities

$$\begin{aligned} W_{\max}(A', \mathbf{p}') &= K(w_{\max}(A', \mathbf{p}')) \\ &< 2^{2s}\, s \cdot (s-1) \cdots 3 \cdot 2 \prod_{j=1}^{s} K(j^{m-1}) \\ &= 2^{2s}\, s! \prod_{j=1}^{s} Q_{j,m-1} \\ &< 2^{2s}\, s! \prod_{j=1}^{s} (j+1)^m \\ &= 2^{2s}\, s!((s+1)!)^m \end{aligned}$$

and hence

$$P(A, \mathbf{p}) < 2^{2s}\, s!\, ((s+1)!)^m. \tag{4}$$

For each $s \geq 2$ we define $m_0 = m_0(s)$ to be the smallest positive integer such that

$$2^{2s} \, s! \leq \left(\frac{100}{99}\right)^{m_0}.$$

Then

$$P(A, \mathbf{p}) < \left(\frac{100}{99} \, ((s+1)!)\right)^m \quad \text{for all } m \geq m_0(s). \tag{5}$$

On the other hand, we have the following lower bound for the number of different words in $\mathcal{X}(w)$:

$$N(A, \mathbf{p}) \geq \frac{((s-l+t)\,m)!}{2(m!)^{s-l+t}} \tag{6}$$

Based on the condition (7) below, we will later make a choice of $s = s'(t, l)$ depending on the parameters t, l. We apply the estimates provided by Sterling's formula to the factorial terms occurring in relations (5) and (6) to obtain

$$(P(A, \mathbf{p}))^{1/m} < \frac{100}{99}(s+1)! < \frac{100}{99}\frac{12}{11} \, e^{-(s+1)}(s+1)^{s+1}\sqrt{2\pi(s+1)}.$$

$$(((s-l+t)\,m)!)^{1/m} > e^{-(s-l+t)}\left((s-l+t)\,m)^{s-l+t}\sqrt{2\pi(s-l+t)\,m}\right)^{1/m}.$$

$$\left(2\,(m!)^{s-l+t}\right)^{1/m} < e^{-(s-l+t)}m^{s-l+t}\left(2\,\frac{12}{11}\,\left(\sqrt{2\pi m}\right)^{s-l+t}\right)^{1/m}.$$

When we put the right hand sides of the last two inequalities together, the terms e^{s-l+t} and m^{s-l+t} cancel out, and if we keep the parameters t, l fixed for the moment, the terms of the form $\sqrt{\cdot}^{\,1/m}$ tend to 1 as $m \to \infty$. Letting $m \to \infty$ we get

$$\lim_{m\to\infty}\left(\frac{N(A, \mathbf{p})}{P(A, \mathbf{p})}\right)^{1/m} \geq \frac{99}{100}\frac{11}{12}\frac{e^{s+1}(s-l+t)^{s-l+t}}{\sqrt{2\pi(s+1)}(s+1)^{s+1}}$$

$$= \frac{363}{400}\frac{e^{s+1}(s+1-l+t-1)^{\,s+1-l+t-1}}{\sqrt{2\pi(s+1)}(s+1)^{s+1}}$$

$$= \frac{363}{400}\frac{e^{s+1}}{\sqrt{2\pi(s+1)}\,(s-l+t)^{l-t+1}}\left(1-\frac{l-t+1}{s+1}\right)^{s+1}.$$

For fixed t, l $(l - t \geq 1)$ the function $f(t, l, s) = \left(1 - \frac{l-t+1}{s+1}\right)^{s+1}$ in the variable s is strictly increasing on the interval $[l - t + 1, \infty)$ with $f(t, l, s) \nearrow e^{-(l-t)-1}$ as $s \to \infty$. We define s_0 to be the lowest integer such that $f(t, l, s_0) \geq \frac{1}{2}\,e^{-(l-t)-1}$. Then

$$\lim_{m\to\infty}\left(\frac{N(A, \mathbf{p})}{P(A, \mathbf{p})}\right)^{1/m} \geq \frac{363}{400}\frac{e^{s+1}}{\sqrt{2\pi(s+1)}\,(s-l+t)^{l-t+1}}\,\frac{1}{2}\,e^{-(l-t)-1} =: H(t, l, s)$$

for all $s \geq s_0$. Obviously there exists some sufficiently large $s' = s'(t, l) \geq s_0$ such that

$$H(t, l, s') > 1. \tag{7}$$

Therefore the right hand side of

$$\left(\frac{N(A, \mathbf{p})}{P(A, \mathbf{p})} \right) > (H(t, l, s'))^m \tag{8}$$

can be made arbitrarily large by letting $m \to \infty$. We call an $(s'-l+t)$-digit alphabet $A = \{(1 \leq) b_1 < \cdots < b_t < \cdots < s'\}$ *admissible* if $s' = s'(t, l)$ fulfills condition (7) We consider the word $u(A, \mathbf{p}_1) = (b_1)^{m_1} \cdots (b_t)^{m_1} (l+1)^{m_1} \cdots (s')^{m_1}$ of length $n = (s' - l + t) m_1$ with Parikh vector $\mathbf{p}_1 = ((m_1)^{s'-l+t})$ where we choose $m_1 \geq m_0$ such that $\left(\frac{N(A, \mathbf{p}_1)}{P(A, \mathbf{p}_1)} \right) > (H(t, l, s'))^{m_1}$. The multi-set $\mathcal{X}_1 = \mathcal{X}(A, \mathbf{p}_1)$ is made up of the $N(A, \mathbf{p}_1) = \#\mathcal{X}_1$ permuted arrangements of u. There exists at least one word $w_1 \in \mathcal{X}_1$ with multiplicity $\mu \geq 2$ because otherwise we would have $N(A, \mathbf{p}_1) = P(A, \mathbf{p}_1)$ which contradicts (8) with $m = m_1$. Let $\mu_1 (\geq 2)$ be the maximal multiplicity attained by words $w \in \mathcal{X}_1$. Next choose $m_2 > m_1(s')$ such that $H(t, l, s')^{m_2} > \mu_1$. We claim that at least one word w_2 from $\mathcal{X}_2 = \mathcal{X}(A, \mathbf{p}_2)$, $\mathbf{p}_2 = ((m_2)^{s'-l+t})$ has multiplicity $\mu > \mu_1$. Otherwise we would have $N(A, \mathbf{p}_2) \leq \mu_1 P(A, \mathbf{p}_2)$ which contradicts (8) with $m = m_2$. Next let $\mu_2 (\geq \mu_1)$ be the maximal multiplicity attained by words $w \in \mathcal{X}_2$. Proceeding with this construction step by step we end up with a sequence of words w_k on A with multiplicities $\mu_k \to \infty$ as $k \to \infty$. The construction can be carried out for infinitely many different admissible alphabets. This completes the proof of Theorem 1.

The question remains largely unsolved in the case of semi-regular continuants though it seems certain that the behavior is quite similar to the regular case.

There is some evidence supporting the following:

Conjecture. Given any ordered alphabet $A = \{a_1 < \cdots < a_s\}$ $(a_j \in \mathbb{N}, s \geq 2)$, let $\mu \geq 2$ be a positive integer. Then there exist infinitely many words on A whose multiplicity is precisely μ. The problem appears to require a deeper investigation into the values of continuants. Most likely our theorem and the conjecture also hold for continuants of semi-regular continued fractions. Unfortunately no higher-dimensional analogue of the theorem is available at present for $s \geq 4$ due to the fact that very little is known about the maximizing arrangements $w'_{\max}$ for $s \geq 4$ (see [8]).

References

1. Baxa, C.: Extremal values of continuants and transcendence of certain continued fractions. Adv. Appl. Math. **32**, 754–790 (2004)
2. Perron, O.: Die Lehre von den Kettenbrüchen, Bd. 1. Vieweg and Teubner, Leipzig (1977)

3. Ramharter, G.: Extremal values of continuants. Proc. Amer. Math. Soc. **89**(2), 189–201 (1983)
4. Ramharter, G.: Some metrical properties of continued fractions. Mathematica **30**(2), 117–132 (1983)
5. Ramharter, G.: Maximal continuants and the Fine-Wilf theorem. J. Comb. Th. A **111**, 59–77 (2005)
6. Ramharter, G.: Maximal continuants and periodicity. Integers, January 2006. 10 p. (2005)
7. Shallit, J., Sorensen, J.: Analysis of a left shift binary GCD algorithm. J. Symb. Comput. **17**(6), 473–486 (1994)
8. De Luca, A., Edson, M., Zamboni, L.Q.: Extremal values of semi-regular continuants and codings of interval exchange transformations, preprint (2021). arXiv:2105.00496

Quaternary n-cubes and Isometric Words

Marcella Anselmo[1], Manuela Flores[1], and Maria Madonia[2(✉)]

[1] Dipartimento di Informatica, Università di Salerno,
Via Giovanni Paolo II, 132-84084 Fisciano, SA, Italy
manselmo@unisa.it
[2] Dipartimento di Matematica e Informatica, Università di Catania,
Viale Andrea Doria 6/a, 95125 Catania, Italy
madonia@dmi.unict.it

Abstract. A k-ary n-cube is a graph with k^n vertices, each associated to a word of length n over an alphabet of cardinality k. The subgraph obtained deleting those vertices which contain a given k-ary word f as a factor is here introduced and called the *k-ary n-cube avoiding f*. When, for any n, such a subgraph is isometric to the cube, the word f is said *isometric*. In the binary case, isometric words can be equivalently defined, independently from hypercubes. A binary word f is isometric if and only if it is *good*, i.e., for any pair of f-free words u and v, u can be transformed in v by exchanging one by one the bits on which they differ and generating only f-free words. These two approaches are here considered in the case of a k-ary alphabet, showing that they are still coincident for $k = 3$, but they are not from $k = 4$ on. Bad words are then characterized in terms of their overlaps with errors. Further properties are obtained on non-isometric words and their index, in the case of a quaternary alphabet.

Keywords: Hypercubes · Words avoiding factors · Index of a word · Overlap · Hamming and Lee distance

1 Introduction

Graphs and words, or strings, are two of the most central notions in computer science, especially in interconnection networks and combinatorics on words areas. Many parallel processing applications have communication patterns that can be viewed as graphs called k-ary n-cubes [10]. The k-ary n-cube, Q_n^k, is a graph with k^n vertices, each associated to a word of length n over a k-ary alphabet identified with $\mathbb{Z}_k = \{0, 1, \ldots, k-1\}$. Two vertices in Q_n^k are adjacent whenever their associated words differ in exactly one position, and the mismatch is given by two symbols x and y, with $x \equiv y \pm 1 \mod k$. Special cases include rings (when $n = 1$), hypercubes Q_n (when $k = 2$) and tori.

The binary case has been extensively investigated [4]. In order to obtain some variants of hypercubes such that the number of vertices increases slower than in a hypercube, Hsu [5] introduced Fibonacci cubes. They received a lot of attention afterwards

Partially supported by INdAM-GNCS Projects 2020-2021, FARB Project ORSA203187 of University of Salerno and TEAMS Project of University of Catania.

T. Lecroq and S. Puzynina (Eds.): WORDS 2021, LNCS 12847, pp. 27–39, 2021.
https://doi.org/10.1007/978-3-030-85088-3_3

and found application in theoretical chemistry (see [8] for a survey). These notions have been then extended to define the generalized Fibonacci cube $Q_n(f)$ [6]. It is the subgraph of Q_n obtained by considering only vertices associated to binary words that do not contain a given word f as a factor, i.e. f-free binary words. Using this notation, and binary alphabet $\{0, 1\}$, Fibonacci cubes are the ones avoiding $f = 11$. In this framework, a binary word f is said *isometric* when, for any $n \geq 1$, $Q_n(f)$ can be isometrically embedded into Q_n, and *non-isometric*, otherwise [9]. Isometric words have been recently investigated from many points of view (see [2, 15]).

Observe that, in the binary case, the distance of two vertices in the hypercube coincides with their Hamming distance. Hence, the definition of isometric binary word can be equivalently given ignoring hypercubes and adopting a point of view closer to combinatorics on words. A binary word f is *d-good* if for any pair of f-free words u and v of length d, u can be transformed in v by exchanging òne by one the bits on which they differ and generating only f-free words. It is *good* if it is d-good for all d. A binary word f is *bad* if it is not good. The index of a binary bad word is the threshold d from which the word is no longer d-good. The structure of binary bad words has been characterized in [7, 9, 14, 16, 17]. In particular, a binary word is good if and only if it is isometric. Recently, binary bad words have been considered in the two-dimensional setting, and bad pictures have been investigated [1].

In this paper, we extend these definitions to the case of a generic k-ary alphabet. We introduce the *k-ary n-cube avoiding* f, $Q_n^k(f)$, for any k-ary word f. It is obtained from Q_n^k by elimination of the vertices containing f as a factor. In other words, only vertices avoiding factor f are kept. Then, we define isometric/non-isometric, and good/bad k-ary words, keeping the terminology of the binary case. While in the binary and ternary cases the two approaches coincide, this is no more true for k-ary alphabets, with $k \geq 4$. Then, we introduce two distinct definitions of index of a word and show lower and upper bounds in relation to the length of the word.

Binary bad words are characterized in [9] and [14] as the ones having a 2-error overlap, i.e., a prefix that differs from the suffix of the same length in exactly 2 positions. We generalize this nice characterization to the case of an alphabet of any cardinality $k \geq 2$. Recall that the notions of bad and non-isometric word do not coincide for k-ary alphabets with $k \geq 4$. It turns out that the characterization still holds for bad (not for non-isometric) k-ary words.

Particular interest is devoted to the quaternary alphabet, as the first value of k such that the definitions of bad and non-isometric word no longer coincide, and taking in mind the importance of quaternary alphabet, for example to model DNA sequences. Hence, our preferred quaternary alphabet will be the *genetic alphabet* $\Delta = \{A, C, T, G\}$. Some of the results on quaternary (non-) isometric words will be achieved applying known properties of binary ones. In fact, for all n, the cube Q_n^4 is isomorphic to the hypercube Q_{2n}^2 [3]. The isomorphism is obtained by the Gray map, g from Δ^* to the set of binary words of even length given by $g(A) = 00$, $g(C) = 01$, $g(T) = 11$, and $g(G) = 10$. Observe that the distance of two vertices in Q_n^4 is no longer their Hamming distance, but it is their Lee distance. Indeed, the Hamming and the Lee distance are the same for binary (and ternary) words, while they differ when dealing with k-ary words, with $k \geq 4$. We will characterize non-isometric quaternary words as the ones having a prefix and a suffix of same length whose Lee distance is 2.

Another question that naturally arises on quaternary words is whether there is a relation between the isometricness of a quaternary word f and of its binary representation by the Gray map, $g(f)$. We obtain that if f is non-isometric then so is $g(f)$. The vice versa holds if and only if $g(f)$ has an overlap of even length with exactly 2 errors.

As a conclusion, let us point out that the topics treated in this paper, for their interdisciplinary feature, promise to contribute to many areas of computer science. For instance, the property of avoiding a factor intervenes in combinatorics of words, as well as in the investigation of DNA sequences, where the avoided factor is referred to as an absent word [13]. Finally, overlaps with errors can play an important role in approximate pattern matching [11].

2 Preliminaires

Let Σ be an alphabet and $|\Sigma| = k$. Throughout the paper, Σ will be identified with $\mathbb{Z}_k = \{0, 1, \ldots, k - 1\}$. A word (or string) $f \in \Sigma^*$ of length n is $f = x_1 x_2 \cdots x_n$, where $x_1, x_2, \cdots, x_n$ are symbols in Σ. The set of words over Σ of length n is denoted Σ^n. Let $f[i]$ denote the symbol of f in position i, i.e. $f[i] = x_i$. Then, $f[i \ldots j] = x_i \cdots x_j$, for $1 \leq i \leq j \leq n$, is a factor of f. The prefix of f of length l is $pre_l(f) = f[1 \ldots l]$; while the suffix of f of length l is $suf_l(f) = f[n - l + 1 \ldots n]$. When $pre_l(f) = suf_l(f)$ then $pre_l(f)$ is referred to as an overlap of f of length l.

Let $u, v \in \Sigma^*$ be two words of the same length. Then, the *Hamming distance* $d_H(u, v)$ between u and v is the number of positions at which u and v differ. The *Lee distance* between two words $u, v \in \mathbb{Z}_k^n$, $u = x_1 \cdots x_n$ and $v = y_1 \cdots y_n$ is

$$d_L(u, v) = \sum_{i=1}^{n} min(|x_i - y_i|, k - |x_i - y_i|).$$

The distance $d_G(u, v)$ between two vertices u and v of a graph G is the length of any shortest path in G from u to v (zero, if $u = v$). A subgraph G' of G is called an *isometric subgraph* of G if for any two vertices $u, v \in V(G')$, $d_{G'}(u, v) = d_G(u, v)$.

A k-ary n-cube, denoted Q_n^k, is a graph with k^n vertices, each associated to a word of length n over an alphabet of cardinality k identified with $\mathbb{Z}_k = \{0, 1, \ldots, k-1\}$. Two vertices u and v in Q_n^k are adjacent if and only if there exists an integer $j \in \{1, \ldots, n\}$ such that $u[j] \equiv v[j] \pm 1 \mod k$, and $u[i] = v[i]$ for all $i \in \{1, \ldots, n\} \backslash \{j\}$ [12]. A k-ary n-cube can be recursively defined as in [10], or in the following way. A k-ary 1-cube, Q_1^k, is a ring of k nodes labelled $0, 1, \ldots, k-1$, with edges connecting i and $(i+1)$ mod k, for any $i \in \mathbb{Z}_k$, while a k-ary n-cube can be obtained as $Q_n^k = Q_1^k \times Q_{n-1}^k$.

In the sequel, Σ will denote a generic alphabet of cardinality k, while B is used to denote the binary alphabet $B = \{0, 1\}$, and Δ to denote the quaternary alphabet $\Delta = \{A, C, T, G\}$, referred to as the *genetic alphabet*. Symbols A and T (C and G, resp.) will be called *complementary symbols*, in analogy to the Watson-Crick complementary bases they represent. The alphabet Δ will be identified with $\mathbb{Z}_4$, in such a way that A, C, T, and G will be identified with 0, 1, 2, and 3, respectively. Therefore, pairs of complementary symbols have Lee distance 2, whereas pairs of non-complementary symbols have Lee distance 1. Moreover, Q_n^4 is the n-cube defined starting from the ring (A, C, T, G) (see Fig. 1 for $n = 1$ and $n = 2$). Thus, $d_{Q_1^4}(A, T) = d_{Q_1^4}(C, G) = 2$, i.e. the complementary symbols have distance 2.

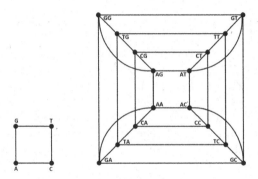

Fig. 1. Q_1^4 and Q_2^4 with alphabet $\Delta = \{A, C, T, G\}$

3   Non-isometric and Bad Words

The notions of *non-isometric* and *bad* binary words have been introduced in [7,9], where the authors observe that they are equivalent definitions. In this section, we generalize these definitions to words over an alphabet Σ with k symbols, $k \geq 2$. It will turn out that the definitions are equivalent for $k = 2, 3$, while they differ from $k = 4$ onwards. That is why we introduce two distinguished definitions. Then, we present a characterization of bad words which holds for any cardinality k of the alphabet.

Let Σ be an alphabet with $|\Sigma| = k$ and f be a word over Σ. Then, $s \in \Sigma^*$ is said f-*free* if it does not contain f as a factor. Given two f-free words, u and v of the same length, we can transform u in v by replacing one-by-one all the symbols in which they differ. This results in a sequence of words $w_0, w_1, \ldots, w_h$ such that $w_0 = u$, $w_h = v$ and for $i = 0, 1, \cdots, h - 1$, w_i differs from w_{i+1} only in one position, i.e., $d_H(w_i, w_{i+1}) = 1$. Its length is h. When each word w_i, for $i = 0, 1, \cdots, h$, is f-free the sequence is called an f-*free transformation* from u to v. Note that the next definitions are referred to a given map which identifies Σ with $\mathbb{Z}_k = \{0, 1, \ldots, k - 1\}$.

Example 1. Let $f = 010$, $u = 0011$, $v = 0110$ be in B^*. Two transformations from u to v are the following ones, $\tau: w_0 = 0011, w_1 = 0111, w_2 = 0110$, and $\tau': w_0' = 0011, w_1' = 0010, w_2' = 0110$. Then, τ is f-free, while τ' is not f-free.

Definition 2. *Let Σ be a k-ary alphabet and d a positive integer. A word $f \in \Sigma^*$ is d-good if for any pair of f-free words u and v of length d, there exists an f-free transformation from u to v of length $d_H(u, v)$. A word is good if it is d-good for any $d \geq 1$. A word is bad if it is not good.*

An example of bad word is given in Example 11. On the other hand, one can show that the words 10^s, 11010 are examples of good words (see [6,9]).

Definition 3. *Let Σ be a k-ary alphabet and let $f \in \Sigma^*$. The subgraph of Q_n^k obtained by deleting those vertices which contain f as a factor is called k-ary n-cube avoiding f, and is denoted $Q_n^k(f)$.*

Definition 4. *Let Σ be a k-ary alphabet. A word $f \in \Sigma^*$ is* isometric *if for all $d \geq 1$, $Q_d^k(f)$ is an isometric sub-graph of Q_d^k. A word $f \in \Sigma^*$ is* non-isometric *if it is not isometric.*

In analogy to the definition of good word, the definition of isometric word can be equivalently given by ignoring the notion of n-cube and referring to some kind of f-free transformation, as defined below.

Definition 5. *Let Σ be a k-ary alphabet and $f \in \Sigma^*$. Let $u, v \in \Sigma^n$ be f-free words. An* L-transformation *of length h from u to v is a sequence of words $w_0, w_1, \ldots, w_h$ such that $w_0 = u$, $w_h = v$, and for any $i = 0, 1, \cdots, h-1$, $d_L(w_i, w_{i+1}) = 1$. The L-transformation is f-free if for any $i = 0, 1, \cdots, h$, word w_i is f-free.*

Note that the definition of f-free L-transformation differs from the one of f-free transformation only in the distance that is used, the Lee or the Hamming distance. The definitions coincide when $k = 2$, since the Lee distance coincides with the Hamming distance when $k = 2$. Observing that for any $u, v \in \Sigma^n$, $d_{Q_n^k}(u, v) = d_L(u, v)$ [3], the following result holds.

Lemma 6. *Let Σ be a k-ary alphabet. A word $f \in \Sigma^*$ is isometric if and only if for all $n \geq 1$, and f-free words $u, v \in \Sigma^n$, there is an f-free L-transformation from u to v of length equal to $d_L(u, v)$.*

Remark 7. Let $f \in \Sigma^*$ and let $u, v \in \Sigma^*$ be two f-free words. Consider any f-free L-transformation from u to v of length equal to $d_L(u, v)$. Then, only symbols in the positions where u and v differ are modified in this transformation. Moreover, at each step of the transformation, a symbol x can be replaced by y only if $d_L(x, y) = 1$. Hence, each position i such that $d_L(u[i], v[i]) = d$ is replaced exactly d times.

Remark 8. Let $f \in \Sigma^*$ and let $u, v \in \Sigma^*$ be f-free words. If there exists an f-free L-transformation from u to v, then there exists an f-free L-transformation from v to u.

Remark 9. Consider a non-isometric word $f \in \Sigma^*$ and two f-free words $u, v \in \Sigma^n$, for some $n \geq 1$, such that no f-free L-transformation from u to v of length equal to $d = d_L(u, v)$ exists; suppose d to be minimal. Let $V = \{i_1, i_2, \ldots, i_m\}$, with $1 \leq i_1 < i_2 < \cdots < i_m \leq n$ be the set of all the positions where u and v differ; $m \leq d$.

The minimality of d implies that when, in any L-transformation from u to v of length d, $u[i]$, for $i \in V$, is replaced, then the resulting word, say w_i, has an occurrence of f including position i. Otherwise, let $\bar{\imath} \in V$ such that $w_{\bar{\imath}}$ is f-free. Then, $w_{\bar{\imath}}$ and v are f-free words, with $d_L(w_{\bar{\imath}}, v) = d-1 < d$. Further, there is no f-free L-transformation from $w_{\bar{\imath}}$ to v of length $d - 1$ (otherwise it would provide an f-free L-transformation from u to v of length d).

Let us show a characterization of k-ary bad words. It involves the overlaps with errors of a word. Note that in other frameworks, a prefix of a word that is equal to the suffix of the same length is (commonly) called a border or a bifix instead of an overlap. Even though, we have preferred to maintain the original term (in the paper where it is defined). A word f of length n has a *q-error overlap* of length l, $0 \leq l, q < n$, if

$d_H(pre_l(f), suf_l(f)) = q$. Differently saying, the prefix and the suffix of length l of f differ in exactly q positions [16]. The characterization of bad words is proved in [9,14] for the binary case. Let us emphasize that the generalization holds true for *bad* words (and not for non-isometric). Please, note that the proof of Theorem 10 closely follows the proof of Theorem 5.1 in [9] and Lemma 2.2 in [14], unless for the definition of a new $Condition^+$ for 2-error overlaps. It is $Condition^*$ of [14] plus a further property of f. Nevertheless, we sketch the proof in order to give some details which will be used in the following proofs.

Theorem 10. *Let Σ be a k-ary alphabet and $f \in \Sigma^*$. Then, f is bad if and only if f has a 2-error overlap.*

Proof (Sketch). The proof that, if $f \in \Sigma^*$ is bad then f has a 2-error overlap, can be given as in the case of a binary alphabet (see Theorem 5.1 in [9]).

From now on, readers can find more details, to be adapted to the k-ary case, in the proof of Lemma 2.2 in [14].

Suppose that $|f| = n$ and f has a 2-error overlap of length $l = n - r$. We show that f is bad, by constructing two f-free words over Σ, of same length, for which no f-free transformation between them, of length equal to their Hamming distance, exists.

Let $pre_l(f)$ disagree from $suf_l(f)$ in positions i and j, with $i < j$, and $f[i] = x$, $f[j] = y$, $f[r + i] = x'$ and $f[r + j] = y'$, for some $x, y, x', y' \in \Sigma$, with $x \neq x'$ and $y \neq y'$. Let $f^{(i)}$ ($f^{(j)}$, resp.) be obtained from f by replacing symbol x (y, resp.) in position i (j, resp.) with x' (y', resp.). Consider $\alpha = pre_r(f)f^{(i)}$ and $\beta = pre_r(f)f^{(j)}$.

Let us show that α is f-free. Indeed, assume, on the contrary, that α is not f-free and let f^α a copy of f that occurs in α starting from position $r_1 + 1$, with $r_1 < r$. Since $f[i] = x = f^\alpha[i]$, then $\alpha[r_1+i] = x$. Note that $r_1+i < r+i$ and, hence, $\alpha[1 \dots r_1+i] = f[1 \dots r_1 + i]$. This implies $x = \alpha[r_1 + i] = f[r_1 + i] = \alpha[r + r_1 + i] = f^\alpha[r + i]$, against the fact that $f^\alpha[r + i] = x'$.

Moreover, one can show that β has a factor f if and only if the 2-error overlap of f satisfies

$$Condition^+: \quad \begin{cases} j - i = r/2 \\ f[r + i] = f[r + j] \\ f[i + t] = f[j + t] \quad \text{for all } 0 \le t \le r/2 - 1 \end{cases}$$

Suppose that the 2-error overlap of f does not satisfy $Condition^+$. Therefore, α and β are f-free words and $d_H(\alpha, \beta) = 2$, since they differ only in positions $r + i$ and $r + j$. Moreover, the word obtained from α, by replacing $\alpha[r + i]$ with $\beta[r+i]$, ($\alpha[r+j]$ with $\beta[r + j]$, resp.) contains f as a factor; it occurs in position $r + 1$ (1, resp.). Hence, there exists no f-free transformation from α to β of length equal to $d_H(\alpha, \beta)$.

On the other hand, if the 2-error overlap of f satisfies $Condition^+$, then α is still f-free, but β contains one, and only one, occurrence of f that starts in position $r/2+1$. Now let k_1 and k_2 such that $i - 1 = (k_1r/2) + q$ and $n - j = (k_2r/2) + h$, for some $0 \le q, h \le r/2 - 1$. Then $k_1 \ge 0$ and $k_2 \ge 2$.

In case $k_1 \ge 1$, f has another 2-error overlap, namely the one of length $n - (k_1 + k_2)r/2$, that does not satisfy $Condition^+$. A construction analogous to the one used for α and β, leads to a pair of f-free words μ and ν for which there does not exist any f-free transformation of length equal to their Hamming distance; hence, f is bad.

In case $k_1 = 0$, set $\overline{k} = (r/2) + j$, and denote by $f^{(i)}$ the word obtained from f by replacing $f[i]$ with $f[r+i]$, and by $f^{(j,\overline{k})}$ the word obtained from f by replacing $f[j]$ and $f[\overline{k}]$ with $f[r+j]$ and $f[i]$, respectively. Define $\eta = pre_r(f)f^{(i)}suf_{r/2}(f)$ and $\gamma = pre_r(f)f^{(j,\overline{k})}suf_{r/2}(f)$. Note that η and γ disagree in positions $r+i$, $r+j$ and $(3r/2)+j$. Moreover, the words obtained from η by replacing $\eta[r+i]$ with $\gamma[r+i]$, or $\eta[r+j]$ with $\gamma[r+j]$, or $\eta[(3r/2)+j]$ with $\gamma[(3r/2)+j]$, all contain f as a factor; it occurs in position $r+1$, 1 and $(3r/2)+1$, respectively. Thus, no f-free transformation from η to γ of length equal to their Hamming distance exists and f is bad. Note also that, in this case, since $k_1 = 0$ and $k_2 \geq 2$, we have $3r/2 \leq n = |f|$ and, therefore, $|\eta| = |\gamma| \leq 2n = 2|f|$. $\square$

Example 11. Let $f = ATC \in \Delta^3$. The word f has a 2-error overlap of length 2; here $n = 3, r = 1$, and $n - r = 2$. Hence, ATC is bad from Theorem 10. Let us exhibit two f-free words α and β with no f-free transformation of length equal to $d_H(\alpha, \beta)$, following the proof of Theorem 10. We have that $pre_2(f)$ disagrees from $suf_2(f)$ in positions $i = 1$ and $j = 2$, and $f[i] = A$, $f[j] = T$, $f[r+i] = T$ and $f[r+j] = C$. Then, $\alpha = pre_r(f)f^{(i)} = ATTC$ and $\beta = pre_r(f)f^{(j)} = AACC$. The words α and β are f-free words and $d_H(\alpha, \beta) = 2$. Moreover, there is no f-free transformation from α to β of length $d_H(\alpha, \beta) = 2$. Indeed, if we replace $\alpha[2]$ with $\beta[2] = A$ then f occurs in position 2 of α; if we replace $\alpha[3]$ with $\beta[3] = C$ then f occurs in position 1.

Example 12. Let $f = AACC \in \Delta^4$. The word f has a 2-error overlap of length 2. Hence, $f = AACC$ is bad from Theorem 10. Note that the 2-error overlap of f satisfies $Condition^+$. Indeed, we have $i = 1, j = 2, r = 2, f[r+i] = f[3] = C = f[4] = f[r+j]$ and $f[i] = A = f[j]$. Therefore, following the proof of Theorem 10 in the case that the 2-error overlap of f satisfies $Condition^+$ and $k_1 = 0$, let us set $\overline{k} = (r/2)+j = 3$ and let us consider the two words $\eta = pre_r(f)f^{(i)}suf_{r/2}(f) = AACACCC$ and $\gamma = pre_r(f)f^{(j,\overline{k})}suf_{r/2}(f) = AAACACC$. The words η and γ are f-free words and $d_H(\eta, \gamma) = 3$. Moreover, there is no f-free transformation from η to γ of length $d_H(\eta, \gamma) = 3$. Indeed, if we replace $\eta[3]$ with $\gamma[3] = A$ then f occurs in position 3 of η; if we replace $\eta[4]$ with $\gamma[4] = C$ then f occurs in position 1 and if we replace $\eta[5]$ with $\gamma[5] = A$ then f occurs in position 4.

The definitions of bad and non-isometric words are equivalent for binary words. Let us investigate the case of k-ary words, with a generic $k \geq 2$. Observe that the main reason why the equivalence holds true for binary words is that the ring Q_1^2 is just an edge, hence it is a complete graph. Therefore, $d_{Q_n^2}(u, v) = d_L(u, v) = d_H(u, v)$ for any $u, v \in B^n$. In a similar way, the ring Q_1^3 is a triangle, hence it is a complete graph. Therefore, $d_{Q_n^3}(u, v) = d_L(u, v) = d_H(u, v)$ for any $u, v \in \Sigma^n$. Let us state this result.

Proposition 13. *Let $|\Sigma| \leq 3$ and $f \in \Sigma^*$. Then, f is bad iff f is non-isometric.*

The notions of bad and non-isometric word no longer coincide for words on a quaternary alphabet. The proof is given by Examples 15 and 16 which show a quaternary non-isometric and good word ($AGAC$) and a quaternary isometric and bad word (ATC).

Proposition 14. *There exist a word in Δ^* which is non-isometric and good, and a word in Δ^* which is isometric and bad.*

Example 15. Let $f = AGAC \in \Delta^*$. The word f has no 2-error overlap and thus it is good by Theorem 10. On the other hand, $AGAC$ is non-isometric. In fact, consider vertices $w = AGAAAC$ and $w' = AGATAC$ in Q_6^4. Their distance is 2, because there are two paths of length 2 linking them, namely $(AGAAAC, AGACAC, AGATAC)$ and $(AGAAAC, AGAGAC, AGATAC)$, but no edge, even if they differ only in one position.

Note that $w = AGAAAC$ and $w' = AGATAC$ are f-free, and hence they are vertices of $Q_6^4(f)$, too. To the contrary, $AGACAC$ and $AGAGAC$ both contain f as factor, and thus they are not vertices of $Q_6^4(f)$. Hence, there is no path of length 2 in $Q_6^4(f)$ connecting w and w', and then $d_{Q_6^4(f)}(AGAAAC, AGATAC) > 2$. Therefore, $Q_6^4(f)$ is not an isometric subgraph of Q_6^4, and f is not isometric.

Example 16. Let $f = ATC \in \Delta^*$. The word f has a 2-error overlap, the one of length 2, and thus it is bad by Theorem 10. On the other hand, ATC is isometric.

Suppose by contradiction that it is non-isometric. In view of Lemma 6, there exist $d \geq 1$ and two f-free words $u, v \in \Delta^d$, such that there does not exist any f-free L-transformation from u to v of length equal to $d_L(u, v)$. Without loss of generality, let u and v be of minimal Lee distance $d_L(u, v)$. Let $i_1 < i_2 < \ldots i_m$ be the error positions, i.e., the positions where u and v differ. The minimality of $d_L(u, v)$ implies that, in any L-transformation from u to v of length $d_L(u, v)$, each time a symbol of u in one of these positions, say i, is replaced, then an occurrence of ATC appears, either in position i, or $i - 1$, or $i - 2$ (see Remark 9).

Consider any error position, say i. If $u[i]$ and $v[i]$ are complementary symbols, then the replacement of $u[i]$ by $v[i]$ can be accomplished without factor ATC appearing. In fact, if $u[i]$ and $v[i]$ are A and T, then two replacements are needed on this position (see Remark 7), and going through G, the factor ATC cannot appear. Otherwise, if $u[i]$ and $v[i]$ are C and G, if a factor ATC appears when going through A, then going through T it will not; and vice versa. Therefore, any error position in u and v, does not involve two complementary symbols.

Consider now i_1, the leftmost error position and suppose first that $u[i_1] = A$. The symbol A cannot be (directly) replaced by T (because $d_L(A, T) \neq 1$), nor by G, since this replacement cannot let ATC appear. Let us show that it cannot even be replaced by C. In fact, in this case, an occurrence of ATC appears from position $i_1 - 2$. Then, since neither $i_1 - 2$ nor $i_1 - 1$ is an error position, factor ATC could no longer be removed, against the f-freeness of v. In view of Remark 8, not even $v[i_1] = A$ can hold true.

An analogous reasoning shows that $u[i_1]$ and $v[i_1]$ cannot be T. Finally, $u[i]$ and $v[i]$ cannot be neither C nor G. In fact, if $u[i]=C$ then $v[i]$ cannot be A or T, as just proved, nor G, since the case of complementary symbols has been excluded before.

Observe that the non-isometricness of $AGAC$ and the isometricness of ATC have been proved in Examples 15 and 16 by *ad-hoc* techniques. The proofs will become much more simple, by applying Theorem 23 in the next section.

4   Quaternary Isometric and Non-isometric Words

In this section, we investigate the case of (non-) isometric words over an alphabet of cardinality $k = 4$, as the first value of k such that the definitions of bad and non-isometric word no longer coincide. Some of the results in this section will be achieved applying known properties of binary (non-) isometric words, in view of the following remark.

Remark 17. The graphs Q_n^4 and Q_{2n}^2 are isomorphic, for all n. The isomorphism is obtained by the Gray map g given by $g(A) = 00$, $g(C) = 01$, $g(T) = 11$, and $g(G) = 10$ (see [3]).

The Gray morphism is a bijection from Δ^* to $(B^2)^*$. Hence, in this section we will consider binary words of even length, only. A word $f \in \Delta^*$ will be possibly denoted as $(f)_4$ to stress its belonging to the quaternary alphabet, while the corresponding word $g(f) \in B^*$ will be denoted as $(f)_2$. The following question naturally arises.

Question 18. Does $(f)_2$ non-isometric imply $(f)_4$ non-isometric, or vice versa?

To completely answer the question we will need the characterization of non-isometric quaternary words in Theorem 23. Now, let us show a partial result useful in the sequel.

Proposition 19. *Let $(f)_4 \in \Delta^*$ and $(f)_2 \in (B^2)^*$. If $(f)_2$ is non-isometric and it has a 2-error overlap of even length then $(f)_4$ is non-isometric.*

Proof. Let $(f)_2 \in B^{2n}$ be a non-isometric word with a 2-error overlap of length $2l$.

The sequel of the proof refers to the proof of Theorem 10 for the notations and the construction of a pair of $(f)_2$-free words which witness the non-isometricness of $(f)_2$, starting from a given 2-error overlap. The pair is referred to either as $((\alpha)_2, (\beta)_2)$, or $((\mu)_2, (\nu)_2)$, or $((\eta)_2, (\gamma)_2)$, as appropriate. Let us show that the corresponding pairs in Δ^* witness that $(f)_4$ is non-isometric.

Note that, since $(f)_2$ has a 2-error overlap of even length $2l$ then $(f)_4$ has either a 1-error or a 2-error overlap of length l. More exactly, suppose that $pre_{2l}((f)_2)$ disagrees from $suf_{2l}((f)_2)$ in positions i and j, with $i < j$. If i is odd and $j = i+1$, then $(f)_4$ has a 1-error overlap, where the error is caused by two complementary symbols. Otherwise, $(f)_4$ has a 2-error overlap where any error is given by two non-complementary symbols. Then, $pre_l((f)_4)$ disagrees from $suf_l((f)_4)$ in position $s = \lceil i/2 \rceil$, in the first case, and in positions $s = \lceil i/2 \rceil$ and $t = \lceil j/2 \rceil$, in the second case.

Consider now the words $(\alpha)_4$ and $(\beta)_4$ in Δ^* which correspond to $(\alpha)_2$ and $(\beta)_2$.

If the 2-error overlap of $(f)_2$ does not satisfy *Condition$^+$* then $(\alpha)_2$ and $(\beta)_2$ are $(f)_2$-free words, and therefore, $(\alpha)_4$ and $(\beta)_4$ are $(f)_4$-free words. Moreover, $(\alpha)_4$ and $(\beta)_4$ differ uniquely either in position $l + s$, because of two complementary symbols, or in positions $l + s$ and $l + t$, because of non-complementary symbols. Suppose, by the contrary, that there exists an $(f)_4$-free L-transformation from $(\alpha)_4$ to $(\beta)_4$ of length $d_L((\alpha)_4, (\beta)_4) = 2$, and set $r = 2n - 2l$. The two changes of symbols correspond to complementing $(\alpha)_2[r + i]$ and $(\alpha)_2[r + j]$, yielding two occurrences of $(f)_2$ in $(\alpha)_2$, in positions $r + 1$ and 1, respectively. Since both $r + 1$ and 1 are odd numbers, then $(f)_4$ occurs in $(\alpha)_4$. The contradiction shows that $(f)_4$ is non-isometric in this case.

On the other hand, if the 2-error overlap of $(f)_2$ satisfies $Condition^+$, then $(\alpha)_2$ is still $(f)_2$-free, but $(\beta)_2$ contains a unique occurrence of $(f)_2$; it starts in position $r/2 + 1$. If $r/2$ is an odd integer then $r/2 + 1$ is an even value and, therefore, the occurrence of $(f)_2$ in $(\beta)_2$ starts in an even position. So $(f)_4$ does not occur in $(\beta)_4$ and the conclusion that $(f)_4$ is non-isometric follows as in the previous case. Suppose now that $r/2$ is an even integer. In this case, the occurrence of $(f)_2$ in $(\beta)_2$ starts in an odd position, $r/2 + 1$. So $(\beta)_4$ is not $(f)_4$-free. Let us construct some other pairs of words to prove that $(f)_4$ is non-isometric. Let k_1 and k_2 such that $i - 1 = (k_1 r/2) + q$ and $2n - j = (k_2 r/2) + h$, for some $0 \leq q, h \leq r/2 - 1$.

If $k_1 \geq 1$, then $(f)_2$ has another 2-error overlap of even length $2n - (k_1 + k_2)r/2$, which does not satisfy $Condition^+$. Let $(\mu)_2$ and $(\nu)_2$ be the related words. Hence, $(f)_4$ has either a 2-error or a 1-error overlap. Reasoning as for $(\alpha)_4$ and $(\beta)_4$, we have that $(\mu)_4$ and $(\nu)_4$ are $(f)_4$-free words which witness the non-isometricness of $(f)_4$.

If $k_1 = 0$, consider the words $(\eta)_4, (\gamma)_4 \in \Delta^*$ that correspond to the words $(\eta)_2$ and $(\gamma)_2$. Since $(\eta)_2$ and $(\gamma)_2$ are $(f)_2$-free words, $(\eta)_4$ and $(\gamma)_4$ are $(f)_4$-free words. Moreover, $d_L((\eta)_4, (\gamma)_4) = 3$. In fact, $(\eta)_4$ and $(\gamma)_4$ differ either in two positions (because of a pair of complementary symbols and another pair of non-complementary symbols), or in three positions (because of three pairs of non-complementary symbols). Consider, by contraposition, an L-transformation from $(\eta)_4$ to $(\gamma)_4$ of length equal to $d_L((\eta)_4, (\gamma)_4) = 3$. The three changes of symbols that are necessary to transform $(\eta)_4$ in $(\gamma)_4$, correspond to complementing $(\eta)_2[r+i]$, $(\eta)_2[r+j]$ and $(\eta)_2[(3r/2)+j]$. These changes provide three occurrences of $(f)_2$ in $(\eta)_2$; they start in positions $r + 1$, 1 and $(3r/2) + 1$, respectively. Since they are all odd positions, they yield three occurrences of $(f)_4$ in $(\eta)_4$. Hence, $(f)_4$ is non-isometric. □

Let us introduce a new definition that will be the key of the characterization of quaternary non-isometric words.

Definition 20. Let $f \in \Sigma^*$. The word f has a q-Lee-error overlap of length l if $d_L(pre_l(f), suf_l(f)) = q$.

Note that if, for some $f \in \Delta^*$, $pre_l(f)$ and $suf_l(f)$ differ in exactly one position because of a pair of complementary symbols, then f has a 2-Lee-error overlap of length l. For example, $AGAC \in \Delta^*$ has a 2-Lee-error overlap of length 2. Indeed, $d_L(AG, AC) = 2$.

Proposition 21. Let $f \in \Delta^*$. If f is non-isometric then f has a 2-Lee-error overlap.

Proof. Let $u, v \in \Delta^n$ be f-free words for which there is no f-free L-transformation from u to v of length equal to $d = d_L(u, v)$, and assume that d is as small as possible. Clearly, $d \geq 2$. Let $V = \{i_1, i_2 \ldots, i_m\}$, with $1 \leq i_1 < i_2 \ldots, i_m \leq n$, be the set of all positions in which u differs from v; note that $m \leq d$. Consider an L-transformation τ from u to v of length d; by hypothesis, it is not f-free. Further, for any $i \in V$, whatever is the letter $u[i]$ replaced in the first step of τ, the resulting word, say w_i, contains an occurrence of f (see Remark 9). Two cases are possible: there is no $i \in V$ such that $u[i]$ and $v[i]$ are complementary symbols (case 1), and the opposite one (case 2).

In case 1), we have $m = d$. The same argument used in the case of a binary alphabet (see Theorem 5.1 in [9]) shows that if f is non-isometric then it has an overlap with errors only in two distinct positions. The hypotheses of case 1) imply that f has a 2-Lee-error overlap.

In case 2), let us show the proof for $u[i] = A$ and $v[i] = T$ (the other cases go similarly). If $u[i]$ is replaced by C (G, resp.) then the resulting word contains an occurrence of f. These two occurrences of f start in different positions, but they both include position i of u. So, they have an overlap with one mismatch. This implies that f has a 2-Lee-error overlap (the error is due to a C that does not match a G). $\square$

Proposition 22. *Let $f \in \Delta^*$. If f has a 2-Lee-error overlap then f is non-isometric.*

Proof. Let $f = (f)_4 \in \Delta^n$ be a word with a 2-Lee-error overlap of length l. Consider $pre_l((f)_4)$ and $suf_l((f)_4)$. They differ either in one position, because of two complementary symbols, or in two positions, each one because of two non-complementary symbols. Since $g(A) = 00$, $g(C) = 01$, $g(T) = 11$, and $g(G) = 10$, then, in both cases, it turns out that $g(f) = (f)_2 \in B^{2n}$ has a 2-error overlap of even length $2l$. Moreover, $(f)_2$ is non-isometric and, hence, $(f)_4$ is non-isometric applying Proposition 19. $\square$

Propositions 21 and 22 prove the following characterization of quaternary non-isometric words. Theorem 23 provides a simple tool to test whether a quaternary word is isometric or not. Moreover, it allows us to answer Question 18, about the isometricness of a quaternary word and its representation as a binary word.

Theorem 23. *Let $f \in \Delta^*$. Then, f is non-isometric if and only if it has a 2-Lee-error overlap.*

Proposition 24. *Let $(f)_4 \in \Delta^*$ and $(f)_2 \in (B^2)^*$.*
If $(f)_4$ is non-isometric then $(f)_2$ is non-isometric.
If $(f)_2$ is non-isometric, then $(f)_4$ is non-isometric if and only if $(f)_2$ has a 2-error overlap of even length.

Proof. If $(f)_4$ is non-isometric then it has a 2-Lee-error overlap (Proposition 21) and $(f)_2$ has a 2-error overlap of even length (as shown in the proof of Proposition 21). Hence, $(f)_2$ is bad (Theorem 10) and non-isometric (Proposition 13).

Suppose that $(f)_2$ is non-isometric. If $(f)_2$ has a 2-error overlap of even length then $(f)_4$ is non-isometric (Proposition 19). If $(f)_4$ is non-isometric, then it has a 2-Lee-error overlap and $(f)_2$ has a 2-error overlap of even length, as shown before. $\square$

Example 25. Let $(f)_2 = 000110$ and $(f)_4 = ACG$. Theorem 10 shows that $(f)_2$ is non-isometric (or equivalently bad) because it has a 2-error overlap. Indeed, all the 2-error overlaps of $(f)_2$ are of odd length, namely 3 and 5, and Proposition 19 cannot be applied. Further, $(f)_4$ is isometric, because it has no 2-Lee-error overlap (Theorem 23).

5   The Index of Bad and Non-isometric Words

The index of a binary bad word has been introduced in [7]. In this section, we extend the notion to k-ary alphabets. Two distinct definitions of index are given, accordingly to the definitions of bad and non-isometric k-ary word. We show some bounds on the index of a word in terms of its length. Next proposition warrants the introduction of the index for k-ary words. It can be proved as the analogous result for binary words (Lemma 2.1 in [9]).

Proposition 26. *Let Σ be a k-ary alphabet and $f \in \Sigma^*$.*
If f is not d-good then, for any $d' > d$, f is not d'-good.
If $Q_d^k(f)$ is not an isometric subgraph of Q_d^k then, for any $d' > d$, $Q_{d'}^k(f)$ is not an isometric subgraph of $Q_{d'}^k$.

Definition 27. *The badness-index of a bad word $f \in \Sigma^*$, denoted by $I_{bad}(f)$, is defined as the smallest integer d for which f is not d-good. The badness-index of a good word is ∞.*

Definition 28. *The non-isometricness-index of a non-isometric word $f \in \Sigma^*$, denoted by $I_{iso}(f)$, is defined as the smallest integer d for which $Q_d^k(f)$ is not an isometric subgraph of Q_d^k. The non-isometricness-index of an isometric word is ∞.*

The index of a binary bad word is bounded in [17] as follows. Let $f \in B^n$ be a bad binary word and $I(f)$ be its index. Then $n + 1 \leq I(f) \leq 2n - 1$. Let us generalize the result to k-ary alphabets.

Proposition 29. *Let $f \in \Sigma^n$ be a bad word. Then $n + 1 \leq I_{bad}(f) \leq 2n - 1$.*

Proof. The lower bound directly follows from the definition of $I_{bad}(f)$.

Recall that if f is a bad word then it has a 2-error overlap (Theorem 10). The proof allows to construct a pair of f-free words in Σ^* for which no f-free transformation of length equal to their Hamming distance exists. In some cases, the pair is (α, β), in others (μ, ν) or (η, γ). In any case, one can observe that the length of the words $\alpha, \beta, \mu, \nu, \eta, \gamma$ is strictly less than $2|f|$ and, hence, the bound follows. $\square$

In a similar way, the same bounds can be proved for the non-isometricness-index of a non-isometric word in the case $\Sigma = \Delta$. It is sufficient to use the definition of $I_{iso}(f)$ and the construction in Proposition 19.

Proposition 30. *Let $f \in \Delta^n$ be a non-isometric word. Then $n+1 \leq I_{iso}(f) \leq 2n-1$.*

Example 31. Let $f = ATC$, $\alpha = ATTC$, $\beta = AACC \in \Delta^*$ as in Example 11. Since there is no f-free transformation from α to β, then f is not d-good, where $d = 4$. From the lower bound given in Proposition 29, we have $I_{bad}(f) \geq 4$ and, therefore, $I_{bad}(f) = 4$. On the other hand, as seen in Example 16, $f = ATC$ is isometric and, hence, $I_{iso}(f) = \infty$.

Let $f' = AAT \in \Delta^*$. Consider the f'-free words $\alpha' = AAGT$ and $\beta' = AACT$. We have $d_{Q_4^4}(AAGT, AACT) = 2$, but there is no path of length 2 in $Q_4^4(f')$ connecting α' and β'. Therefore, $Q_4^4(f')$ is not an isometric subgraph of Q_4^4, hence f' is not isometric. Moreover, from the lower bound given in Proposition 30, we have $I_{iso}(f) \geq 4$ and, therefore, $I_{iso}(f) = 4$. On the other hand, $f' = AAT$ has no 2-error overlap, so it is good by Theorem 10 and, hence, $I_{bad}(f') = \infty$.

References

1. Anselmo, M., Giammarresi, D., Madonia, M., Selmi, C.: Bad pictures: some structural properties related to overlaps. In: Jirásková, G., Pighizzini, G. (eds.) DCFS 2020. LNCS, vol. 12442, pp. 13–25. Springer, Cham (2020). https://doi.org/10.1007/978-3-030-62536-8_2

2. Azarija, J., Klavžar, S., Lee, J., Pantone, J., Rho, Y.: On isomorphism classes of generalized Fibonacci cubes. Eur. J. Comb. **51**, 372–379 (2016). https://doi.org/10.1016/j.ejc.2015.05.011

3. Bose, B., Broeg, B., Younggeun Kwon, Ashir, Y.: Lee distance and topological properties of k-ary n-cubes. IEEE Trans. Comput. **44**(8), 1021–1030 (1995). https://doi.org/10.1109/12.403718

4. Harary, F., Hayes, J.P., Wu, H.J.: A survey of the theory of hypercube graphs. Comput. Math. Appl. **15**(4), 277–289 (1988). https://doi.org/10.1016/0898-1221(88)90213-1

5. Hsu, W.: Fibonacci cubes-a new interconnection topology. IEEE Trans. Parallel Distrib. Syst. **4**(1), 3–12 (1993). https://doi.org/10.1109/71.205649

6. Ilić, A., Klavžar, S., Rho, Y.: Generalized Fibonacci cubes. Discrete Math. **312**(1), 2–11 (2012). https://doi.org/10.1016/j.disc.2011.02.015

7. Ilić, A., Klavžar, S., Rho, Y.: The index of a binary word. Theor. Comput. Sci. **452**, 100–106 (2012)

8. Klavžar, S.: Structure of Fibonacci cubes: a survey. J. Comb. Optim. **25**, 505–522 (2013), https://doi.org/10.1007/s10878-011-9433-z

9. Klavžar, S., Shpectorov, S.V.: Asymptotic number of isometric generalized Fibonacci cubes. Eur. J. Comb. **33**(2), 220–226 (2012)

10. Mao, W., Nicol, D.M.: On k-ary n-cubes: theory and applications. Discrete Appl. Math. **129**(1), 171–193 (2003). https://doi.org/10.1016/S0166-218X(02)00238-X

11. Navarro, G.: A guided tour to approximate string matching. ACM Comput. Surv. **33**(1), 31–88 (2001). https://doi.org/10.1145/375360.375365

12. Pai, K.J., Chang, J.M., Wang, Y.L.: Upper bounds on the queue number of k-ary n-cubes. Inf. Process. Lett. **110**(2), 50–56 (2009)

13. Rahman, M., Alatabbi, A., Athar, T., Crochemore, M., Rahman, M.: Absent words and the (dis)similarity analysis of DNA sequences: an experimental study. BMC Res. Notes **9**, 1–8 (2016). https://doi.org/10.1186/s13104-016-1972-z

14. Wei, J.: The structures of bad words. Eur. J. Comb. **59**, 204–214 (2017)

15. Wei, J., Yang, Y., Wang, G.: Circular embeddability of isometric words. Discret. Math. **343**(10), 112024 (2020). https://doi.org/10.1016/j.disc.2020.112024

16. Wei, J., Yang, Y., Zhu, X.: A characterization of non-isometric binary words. Eur. J. Comb. **78**, 121–133 (2019)

17. Wei, J., Zhang, H.: Proofs of two conjectures on generalized Fibonacci cubes. Eur. J. Comb. **51**, 419–432 (2016). https://doi.org/10.1016/j.ejc.2015.07.018

Strings from Linear Recurrences: A Gray Code

Elena Barcucci[ORCID], Antonio Bernini[✉][ORCID], and Renzo Pinzani

Dipartimento di Matematica e Informatica "Ulisse Dini",
Università degli Studi di Firenze, Viale G.B. Morgagni 65, 50134 Firenze, Italy
{elena.barcucci,antonio.bernini,renzo.pinzani}@unifi.it

Abstract. Each strictly increasing sequence of positive integers can be used to define a numeration system so that any non-negative integer can be represented by a suitable and unique string of digits. We consider sequences defined by a two termed linear recurrence with constant coefficients having some particular properties and investigate on the possibility to define a Gray code for the set of the strings arising from them.

Keywords: Gray code · Numeration system · Regular language

1 Introduction

In [4] a very simple and general system of numeration is presented, based on a strictly increasing integer sequence and an iterating division algorithm. Each non-negative integer can then be uniquely represented by a suitable string. These strings form a language over a certain alphabet and, clearly, the language depends on the selected sequence. In some cases interesting and useful representations for the numbers can be obtained. In [4] several application of different systems of numeration can be found, ranging from compressing and partitioning large dictionaries, ranking permutations with repetitions, up to designing error-insensitive codes for data transmission.

In [2] some languages arising from particular sequences (defined by a two termed linear recurrence) are analysed and in [1] a definition af a Gray code for some of them is provided. In the present paper we continue this study and define a new Gray code with Hamming distance equal to 1 for a new group of languages. More precisely, we consider the sequence whose general term is $a_m = ka_{m-1} - ha_{m-2}$, with $k > h > 0$ and initial conditions $a_0 = 1$, $a_1 = k$, and we give a Gray code for the set of the strings representing all the integers $\ell \in \{0, 1, 2, \ldots, a_m - 1\}$ for a fixed value of m, in the case of h being even.

2 Preliminaries

Given a sequence $\{a_m\}_{m \geq 0}$ of positive integers such that $a_0 = 1$ and $a_m < a_{m+1}$ for each $m \in \mathbb{N}$, let N be any non-negative integer. Consider the largest term

© Springer Nature Switzerland AG 2021
T. Lecroq and S. Puzynina (Eds.): WORDS 2021, LNCS 12847, pp. 40–49, 2021.
https://doi.org/10.1007/978-3-030-85088-3_4

a_n of the sequence such that $a_n \leq N$. More precisely, $a_n = \max\{a_m \mid a_m \leq N\}$ (for the particular case $N = 0$, see below). We divide N by a_n obtaining $N = d_n a_n + r_n$. Obviously, for the remainder r_n, it is clear that $r_n < a_n$. If we divide r_n by a_{n-1}, we get $r_n = d_{n-1}a_{n-1} + r_{n-1}$, with $r_{n-1} < a_{n-1}$. Then, iterating this procedure until the division by $a_0 = 1$ (where of course the remainder is 0), we have:

$$N = d_n a_n + r_n \qquad\qquad 0 \leq r_n < a_n ,$$

$$r_n = d_{n-1}a_{n-1} + r_{n-1} \qquad\qquad 0 \leq r_{n-1} < a_{n-1} ,$$

$$r_{n-1} = d_{n-2}a_{n-2} + r_{n-2} \qquad\qquad 0 \leq r_{n-2} < a_{n-2} ,$$

$$\cdots = \cdots\cdots\cdots\cdots \qquad\qquad \cdots\cdots\cdots\cdots$$

$$\cdots = \cdots\cdots\cdots\cdots \qquad\qquad \cdots\cdots\cdots\cdots$$

$$r_3 = d_2 a_2 + r_2 \qquad\qquad 0 \leq r_2 < a_2 ,$$

$$r_2 = d_1 a_1 + r_1 \qquad\qquad 0 \leq r_1 < a_1 ,$$

$$r_1 = d_0 a_0 .$$

The above relations imply that:

$$N = d_n a_n + d_{n-1}a_{n-1} + d_{n-2}a_{n-2} + \cdots\cdots + d_1 a_1 + d_0 a_0 . \qquad (1)$$

Expression (1) is the representation of N in the numeration system $S = \{a_0, a_1, a_2, \ldots\ldots\}$, and the string $d_n d_{n-1} \ldots d_1 d_0$ is associated to the number N (in what follows the term "representation" equivalently refers either to expression (1) or to its associated string). This method [4] can be applied to every non-negative integer and in the case $N = 0$, clearly, all the coefficients d_i are 0 (in other words the representation of 0 is simply the string 0). Moreover, we have

$$r_i = d_{i-1}a_{i-1} + d_{i-2}a_{i-2} + \cdots\cdots + d_1 a_1 + d_0 a_0 < a_i , \qquad (2)$$

for each $i \geq 0$.

It is possible to show [4] that if $N = \sum_{i \geq 0}^{n} d_i a_i$ with

$$d_i a_i + d_{i-1}a_{i-1} + \ldots + d_1 a_1 + d_0 a_0 < a_{i+1} \qquad (3)$$

for each $i \geq 0$, then the representation $N = \sum_{i \geq 0}^{n} d_i a_i$ is unique. For the sake of completeness, we recall the complete theorem:

Theorem 1. *Let $1 = a_0 < a_1 < a_2 < \ldots$ be any finite or infinite sequence of integers. Any non-negative integer N has precisely one representation in the system $S = \{a_0, a_1, a_2, \ldots\}$ of the form $N = \sum_{i \geq 0}^{n} d_i a_i$ where the d_i are non-negative integers satisfying (3).*

As an example, consider the well-known sequence of Pell numbers (sequence M1413 in [5]) $p_m = 1, 2, 5, 12, 29, \ldots$ defined by $p_0 = 1$, $p_1 = 2$, $p_m = 2p_{m-1} + p_{m-2}$. The representation of $N = 16$ is associated to the string 1020.

3   Strings from a Number Sequence

Given a sequence $\{a_m\}_{m \geq 0}$, for a fixed $m > 0$, we consider all the integers $\ell \in \{0, 1, 2, \ldots, a_m - 1\}$. According to the scheme of the previous section, the representations of the integers j with $a_{m-1} \leq j < a_m$ is $j = d_{m-1}a_{m-1} + d_{m-2}a_{m-2} + \ldots + d_0a_0$ (so that the associated string is $d_{m-1}d_{m-2}\ldots d_0$), while, following the same scheme, the remaining integers (i.e. the integers $0 \leq j < a_{m-1}$) have a representation with less than m digits. For example: the representation of $a_{m-1} - 1 = d_{m-2}a_{m-2} + \ldots + d_0a_0$ has $m - 1$ digits. For our purpose (the construction of a Gray code), we require that all the representations of the considered integers $\ell \in \{0, 1, 2, \ldots, a_m - 1\}$ have m digits, so we pad the string on the left with 0's until we have m digits: the representation of $a_{m-1} - 1$ becomes $a_{m-1} - 1 = 0a_{m-1} + d_{m-2}a_{m-2} + \ldots + d_0a_0$ (therefore, the associated string is $0d_{m-2}\ldots d_0$).

With this little adjustment, we now define the following sets:

$$\mathscr{L}_0 = \{\varepsilon\} \, ,$$

$$\mathscr{L}_m = \{d_{m-1}\ldots d_0 \mid \text{the string } d_{m-1}\ldots d_0 \text{ is the representation of an integer}$$
$\ell < a_m$ in the numeration system $\{a_n\}_{n \geq 0}$ }.
Finally, we denote by $\mathscr{L}$ the language obtained by taking the union of all the sets $\mathscr{L}_m$:

$$\mathscr{L} = \bigcup_{m \geq 0} \mathscr{L}_m.$$

We remark that each element of $\mathscr{L}_m$ has precisely m digits, so that some string $d_{m-1}\ldots d_0$ can have a prefix consisting of consecutive zeros. Moreover, each $\mathscr{L}_m$ contains precisely a_m elements (which are the representations of each $\ell \in \{0, 1, \ldots, a_m - 1\}$).

Referring to the sequence of Pell numbers $p_m = \{1, 2, 5, 12, 29, \ldots\}$ defined in Sect. 2, we have:

$$\mathscr{L}_0 = \{\varepsilon\}$$

$$\mathscr{L}_1 = \{0, 1\}$$

$$\mathscr{L}_2 = \{00, 01, 10, 11, 20\}$$

$$\mathscr{L}_3 = \{000, 001, 010, 011, 020, 100, 101, 110, 111, 120, 200, 201\}$$

$$\mathscr{L}_4 = \{0000, 0001, 0010, 0011, 0020, 0100, 0101, 0110, 0111, 0120, 0200, 0201, 1000,$$
$$1001, 1010, 1011, 1020, 1100, 1101, 1110, 1111, 1120, 1200, 1201, 2000,$$
$$2001, 2010, 2011, 2020\}$$

The strings in $\mathscr{L}_2$ are, respectively, the representations of the integers $\ell \in \{0, 1, 2, 3, 4\}$. This corresponds to the case $m = 2$ where $a_m = 5$. Note that $\mathscr{L}_2$ contains exactly $a_2 = 5$ elements.

It is not difficult to realize that the alphabet of the language $\mathscr{L}$ strictly depends on the sequence $\{a_m\}_{m \geq 0}$. In general it is possible to set an upper bound for the digits d_i. From (3), we deduce $d_i a_i < a_{i+1} - \sum_{j=0}^{i-1} d_j a_j$, so that, since the numbers are all integers:

$$d_i a_i \leq a_{i+1} - 1 - \sum_{j=0}^{i-1} d_j a_j \leq a_{i+1} - 1 \,,$$

leading to

$$d_i \leq \left\lfloor \frac{a_{i+1} - 1}{a_i} \right\rfloor . \tag{4}$$

Therefore, the alphabet for $\mathscr{L}_m$ is given by $\{0, 1, \ldots, s\}$ with

$$s = \max_{i=0,1,\ldots,m-1} \left\{ \left\lfloor \frac{a_{i+1} - 1}{a_i} \right\rfloor \right\},$$

and, denoting by Σ the alphabet for $\mathscr{L}$, we have $\Sigma = \{0, 1, \ldots, t\}$ with

$$t = \max_i \left\{ \left\lfloor \frac{a_{i+1} - 1}{a_i} \right\rfloor \right\}.$$

We now investigate on the strings deriving from integer sequences defined by the following recurrences, where $k > h > 0$, with $k, h \in \mathbb{N}$:

$$a_m = \begin{cases} 1 & \text{if } m = 0 \\ k & \text{if } m = 1 \\ k a_{m-1} - h a_{m-2} & \text{if } m \geq 2 \end{cases}$$

It is not difficult to show that $a_m < a_{m+1}$ for each $m \geq 0$, so that Theorem 1 can be applied. Moreover, from [2] we deduce that the alphabet Σ for the language $\mathscr{L}$ is $\Sigma = \{0, 1, \ldots, k-1\}$ and that the language $\mathscr{L}$ is the set constituted by all the words $w = u_r u_{r-1} \ldots u_0 \in \Sigma^*$ (w having length $r+1$, with $r \geq 0$) such that if $u_i = k-1$ then $u_{i-1} \leq k-h-1$, and if $u_{i-1} = u_{i-2} = \ldots = u_j = k-h-1$, $j > 0$, then $u_{j-1} \leq k - h - 1$.

4   A Gray Code for $\mathscr{L}_n$

We introduce some notations (as in [3]) in order to express the language $\mathscr{L}$ in an alternative recursive way.

- If α is a symbol and L is a list of words $L = (v_1, v_2, \ldots, v_s)$, then $\alpha \cdot L = (\alpha v_1, \alpha v_2 \ldots, \alpha v_s)$ is the list obtained by left concatenating α to each string of the list L;

- if i and j are symbols, then $ij \cdot L$ is the list obtained by concatenating i to each string of $j \cdot L$ (or equivalently $ij \cdot L = i \cdot (j \cdot L)$);
- if L is a list of words, $\bar{L}$ is the list in the reverse order;
- if L is a list of words, $(L)^{\underline{i}}$ is L if i is even and $\bar{L}$ if i is odd;
- if L and M are two lists, $L \circ M$ is their positions preserving union. For example, if $L = (v_1, v_2)$ and $M = (w_1, w_2)$, then $L \circ M = (v_1, v_2, w_1, w_2)$ which is considered different from (v_2, w_1, v_1, w_2);
- if $L_j, L_{j+1}, \ldots, L_{j+r}$ are lists, $\bigcirc_{\ell=0}^{r} L_{j+\ell}$ is the list $L_j \circ \ldots \circ L_{j+r}$.
- if L is a list of words, then $first(L)$ is the first element of L and $last(L)$ is the last element of L.

The set $\mathscr{L}_n$ of the of the strings of $\mathscr{L}$ having length n, lexicographically ordered, by using the above notation and operations, can be defined as:

$$
\mathscr{L}_n = \begin{cases}
(\varepsilon) & \text{if } n = 0 \\[2mm]
(0, 1, \ldots, k-1) & \text{if } n = 1 \\[2mm]
\left. \begin{array}{l}
\left(\bigcirc_{i=0}^{k-2} i \cdot \mathscr{L}_{n-1} \right) \\
\circ \left(\bigcirc_{i=0}^{k-h-2} (k-1)i \cdot \mathscr{L}_{n-2} \right) \\
\circ \bigcirc_{\ell=1}^{n-2} \left(\bigcirc_{\alpha=0}^{k-h-2} (k-1)(k-h-1)^{\ell} \alpha \cdot \mathscr{L}_{n-\ell-2} \right) \\
\circ \ (k-1)(k-h-1)^{n-1}
\end{array} \right\} & \text{if } n \geq 2.
\end{cases} \tag{5}
$$

It is rather complicated, but the idea is the following:

- There is nothing to explain for $n = 0$ and $n = 1$.
- For a given $n \geq 2$, any prefix up to $k - 2$ can be appended to any string of length $n - 1$ (this is the first line of the case $n \geq 2$ in the above formula).
- The prefix to append can start also with $k - 1$. In this case we have to pay attention to the second digit since, according to the description of $\mathscr{L}$ at the end of the previous section, the digit following $k - 1$ must be less or equal to $k - h - 1$. So, the possible prefixes of length 2 starting with $k - 1$ are $(k-1)0, (k-1)1, \ldots (k-1)(k-h-2)$, and also $(k-1)(k-h-1)$. If the chosen prefixes are different from $(k-1)(k-h-1)$, there are no restrictions for the following digits, so they can be appended to any string of length $n - 2$ (this is the second line of the case $n \geq 2$ in the above formula).
- If the prefix starting with $(k-1)(k-h-1)$ is chosen, then other digits $k-h-1$ can appear, but after the last one the following digit (if present) must be less than $k - h - 1$. So, the prefix starts with $(k-1)(k-h-1)^{\ell}$ and there are two possibilities:
  - add a digit $\alpha \leq k - h - 2$ followed by any string of length $n - \ell - 2$ (this is the third line of the case $n \geq 2$ in the above formula) or
  - add $\alpha = k - h - 1$ up to the end of the string (this is the fourth line of the case $n \geq 2$ in the above formula).

If h is even, then the strings of $\mathscr{L}_n$ can be suitably arranged so that they form a Gray code with Hamming distance equal to 1. In particular we define

$$
\mathcal{L}_n = \begin{cases}
(\varepsilon) & \text{if } n = 0 \\[2mm]
(0, 1, \ldots, k-1) & \text{if } n = 1 \\[2mm]
\left. \begin{array}{l}
\left(\bigcirc_{r=0}^{k-2} r \cdot \mathcal{L}_{n-1}^{k+r+1} \right) \\[2mm]
\circ \left((k-1) \cdot \bigcirc_{r=0}^{k-h-2} r \cdot \mathcal{L}_{n-2}^{k-h+r+1} \right) \\[2mm]
\circ \left((k-1) \cdot \bigcirc_{\ell=1}^{n-2} (k-h-1)^\ell \cdot \left(\bigcirc_{\alpha=0}^{k-h-2} \alpha \cdot \mathcal{L}_{n-\ell-2}^{k-h+\alpha+1} \right) \right) \\[2mm]
\circ (k-1)(k-h-1)^{n-1}
\end{array} \right\} & \text{if } n \geq 2.
\end{cases}
\tag{6}
$$

We note that the lists $\mathscr{L}_n$ and $\mathcal{L}_n$ contain exactly the same strings. It can be easily seen that the construction of the strings with length $n \geq 2$ is the same but in $\mathcal{L}_n$, sometimes the sublists are read in the reverse order.

We have the following proposition.

Proposition 1. *The list $\mathcal{L}_n$ is a Gray code with Hamming distance equal to one, for each $n \geq 0$.*

Proof. There is nothing to prove if $n = 0, 1$. We have to check that:

1. in lines 1, 2, 3 and 4 in the case $n \geq 2$ of Definition 6 the strings are listed in a way such that in any couple of consecutive strings they differ in only one entry;
2. the last string u of the i-th line and the first string v of the $(i+1)$-th line differ in only one entry, for $i = 1, 2, 3$.

For what the first point is concerned we proceed by induction, starting from the first line $\left(\bigcirc_{r=0}^{k-2} r \cdot \mathcal{L}_{n-1}^{k+r+1} \right)$. The list $\mathcal{L}_0$ and $\mathcal{L}_1$ are, trivially, Gray codes. Suppose that all the lists $\mathcal{L}_j$ are Gray codes up to $j = n-1$. Then, since $\mathcal{L}_{n-1}^{k+r+1}$ is a Gray code for inductive hypothesis, the list $r \cdot \mathcal{L}_{n-1}^{k+r+1}$ is a Gray code, too, as it is obtained by appending the same entry r at each string. We have now to check the Hamming distance between $last \left(r \cdot \mathcal{L}_{n-1}^{k+r+1} \right)$ and $first \left((r+1) \cdot \mathcal{L}_{n-1}^{k+r+2} \right)$, for $r = 0, \ldots, \ldots k-3$. We have:

$$
last \left(r \cdot \mathcal{L}_{n-1}^{k+r+1} \right) = r \cdot last \left(\mathcal{L}_{n-1}^{k+r+1} \right)
$$

and

$$
first \left((r+1) \cdot \mathcal{L}_{n-1}^{k+r+2} \right) = (r+1) \cdot first \left(\mathcal{L}_{n-1}^{k+r+2} \right) .
$$

Since $k + r + 1$ and $k + r + 2$ have opposite parity, it is $\mathcal{L}_{n-1}^{k+r+1} = \mathcal{L}_{n-1}$ and $\mathcal{L}_{n-1}^{k+r+2} = \overline{\mathcal{L}}_{n-1}$, or vice versa. In both cases, clearly, it is $last\left(\mathcal{L}_{n-1}^{k+r+1}\right) = first\left(\mathcal{L}_{n-1}^{k+r+2}\right)$ and the required Hamming distance is 1, since the two strings differ only in the first position.

. A similar argument can be used to show that the strings arising from the second line $\left((k-1) \cdot \bigcirc_{r=0}^{k-h-2} r \cdot \mathcal{L}_{n-2}^{k-h+r+1}\right)$ form a Gray code.

The fourth line is a single string and there is nothing to prove, while in the third line we have to check the Hamming distance between the last string arising with a certain ℓ and the first string arising with $\ell + 1$, for $\ell = 1, 2, \ldots, n - 3$:

$$last\left((k-1)(k-h-1)^\ell \cdot \left(\bigcirc_{\alpha=0}^{k-h-2}\alpha \cdot \mathcal{L}_{n-\ell-2}^{k-h+\alpha+1}\right)\right)$$
$$= (k-1)(k-h-1)^\ell \cdot last\left(\bigcirc_{\alpha=0}^{k-h-2}\alpha \cdot \mathcal{L}_{n-\ell-2}^{k-h+\alpha+1}\right)$$
$$= (k-1)(k-h-1)^\ell(k-h-2) \cdot last\left(\mathcal{L}_{n-\ell-2}^{2k-2h-1}\right)$$

and

$$first\left((k-1)(k-h-1)^{\ell+1} \cdot \left(\bigcirc_{\alpha=0}^{k-h-2}\alpha \cdot \mathcal{L}_{n-\ell-3}^{k-h+\alpha+1}\right)\right)$$
$$= (k-1)(k-h-1)^{\ell+1} \cdot first\left(\bigcirc_{\alpha=0}^{k-h-2}\alpha \cdot \mathcal{L}_{n-\ell-3}^{k-h+\alpha+1}\right)$$
$$= (k-1)(k-h-1)^{\ell+1}0 \cdot first\left(\mathcal{L}_{n-\ell-3}^{k-h+1}\right)$$
$$= (k-1)(k-h-1)^\ell(k-h-1)0 \cdot first\left(\mathcal{L}_{n-\ell-3}^{k-h+1}\right) \ .$$

Since $2k - 2h - 1$ is odd, observe that:

$$last\left(\mathcal{L}_{n-\ell-2}^{2k-2h-1}\right) = last\left(\overline{\mathcal{L}}_{n-\ell-2}\right) = first\left(\mathcal{L}_{n-\ell-2}\right) = 0 \cdot first\left(\mathcal{L}_{n-\ell-3}^{k+1}\right)$$

(the last equality being derived from Definition 6).

Recalling that h is even, we have that $first\left(\mathcal{L}_{n-\ell-3}^{k-h+1}\right) = first\left(\mathcal{L}_{n-\ell-3}^{k+1}\right)$, having $k + 1$ and $k - h + 1$ the same parity. Therefore

$$(k-1)(k-h-1)^\ell(k-h-2) \cdot last\left(\mathcal{L}_{n-\ell-2}^{2k-2h-1}\right)$$

and

$$(k-1)(k-h-1)^\ell(k-h-1)0 \cdot first\left(\mathcal{L}_{n-\ell-3}^{k-h+1}\right)$$

differ only in the $(\ell + 1)$-th position.

We now analyse the last string u of the first line and the first string v of the second line of the case $n \geq 2$ in Definition 6.

We have:

$$u = last\left(\bigcirc_{r=0}^{k-2} r \cdot \mathcal{L}_{n-1}^{k+r+1}\right) = last\left((k-2) \cdot \mathcal{L}_{n-1}^{2k-1}\right)$$

$$= (k-2) \cdot last\left(\mathcal{L}_{n-1}^{2k-1}\right) = (k-2) \cdot last\left(\overline{\mathcal{L}}_{n-1}\right)$$

$$= (k-2) \cdot first\left(\mathcal{L}_{n-1}\right) = (k-2) \cdot first\left(0 \cdot first\left(\mathcal{L}_{n-2}^{k+1}\right)\right)$$

$$= (k-2)0 \cdot first\left(\mathcal{L}_{n-2}^{k+1}\right) ,$$

while

$$v = first\left((k-1) \cdot \bigcirc_{r=0}^{k-h-2} r \cdot \mathcal{L}_{n-2}^{k-h+r+1}\right) = (k-1)0 \cdot first\left(\mathcal{L}_{n-2}^{k-h+1}\right) .$$

The quantities $k+1$ and $k-h+1$ have the same parity since h is even, so that u and v differ only in the first position.

For the second and third line we have:

$$u = last\left((k-1) \cdot \bigcirc_{r=0}^{k-h-2} r \cdot \mathcal{L}_{n-2}^{k-h+r+1}\right)$$

$$= (k-1)(k-h-2) \cdot last\left(\mathcal{L}_{n-2}^{2k-2h-1}\right)$$

$$= (k-1)(k-h-2) \cdot last\left(\overline{\mathcal{L}}_{n-2}\right) = (k-1)(k-h-2) \cdot first\left(\mathcal{L}_{n-2}\right)$$

$$= (k-1)(k-h-2)0 \cdot first\left(\mathcal{L}_{n-3}^{k+1}\right)$$

and

$$v = first\left((k-1) \cdot \bigcirc_{\ell=1}^{n-2}(k-h-1)^{\ell} \cdot \left(\bigcirc_{\alpha=0}^{k-h-2}\alpha \cdot \mathcal{L}_{n-\ell-2}^{k-h+\alpha+1}\right)\right)$$

$$= (k-1)(k-h-1)0 \cdot first\left(\mathcal{L}_{n-3}^{k-h+1}\right) .$$

Again, since h is even, the parity of $k+1$ and $k-h+1$ is the same and the strings u and v differ in the second position.

Finally, we have:

$$u = last\left((k-1) \cdot \bigcirc_{\ell=1}^{n-2}(k-h-1)^{\ell} \cdot \left(\bigcirc_{\alpha=0}^{k-h-2}\alpha \cdot \mathcal{L}_{n-\ell-2}^{k-h+\alpha+1}\right)\right)$$

$$= (k-1)(k-h-1)^{n-2}(k-h-2) \cdot last\left(\mathcal{L}_{0}^{2k-2h+1}\right)$$

$$= (k-1)(k-h-1)^{n-2}(k-h-2) ,$$

which differ only in the last position from the string $v = (k-1)(k-h-1)^{n-1}$. The proof is completed.

■

As an example, we consider the case $k = 4$, $h = 2$ and we give the Gray codes for the strings of length up to 4. In each line of the lists $\mathcal{L}_2$, $\mathcal{L}_3$, and $\mathcal{L}_4$ we grouped together the strings corresponding to each line of the case $n \geq 2$ of Definition 6. The first terms of the sequence are $1, 4, 14, 48, 164, \ldots$ which correspond to the cardinality of the showed Gray codes.

$$\mathcal{L}_0 = (\varepsilon)$$

$$\mathcal{L}_1 = (0, 1, 2, 3)$$

$$\mathcal{L}_2 = (03, 02, 01, 00, 10, 11, 12, 13, 23, 22, 21, 20,$$
$$30,$$
$$empty\ldots$$
$$31)$$

$$\mathcal{L}_3 = (0 \cdot \overline{\mathcal{L}_2}, 1 \cdot \mathcal{L}_2, 2 \cdot \overline{\mathcal{L}_2},$$
$$30 \cdot \overline{\mathcal{L}_1},$$
$$310,$$
$$311)$$
$$= (031, \cdots, 003, 103, \cdots 131, 231 \cdots, 203,$$
$$303, 302, 301, 300,$$
$$310,$$
$$311)$$

$$\mathcal{L}_4 = (0 \cdot \overline{\mathcal{L}_3}, 1 \cdot \mathcal{L}_3, 2 \cdot \overline{\mathcal{L}_3},$$
$$30 \cdot \overline{\mathcal{L}_2},$$
$$310\overline{\mathcal{L}_1}, 3110,$$
$$3111)$$
$$= (0311, \cdots, 0031, 1031, \cdots, 1311, 2311, \cdots, 2031,$$
$$3031, \cdots, 3003,$$
$$3103, 3102, 3101, 3100, 3110,$$
$$3111)$$

5   Conclusion and Further Developments

In the present paper we considered the recurrence relation defined by $a_m = ka_{m-1} - ha_{m-2}$ with $k > h > 0$, which leads to a strictly increasing sequence defining a language $\mathcal{L}$ over the alphabet $\{0, 1, 2, \ldots, k - 1\}$. In the case h even, a general recursive method defining a Gray code $\mathcal{L}$ with Hamming distance 1 for the language $\mathcal{L}$ is given, so extending the results obtained in [1]. The case h odd is still open, indeed we were not able to find a general recursive method to generate a Gray code with Hamming distance 1.

References

1. Barcucci, E., Bernini, A., Pinzani, R.: A gray code for a regular language. In: Ferrari, L., Vamvakari, M. (eds.) GASCom 2018, CEUR Workshop Proceedings, vol. 2113, pp. 87–93 (2018). http://ceur-ws.org/Vol-2113/
2. Barcucci, E., Rinaldi, S.: Some linear recurrences and their combinatorial interpretation by means of regular languages. Theoret. Comput. Sci. **255**, 679–686 (2001)
3. Bernini, A., Bilotta, S., Pinzani, R., Sabri, A., Vajnovszki, V.: Gray code orders for q-ary words avoiding a given factor. Acta Inform. **52**, 573–592 (2015)
4. Fraenkel, A.S.: Systems of numeration. Amer. Math. Monthly **92**, 105–114 (1985)
5. Sloane, N.J.A.: The online encyclopedia of integer sequences (2003). http://oeis.org

String Theories Involving Regular Membership Predicates: From Practice to Theory and Back

Murphy Berzish[1], Joel D. Day[2], Vijay Ganesh[1], Mitja Kulczynski[3(✉)],
Florin Manea[4], and Federico Mora[5], and Dirk Nowotka[3]

[1] University of Waterloo, Waterloo, Canada
[2] Loughborough University, Loughborough, UK
[3] Kiel University, Kiel, Germany
`mku@informatik.uni-kiel.de`
[4] University of Göttingen and Campus-Institute Data Science, Göttingen, Germany
[5] University of California, Berkeley, USA

Abstract. Widespread use of string solvers in formal analysis of string-heavy programs has led to a growing demand for more efficient and reliable techniques which can be applied in this context, especially for real-world cases. Designing an algorithm for the (generally undecidable) satisfiability problem for systems of string constraints requires a thorough understanding of the structure of constraints present in the targeted cases. In this paper, we investigate benchmarks presented in the literature containing regular expression membership predicates, extract different first order logic theories, and prove their decidability, resp. undecidability. Notably, the most common theories in real-world benchmarks are PSPACE-complete and directly lead to the implementation of a more efficient algorithm to solving string constraints.

1 Introduction

String constraint solving (for short, *string solving*) is a topic within the more general constraint solving area, where one is interested in checking the satisfiability of particular quantifier-free first order logic formulae over a structure involving string equalities, linear arithmetic over string length, and regular language membership, all built on top of string variables. While deeply rooted in algebra and combinatorics on words (more precisely, in the theory of word equations [15]), in recent years, string solving has also attained widespread interest in the formal methods community. Indeed, this model arises naturally in, e.g., tasks related to formal analysis of string-heavy programs such as sanitization and validation of inputs (cf. [27]), leading to the development of multiple string solvers such as CVC4 [4], Z3SEQ [8], Z3STR3 [5], and WOORPJE [13]. Even though these solvers are quite efficient for certain practical use cases, novel applications demand even more *efficient* and *reliable techniques*, especially for real-world inputs. Taking

© Springer Nature Switzerland AG 2021
T. Lecroq and S. Puzynina (Eds.): WORDS 2021, LNCS 12847, pp. 50–64, 2021.
https://doi.org/10.1007/978-3-030-85088-3_5

a closer look at all reported security-related vulnerabilities listed in the Common Vulnerabilities and Exposures Repositories [12], the most frequent issues are related to strings, e.g. *Cross-site Scripting*. For such an attack, an attacker inserts malicious data into an HTML document, which is usually countered by input sanitization using regular expressions. However, due to the complex requirements, coming up with correct regular expressions is error-prone. Consider for example the regular expression /[^A-Za-z0-9 .-@:/]/ taken from [10] which was used inside the PHP web application MyEasyMarket [3] to sanitize a user's input. The intention of the developer was to remove everything other than alphanumeric characters and the symbols ., -, @, :, and /. Unfortunately, this expression overlooks the special semantics of - within a regular expression. Instead of listing all unwanted symbols individually (.-@), this regex specifies the union of all characters between . and @. Since < is within this range, an attacker can inject HTML elements which bypass the sanitizer. The correct regular expression using proper escaping has the form /[^A-Za-z0-9 .\-@:/]/. Detecting these kinds of errors is extremely hard. This is where modern string solvers come into play. Based on a proper specification, a string solver that handles regular expression membership predicates is able to reveal human mistakes as seen above.

Theories containing string constraints have been studied for decades. In [25] Makanin proved that the satisfiability of word equations is decidable. Recently, Jeż [21] showed that word equations can be solved in non-deterministic linear space. In [26] a reduction from the more powerful theory of word equations with linear length constraints (i.e., linear relations between word lengths) to Diophantine equations is shown. Whether this extended theory of word equations is decidable remains a major open problem. Solely considering the theory of regular expression membership predicates, an elegant proof of their decidability is given in [1]. The theory of word equations and regular expression membership predicates is known to be decidable [24]. It is not known whether the satisfiability problem for string constraints involving all aforementioned theories is decidable or not. However, already in the presence of other simple and natural constraints, like string-number conversion, this problem becomes undecidable (cf. [14]).

Driven by practical relevance and the need of more efficient algorithms, we analysed 56993 string solving instances from industrial applications and solver developers containing regular expression membership predicates, gathered in [22], and identified numerous relevant sub-theories based around regular membership predicates. In particular, we identified theories which may have a string-number conversion predicate numstr (contains pairs of integers and their string representation), a string length function and/or string concatenation, and prove decidability resp. undecidability for certain sub-theories. One benefit arising from this analysis is the observation that the sub-theory occurring most frequently within the benchmarks is actually PSPACE-complete. Most notably, these results lead to an algorithm implemented within Z3STR3 showing superior performance compared to its competitors [7]. The algorithm itself was directly informed by the ideas we used in proofs of the theorems presented in this work. Within this paper we show that the theory of complement-free-regular expression membership predicates, with linear length constraints and concatenation is PSPACE-complete. Furthermore, if we additionally allow complement, we prove decidability and a $\mathsf{NSPACE}(f(n))$ lower bound, where $f(n)$ is a tetration $2 \uparrow^h (cn)$ whose height h depends on the number of stacked complements (and c is a constant). Continuing this trail, we prove PSPACE-completeness for the theory of complement-free regular expression membership predicates and a string-number conversion predicate, which naturally leads to decidability when considering complements. We show corresponding lower bounds in this case too. At the opposite end of our spectrum, we show that the theory of regular expression membership predicates, linear length constraints, concatenation and string-number conversion is in fact undecidable.

To summarize, our analysis of the benchmarks not only revealed these theories, but also shows that most considered real-world string constraints actually fall into a decidable fragment. Out of 56993, about 51% lay in a decidable fragment. Only considering string constraints without word equations (30540 of 56993 instances), 26140 of these instances (85%) fall into a decidable fragment. Therefore, our theoretical analysis gives an intuition wrt. the performance of our solver.

2   Preliminaries

Let $\mathbb{N}$ be the set of natural numbers (including 0). By $\mathrm{dom}(r)$ we denote the domain of a function r. An *alphabet* Δ is a set of symbols, whereas $a \in \Delta$ are called *letters*. By Δ^* we denote the set of all finite words over Δ and let $\varepsilon \in \Delta^*$ denote the *empty word*. For $n \in \mathbb{N}$ let $w = a_1 \dots a_n \in \Delta^*$ be a word, i.e. a finite sequence. By $w[i] = a_i$ we refer to the letter in the ith position of w. Let $|w| = n$ denote the *length* of a word w. Let Δ' be an alphabet. A mapping $h : \Delta^* \to \Delta'^*$ satisfying $h(uv) = h(u)h(v)$ for all $u, v \in \Delta^*$ is called a *morphism*. In particular, for a morphism h we have $h(\varepsilon) = \varepsilon$ and by defining h for each $a \in \Delta$ the mapping is completely specified.

A finite automaton is a structure $A = (Q, \Delta, \delta, q_0, F)$ where Q is the set of states, Δ an alphabet, $\delta : Q \times \Delta \to 2^Q$ a transition function, $q_0 \in Q$ the initial state, and $F \subseteq Q$ a set of accepting states. We call A a *deterministic finite automaton* (DFA) if for all $q \in Q$ and $a \in \Delta$ we have $(q, a) \in \mathrm{dom}(\delta)$ and $|\delta(q, a)| = 1$. Otherwise, A is a *non-deterministic finite automaton* (NFA). We say A *accepts* a word $w \in \Delta$ if there is a path via δ leading from q_0 to some $f \in F$ (shortly $w \in L(A)$). We define regular expressions RegEx_Δ over three operations, namely concatenation $\cdot : \mathsf{RegEx}_\Delta \times \mathsf{RegEx}_\Delta \to \mathsf{RegEx}_\Delta$, union $\cup : \mathsf{RegEx}_\Delta \times \mathsf{RegEx}_\Delta \to \mathsf{RegEx}_\Delta$, and Kleene star$^* : \mathsf{RegEx}_\Delta \to \mathsf{RegEx}_\Delta$. On top of these operations we define the set of regular expressions RegEx_Δ inductively as follows: we have $\varepsilon, \emptyset, a \in \mathsf{RegEx}_\Delta$ for $a \in \Delta$. Given $R_1, R_2 \in \mathsf{RegEx}_\Delta$ we have $R_1 \cdot R_2, R_1 \cup R_2, R_1^* \in \mathsf{RegEx}_\Delta$. The semantics $L : \mathsf{RegEx}_\Delta \to 2^{\Delta^*}$ are given by $L(a) = \{ a \}$ for $a \in \Delta \cup \{ \varepsilon \}$, $L(\emptyset) = \emptyset$. For $R_1, R_2 \in \mathsf{RegEx}_\Delta$, let $R_1 \cdot R_2 = \{\, \alpha \cdot \beta \mid \alpha \in L(R_1), \beta \in L(R_2) \,\}$, $R_1 \cup R_2 = L(R_1) \cup L(R_2)$, and $L(R_1^*) = L(R_1)^*$.

We shall generally distinguish between two alphabets, namely a finite set $A = \{\, a, b, c, \dots \,\}$ called *terminals* or *constants* and a possibly infinite set $\mathcal{X} = \{\, x_1, x_2, \dots \,\}$ called *variables* such that $A \cap \mathcal{X} = \emptyset$. We call a word $\alpha \in (\mathcal{X} \cup A)^*$ a *pattern*. Let $\mathtt{Pat}_A = (\mathcal{X} \cup A)^*$ denote the set of all patterns and $\mathtt{vars}\,(\alpha) \subseteq \mathcal{X}$ denotes the set of all variable occurring in α.

Where not specified otherwise, we shall rely on the basic logical definitions and notations as presented in [16]. We consider first-order logical theories of the Σ_1 fragment. Whenever the connection of constants c^A, functions f^A, or relations R^A to a $\mathcal{V}$-structure is clear from context we omit the superscript^A and simply write c, f, and R, instead of c^A, f^A, and R^A, respectively. Let $\mathcal{A}$ be a $\mathcal{V}$-structure having the domain A. An *assignment* $h : A \cup \mathcal{X} \to A$ is a morphism such that $h(\mathsf{x}) \in A^*$ and $h(\mathsf{c}) = c^A$ holds. The morphism naturally extends to $\mathtt{Pat}_A$. Let $\mathcal{H}_\mathcal{A} = \{\, h \mid h : \mathtt{Pat}_A \to A \text{ morphism}, \forall\, c \in A : h(c) = c^A \,\}$ denote the set of all assignments. We call a $\mathcal{V}$ formula φ in a $\mathcal{V}$-structure $\mathcal{A}$ *satisfiable* if there exists an assignment $h \in \mathcal{H}_\mathcal{A}$ such that $\mathcal{A}, h \models \varphi$ holds and use $\mathcal{A} \models \varphi$ as a short form. In this case we also call h a *solution* to φ. Consequently, we call φ *unsatisfiable* if there does not exist an assignment $h \in \mathcal{H}_\mathcal{A}$ such that $\mathcal{A}, h \models \varphi$ holds and shortly write $\mathcal{A} \not\models \varphi$. A set $\Phi \subseteq \mathsf{FO}(\mathcal{V})$ of $\mathcal{V}$ formulae is satisfiable within a $\mathcal{V}$-structure $\mathcal{A}$ if there exists an assignment $h \in \mathcal{H}_\mathcal{A}$ such that $\mathcal{A}, h \models \varphi$ holds for all $\varphi \in \Phi$ and we denote this by $\mathcal{A} \models \Phi$. Otherwise, the set of formulae Φ is unsatisfiable within the $\mathcal{V}$-structure $\mathcal{A}$ ($\mathcal{A} \not\models \Phi$). As commonly known, the Σ_1 fragment is as expressive as the quantifier-free fragment of the corresponding theory, and we refer to the quantifier-free fragment whenever we are talking about a specific assignment.

The theory of *word equations* is built on top of the vocabulary $\mathcal{W} = \{\, \cdot /\!\!/ 2, \dot{\varepsilon} \,\}$ having the axioms of $(\mathtt{Pat}_A, \cdot^A, \varepsilon)$ forming a monoid. We consider the $\mathcal{W}$-structure $\mathcal{A}^{\doteq} = \{\, A^*, \cdot^A, \dot{\varepsilon}^A \,\}$, whereas $\cdot^A$ is defined as the concatenation of words. For $\mathcal{W}$ terms $\alpha, \beta \in \mathtt{Pat}_A$ the atom $\alpha \doteq \beta$ is called a *word equation*. Let $h \in \mathcal{H}$. The semantics of a word equation $\alpha \doteq \beta$ are induced through h by $h(\alpha) = h(\beta)$, meaning h unifies both sides of the word equation.

The basis theory involving a regular expression membership predicate called *simple regular expressions* is defined on top of the vocabulary $\mathcal{R}_s = \{\, \cdot /\!\!/ 2, \cup /\!\!/ 2,$ $^*/\!\!/ 1, \dot{\in} /2, \dot{\emptyset}, \dot{\varepsilon} \,\}$ being axiomatized as 1. the existence and associativity of a neutral element $\dot{\varepsilon}$ of $\cdot /\!\!/ 2$, 2. the existence, associativity, and commutativity of a neutral element $\dot{\emptyset}$ and idempotents, 3. the distributivity, 4. the annihilation by $\dot{\emptyset}$, We consider the many-sorted $\mathcal{R}_s$-structure $\mathcal{A}_s = \{\, \mathtt{RegEx}_A, A^*, \cdot^A, \cdot^{\mathtt{RegEx}_A}, \cup^{\mathtt{RegEx}_A},$ $^{*\mathtt{RegEx}_A}, \dot{\emptyset}^{\mathtt{RegEx}_A}, \dot{\varepsilon}^{\mathtt{RegEx}_A}, \dot{\in}^{A\,\mathtt{RegEx}_A} \,\}$. Our regular expression operations and constants over $\mathtt{RegEx}_A$ are defined as given before. The semantics of our relation $\dot{\in}^{A\,\mathtt{RegEx}_A}$ is defined by $\alpha \dot{\in}^{A\,\mathtt{RegEx}_A} R$ iff there exists a solution $h \in \mathcal{H}_A$ s.t. $h(\alpha) \in L(R)$ for $\alpha \in A^* \cup \mathcal{X}$ and $R \in \mathtt{RegEx}_A$. Both theories can be combined by considering the union of their components and denote the structure by $\mathcal{A}_s^{\doteq}$.

3 From Practice to Theory

During the development of an extension to cope with regular membership constraints within our SMT solver Z3str3RE [7] we analysed a huge set of over 100,000 industrial influenced benchmarks gathered by the authors of ZaligVinder [22] and identified 22425 instances containing at least one regular expression membership constraint. This set includes instances from the AppScan [31], BanditFuzz,[1] JOACO [29], Kaluza [27], Norn [1], Sloth [19], Stranger [30], and Z3str3-regression [5] benchmarks. Additionally we generated 19979 benchmarks based on a collection of real-world regex queries collected by Loris D'Antoni from the University of Wisconsin, Madison, USA. Thirdly, we applied StringFuzz's [9] transformers to instances supplied by Amazon Web Services related to security policy validation to obtain roughly 15000 instances. All benchmarks follow the widely used SMT-LIB Standard [8] commonly used by SMT string solvers. More details on the selected benchmarks are available in Section 5.2 in [7].

We analysed the benchmarks according to their structure, as well as predicates and functions, by using a small parser for the input formulae, allowing us to observe used operations and structural properties of the regular expressions. We identified sets which contain string-number conversion, string concatenation, and/or linear length constraints over variables used within the regular expression membership predicate. The benchmarks contained combinations of these operations. The goal was now to group them into different first order logic theories, which will be introduced in the next section.

The Resulting First Order Logic Theories. The basis of the following theories is built by $\mathcal{A}_s$, the theory of simple regular expressions. While categorising the benchmarks, we identified four important, (partially) disjoint theories, forming extensions of the aforementioned theory. The vocabulary of *extended regular expressions* is given by $\mathcal{R}_e = \mathcal{R}_s \cup \{ \overline{}/\!/1 \}$. In principle, it adds the complement to our basis. The many-sorted $\mathcal{R}_e$-structure $\mathcal{A}_e = \mathcal{A}_e \cup \{ \overline{}^{\text{RegEx}_A} \}$ therefore simply adds the complement having the semantics $L(\overline{R_1}) = L(R_1^*) \setminus L(R_1)$ to the theory $\mathcal{A}_s$. Let RegExC_A denote the set of all regular expressions including complement, inductively defined as seen above.

Furthermore, in practice solutions to variables are often restricted by linear inequalities ranging over the length of potential solutions. Therefore a natural extension is adding a function to our vocabularies allowing us to reason about length. Let $\mathcal{R}_{il} = \mathcal{R}_i \cup \{ \mathbb{Z}, +/\!/2, \leq/\!/2, \dot{0}, \text{len}/\!/1 \}$ be a vocabulary where $i \in \{ e, s \}$, being characterised by previously defined axioms and additionally the associativity and commutativity of $+/\!/2$, the existence of a neutral element, and the requirement that $\leq$ be a total ordering and monotonic on our domain. The many-sorted $\mathcal{R}_{il}$-structure of *regular expressions with length* is defined by $\mathcal{A}_{il} = \mathcal{A}_i \cup \{ +^{\mathbb{Z}}, \leq^{\mathbb{Z}}, \dot{0}^{\mathbb{Z}}, \text{len}^{A \to \mathbb{Z}} \}$, where $+^{\mathbb{Z}}, \leq^{\mathbb{Z}}$ are defined as commonly used

---

[1] The BanditFuzz benchmark was obtained via private communication with the authors.

operations over $\mathbb{Z}$, $\dot{0}^{\mathbb{Z}} = 0 \in \mathbb{Z}$, and the length function $\mathsf{len}^{A \to \mathbb{Z}}$ for a pattern $\alpha \in \mathsf{Pat}_A$ and an assignment $h \in \mathcal{H}_A$ by $\mathsf{len}^{A \to \mathbb{Z}}(\alpha) = |h(\alpha)|$.

A third addition often occurring in real-world program analysis is a string-number conversion predicate. To this extend let $\mathcal{R}_{in} = \mathcal{R}_i \cup \{\, \mathsf{numstr}/2 \,\}$ whereas $i \in \{\, e, s, el, sl \,\}$ be a vocabulary. The axioms are derived from the corresponding base theory. The many-sorted $\mathcal{R}_{in}$-structure of *regular expressions with number conversation* is defined by $\mathcal{A}_{in} = \mathcal{A}_i \cup \{\, \mathbb{N}, \mathsf{numstr}^{\mathbb{N}\, A^*} \,\}$, whereas $\mathsf{numstr}^{\mathbb{N}\, A^*}$ is a relation, which holds for all positive integers $i \in \mathbb{N}$ and words $w \in \{\, 0, 1 \,\}^*$ where w – possibly having leading zeros – is the binary representation of i, formally defined by $\mathsf{numstr}(n, w)$ iff $w \dot{\in} (0 \cup 1)^* \wedge n \geq 0 \wedge \sum_{j \in \{\, 1, \ldots,\, |\, w\, |\,\}} w[j] \cdot 2^{|w| - j}$.

Naturally, not only in real-world applications, it is interesting to ask whether a pattern $\alpha \in \mathsf{Pat}_A$ possibly containing variables is bound by a regular language. This leads to the last extension we are considering in this work. Let $\mathcal{R}_{ic} = \mathcal{R}_i \cup \{\, \cdot /\!/ 2 \,\}$ whereas $i \in \{\, e, s, el, sl, eln, sln, en, sn \,\}$ be a vocabulary, having the additional axioms induced by $(\mathsf{Pat}_A, \cdot^A, \varepsilon)$ forming a monoid. The many-sorted $\mathcal{R}_{ic}$-structure of *regular expressions with concatenation* is defined by $\mathcal{A}_{ic} = \mathcal{A}_i \cup \{\, \cdot^A, \dot{\varepsilon}^A \,\}$, whereas $\cdot^A$ is defined as the classical concatenation over Pat_A and $\dot{\varepsilon}^A = \varepsilon \in A^*$. These theories are again naturally combined with the theory of word equations by simply considering the union of their components.

As an example, consider the string constraint $C = \mathsf{x}_1 \dot{\in} 1^* \wedge \mathsf{numstr}(15, \mathsf{x}_1) \wedge \mathsf{len}(\mathsf{x}_1) \geq 3$ where $x_1 \in \mathcal{X}$ and $1 \in A$. A solution $h \in \mathcal{H}_A$ is given by $h(\mathsf{x}_1) = 1111$, since $h(\mathsf{x}_1) = 1111 \in L(1^*)$, $\mathsf{numstr}(15, 1111)$ because 1111 is the binary representation of 15, and $h(\mathsf{x}_1) \geq 3$. Therefore $\mathcal{A}_{sln}, h \models C$.

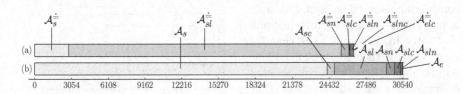

Fig. 1. Distribution of instances among their theories. (a) instances with word equations (b) instances without word equations.

Benchmark Analysis. The analysis of the 56993 instances reveals that 30540 instances are solely a member of one of our regular expression theories, while 26453 additionally contained word equations. In Fig. 1 we plot the distribution of all instances w.r.t. their theory. We display the instances according to the presence of word equations into two bars (a) and (b). The width of a single block within a bar corresponds to the instance count of the smallest theory. Since some of the theories are disjoint (e.g. $\mathcal{A}_{sl}$ and $\mathcal{A}_{sn}$) the diagram does not visualise inclusions.

Within the pure regex formulae, the most frequented theory is $\mathcal{A}_s$ holding 24256 instances. As we will see in this work, this theory and also its successor $\mathcal{A}_{sl}$ with 4327 instances are PSPACE-complete and raises hope for practically viable solving strategies. The theories $\mathcal{A}_{elnc}$ and $\mathcal{A}_{slnc}$, for which we prove undecidability within this work, do not seem to have a high relevance in application since they do not occur at all within our analysed set of benchmarks.

On the other hand, the instances containing word equations are also based around simple regular expressions. The most prominent theory is $\mathcal{A}_{sl}^{\doteq}$ holding 22604 instances, followed by $\mathcal{A}_{s}^{\doteq}$ containing 2813 instances. Unfortunately, the decidability of the largest set of instances is not known. Notably, the total set only contains 9 instances based on the theory $\mathcal{A}_e$ where the complement is actually needed. All other instances can be rewritten to simply avoid the complement.

4   Decidability of the Theories

In this section[2], we characterise the related quantifier-free first-order theories introduced in Sect. 2 according to their decidability. The contributions are summarized in Fig. 2. The arrows lead from stronger and more expressive theories to weaker ones. Theories in the upper box are undecidable, while those in the lower box are decidable (similarly, the theories within the inner dashed box are PSPACE-complete). We proceed with a summary of the theorems we prove and some discussion of the motivation and intuition for the proofs.

In an attempt to move from simpler to more complicated theories, we will begin our journey with the theory without complement operation for regular expressions. We will start by considering $\mathcal{A}_{slc}$. The motivation in approaching this theory first (formalized later in Theorem 3) is that for more general theories, which include regular expressions with complement operations, even simple tasks (like checking whether there exists a common string in the languages of two given expressions) require an exponential amount of space. One way to understand this is that the exponential blow-up with respect to the size of the regular expressions comes from transforming this expression into an NFA, determinising it, and then computing its complement. In fact, we will see that any other approach inherently leads to such an exponential blow-up. We can state the following result.

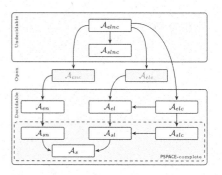

Fig. 2. Visualization of relationship and decidability of various extensions of $\mathcal{A}_s$, with arrows leading from stronger theories to theories which they contain.

Theorem 1. *The satisfiability problems for $\mathcal{A}_{slc}$ and $\mathcal{A}_{sl}$ of simple regexes, linear integer arithmetic, string length, and concatenation are PSPACE-complete.*

---

[2] All proofs can be found in our companion paper [6].

In the following we sketch the proof of this theorem. It is enough to show the statement for $\mathcal{A}_{slc}$. Assume that the input is a formula φ. We first propagate all negations top-down in the formula, so that we obtain an equivalent formula φ' which consists of a Boolean combination of atoms of the form $\alpha \doteq R$ or $\neg(\alpha \doteq R)$, where $\alpha \in \mathtt{Pat}_A$ and $R \in \mathtt{RegEx}_A$, as well as atoms encoding arithmetic constraints. Clearly, $|\varphi'| \in \mathcal{O}(|\varphi|)$. Then, we non-deterministically choose an assignment of truth values for all atoms such that φ' evaluates to true. As such, we get from our formula a list $\mathcal{L}_r$ of atoms of the form $\alpha \doteq R$ or $\neg(\alpha \doteq R)$, where $\alpha \in \mathtt{Pat}_A$ and $R \in \mathtt{RegEx}_A$, that have to be true. If an atom $\alpha \in R$ is false in our assignment, then $\mathcal{L}_r$ will contain $\neg(\alpha \doteq R)$, and if $\neg(\alpha \doteq R)$ is false in the assignment, then $\mathcal{L}_r$ will contain $\alpha \doteq R$. We similarly construct a second list $\mathcal{L}_n$ with the arithmetic linear constraints that should be true. Clearly, an assignment of the variables occurring in these two lists such that all the atoms they contain are evaluated to true exists if and only if φ is satisfiable.

Let us first neglect the polynomial space requirement. We construct the NFA M_R for each regex $R \in \mathtt{RegEx}_A$ occurring in $\mathcal{L}_r$. Following a folklore automata-theoretical approach (reminiscent of classical algorithms converting finite automata into regexes and vice versa, and also used in string solving in, e.g., [1,2]), each occurrence of a variable $x \in \mathcal{X}$ should be assigned a path in one of the NFAs M_R (if $x \in \mathtt{vars}(\alpha)$ for an atom $\alpha \doteq R$) or in the NFA $\overline{M_R}$, accepting the complement of the language accepted by M_R (if $x \in \mathtt{vars}(\alpha)$ for an atom $\neg(\alpha \doteq R)$). This assignment should be *correct*: for each atom $\alpha \doteq R$ (resp., $\neg(\alpha \doteq R)$), concatenating the paths assigned to the occurrences of the variables of α, in the order in which they occur in α, we should get an accepting path in M_R (resp., $\overline{M_R}$). Hence, it is enough to associate to each occurrence of each variable the starting and ending state of the respective paths, and then ensure that we connect these states by the same word for all occurrences of the same variable. That is, we associate to an occurrence of a string variable $x \in \mathcal{X}$ occurring in $\alpha \doteq R$ a copy of the automaton M_R with the initial and final state changed, so that they correspond to the starting and ending state on the path of M_R associated to the respective occurrence of x (and similarly for $\overline{M_R}$). So, if x_i is the ith occurrence of x in φ, then we associate an NFA $M_{x,i}$ to it. We intersect all the automata $M_{x,i}$ to obtain an NFA A_x which accepts exactly those strings which are a *correct* assignment for the variable x.

Observe now that if a word is accepted in A_x then its length is part of an arithmetic progression, from a finite set of arithmetic progressions [11,18]. Conversely, each element of these arithmetic progressions is the length of a word accepted by A_x, and the set of progressions can be computed based only on the underlying graph of the NFA A_x. Hence, we get several new linear arithmetic constraints on the length of our variables, which are satisfied if and only if there exists a *correct* assignment for the variables. We add this new set of constraints to $\mathcal{L}_n$ and then solve the resulting linear integer system with standard methods.

Finally, if, and only if, the final set of linear constraints we defined is satisfiable, then φ' and, consequently, φ are also satisfiable.

This ends the description of our decision procedure, which is based on relatively standard automata-theory techniques. To show the PSPACE-membership we use the fact that the regexes of $\mathcal{A}_{slc}$ do not have complements. Firstly, note that we can just build the NFAs for all the regexes occurring in the positive or negative atoms φ' (and not complement any of them). Once these automata are built, we do not have to explicitly construct the automata $M_{\mathsf{x},i}$ or A_{x}: we implicitly know their states and the transitions that may occur between them. Indeed, the states are tuples of states of the original NFAs M_R, and, as we do not have complements in any expression R, the number of components in each tuple is bounded by a polynomial in the size of φ; the transitions between such states can be simulated by looking at the transitions of the original NFAs. Computing (and storing) the linear constraints on the length of the *correct* assignments for x from A_{x} can also be done in polynomial space (because of the bounds on the number of states of the automata A_{x}). We obtain, as such, a system of linear arithmetic constraints with coefficients of polynomial size (w.r.t. the size of φ). Thus, solving the derived system can be done in polynomial space.

The lower bounds stated in Theorem 1 follow immediately from the PSPACE-completeness of the intersection problem for NFAs.

When we allow arbitrary complements in the regular expressions, we can still prove the decidability of the respective theories but the complexity increases.

Theorem 2. *The satisfiability problems for $\mathcal{A}_{elc}$ and $\mathcal{A}_{el}$ of regular expressions, linear integer arithmetic, concatenation, and string length are decidable.*

The idea is to use the same strategy as explained above for $\mathcal{A}_{slc}$. Since regular expressions may now contain complements, when constructing the automaton M_R associated with a regex $R \in \mathsf{RegEx}_A$ we might have an exponential blow-up in size, even if the alphabet of the regex (resp. NFA) is binary and only one complement is used (as shown, for instance, in [20]). We can no longer guarantee the polynomial space complexity of our approach, but the decidability result holds.

This theorem is supplemented by the following remark, which shows upper and, more interestingly, lower bounds for the space needed to decide the satisfiability problem for a formula in the quantifier-free theories $\mathcal{A}_{el}$ and $\mathcal{A}_{elc}$.

Remark 1. Let $g : \mathbb{N}_{>0} \times \mathbb{Q} \to \mathbb{Q}$ recursively defined by $g(1,c) = 2^c$ and $g(k+1,c) = 2^{g(k,c)}$ for $k \in \mathbb{N}_{>0}$ and $c \in \mathbb{Q}$. Informally this mapping corresponds to the following tower of powers (a.k.a. tetration) $g(k,c) = \underbrace{2^{2^{2^{\cdots^{2^c}}}}}_{k \text{ times}} = 2 \uparrow^k c$.

For a regex $R \in \mathsf{RegExC}_A$, define the complement-depth $\mathsf{cDepth} : \mathsf{RegExC}_A \to \mathbb{N}$ recursively as follows. If $R \in \{\emptyset, \varepsilon, a\}$ for $a \in A$ let $\mathsf{cDepth}(R) = 0$. Otherwise if $R \in \{R_1 \cup R_2, R_1 \cdot R_2\}$ let $\mathsf{cDepth}(R) = \mathsf{cDepth}(R_1) + \mathsf{cDepth}(R_2)$, if $R = R_1^*$ let $\mathsf{cDepth}(R) = \mathsf{cDepth}(R_1)$, and if $R = \overline{R_1}$ let $\mathsf{cDepth}(R) = 1 + \mathsf{cDepth}(R_1)$ for appropriate $R_1, R_2 \in \mathsf{RegExC}_A$. For a formula φ in the quantifier-free theory $\mathcal{A}_{elc}$ (as well as $\mathcal{A}_{el}$) we let $\mathsf{cDepth}(\varphi)$ be the maximum depth of a regex in φ.

One can show, using for instance our approach from the proofs of Theorems 1 and 2, that the satisfiability problem for formulae φ from the quantifier-free theory $\mathcal{A}_{elc}$ (and $\mathcal{A}_{el}$ as well), with size $n \in \mathbb{N}$ and $\mathsf{cDepth}(\varphi) = k \in \mathbb{N}$, is in $\mathsf{NSPACE}(f(g(k-1,2n)))$, where f is a polynomial function. However, there exists a positive rational number $c \in \mathbb{Q}$ such that the respective problem is not contained in $\mathsf{NSPACE}(g(k-1,cn))$. This lower bound follows from [28]. There, the following problem is considered: Given a regex $R \in \mathsf{RegExC}_A$, of length n, with $\mathsf{cDepth}(R) = k \in \mathbb{N}$ over an alphabet A, decide whether $L(R) = A^*$. It is shown that there exists a positive rational number c such that the respective problem cannot be solved in $\mathsf{NSPACE}(g(k,cn))$. So, deciding whether a formula φ of $\mathcal{A}_{el}$ consisting of the atoms $\alpha \,\dot{\in}\, \overline{R}$ and $\alpha \in A^*$, where $R \in \mathsf{RegExC}_A$ is a regex of length n with $\mathsf{cDepth}(R) = k - 1$, is not contained in $\mathsf{NSPACE}(g(k-1,cn))$ (note that, in this case, the length of the formula φ is also $O(n)$).

Intuitively, this lower bound shows that if the complement-depth of a formula of length n is k, then checking its satisfiability inherently requires an amount of space proportional to the value of the exponentiation tower of height $k-1$, and with the highest exponent cn. ◁

Clearly, the satisfiability problem for the quantifier-free theory $\mathcal{A}_{el}$ is also decidable according to the theorem above. Let g be defined as given in Remark 1. Based on the classical results from [28], we can derive the following theorem.

Theorem 3. *There exists a positive rational number c such that the satisfiability problem for the fragments of $\mathcal{A}_s$ and $\mathcal{A}_{sc}$ allowing only formulae of complement-depth at least k is not in* $\mathsf{NSPACE}(g(k-1,cn))$.

This theorem shows that, in fact, when deciding the satisfiability problem for the quantifier-free theories $\mathcal{A}_{elc}$ and $\mathcal{A}_{el}$ the automata-based proof we presented is relatively close to the space-complexity lower bound for this problem. Any other approach, automata-based or otherwise, would still meet the same obstacle: the space complexity of any algorithm deciding the satisfiability of formulae of complement-depth k cannot go under the $\mathsf{NSPACE}(g(k-1,cn))$ bound. This, on the one hand, explains our interest in analysing the theory $\mathcal{A}_{sl}$ (and its variants): as soon as we consider stacked complements, we are out of the PSPACE complexity class. On the other hand, this also explains the reason why in developing a practical solution for the satisfiability problem of $\mathcal{A}_{el}$ formulae within our tool Z3str3RE we use many heuristics. While the result of Theorem 2 was known from [23], our approach seems to provide a deeper understanding of the hardness of this problem (and where this stems from) and of the ways we can deal with it.

Next we consider the case when we replace the length function by the numstr predicate. The lower bound of Theorem 3 applies also to the case of $\mathcal{A}_{en}$. So one cannot hope to solve the satisfiability problem for this theory in polynomial space, as soon as we allow arbitrary complements in our regular expressions. However, we can show that the satisfiability problem for $\mathcal{A}_{en}$ is decidable, and in PSPACE when only simple regular expressions are allowed.

Theorem 4. *The satisfiability problem for $\mathcal{A}_{sn}$ (resp. for $\mathcal{A}_{en}$) of (simple) regexes and a string-number predicate is* PSPACE-*complete (resp. decidable).*

While the general idea to prove the above result is based on a similar construction to that in Theorem 1, in this case we need to use a different strategy to work with the linear arithmetic constraints (due to the fact that numstr predicates are involved, and their fundamentally different nature w.r.t. the length function). Assume that the input is a formula φ. Similar to the strategy seen in the proof of Theorem 1 we split the atoms of the formula this time into three disjoint lists. We construct the list $\mathcal{L}_r$ of atoms of the form $\alpha \,\dot{\in}\, R$ or $\neg(\alpha \in R)$, where α is a string term and R is a simple regular expression and the second list $\mathcal{L}_n$ containing a set of arithmetic linear constraints as seen as well in the proof of Theorem 1. Each atom of the form numstr(m, α) and $\neg$numstr(m, α), where m is an integer term and $\alpha \in A^* \cup \mathcal{X}$. Note, since we do not allow concatenation, α can only be a word consisting of constants or a single variable. If m is neither a variable nor a constant, we add a new integer variable x_m and replace numstr(m, α) (respectively, $\neg$numstr(m, α)) by the predicate numstr(x_m, α) (respectively, $\neg$numstr(x_m, α)) and the arithmetic atom $x_m = m$. A similar processing can be done to replace the constant strings from numstr predicates by variables. We obtain in this way a new formula φ'', still of size $O(|\varphi|)$. After this, each term in every numstr predicate is either a constant or variable of the appropriate sort.

Now, in φ'', if we have a predicate numstr(m, α) (respectively, $\neg$numstr(m, α)) where $m \in \mathbb{Z}$ is a constant, we let M be the constant string consisting of the shortest binary representation of m. We add $\alpha \,\dot{\in}\, 0^*M$ (respectively, $\neg(\alpha \,\dot{\in}\, 0^*M)$) to the list of regular constraints $\mathcal{L}_r$. We remove numstr(m, α) (respectively, $\neg$numstr(m, α)) from φ''. If we have numstr(x, α) (respectively, $\neg$numstr(x, α)) where x is an integer variable, we add $\alpha \,\dot{\in}\, 0^*\{0, 1\}^*$ (respectively, $\neg(\alpha \,\dot{\in}\, 0^*\{0, 1\}^*)$) to the regular constraints $\mathcal{L}_r$. We remove numstr(x, α) (respectively, $\neg$numstr(x, α)) from φ'', but store in a new list $\mathcal{L}_b$ the information that the binary representation of x fulfils the same regular constraints as α (e.g., if we have $\alpha \,\dot{\in}\, R$ we add $x \,\dot{\in}\, R$ as well), or, respectively, the complement of the regular constraints of α. In the latter case, it is worth nothing that if we have a restriction $\neg(\alpha \,\dot{\in}\, R)$, the binary representation of x must be in the language defined by R, so we will not obtain regular expressions with stacked complements. In this way we obtain a list of regular constraints that need to be true, a list of arithmetic linear constraints that need to be true, as well as a list of constraints stating the binary representation of certain integer variables must also fulfil the same regular constraints as certain variables.

We afterwards use the fact that deciding whether the set of linear constraints is satisfiable is equivalent to checking whether the language accepted by a finite synchronized multi-tape automaton is empty or not (see [17]) where each tape of the automaton corresponds to a variable. The entire approach is now automata-based and, once again, the key to showing the PSPACE membership is the fact that these automata can be simulated in polynomial space.

Dropping the polynomial upper bound on the space we use gives us the decidability of $\mathcal{A}_{en}$.

It is natural to ask whether the decidability result extends to the theories $\mathcal{A}_{enc}$ (and $\mathcal{A}_{snc}$), which also allow concatenation. While we leave this open, one can make interesting observations which we will highlight in the proof of the following theorem.

Theorem 5. *The satisfiability problem for $\mathcal{A}_{slnc}$ of regular expressions, linear integer arithmetic, a string-number predicate and concatenation is undecidable.*

We begin by looking at the theory $\mathcal{A}_{snc}$ and define a predicate $eqLen \subseteq \text{Pat}_A \times \text{Pat}_A$ defined by $eqLen(\alpha, \beta)$ iff $\text{len}(\alpha) = \text{len}(\beta)$ for $\alpha, \beta \in \text{Pat}_A$. We can express $eqLen(\alpha, \beta)$ as:

$$eqLen(\alpha, \beta) = (\mathsf{z} \in 1\{0\}^*)$$
$$\wedge \; \mathsf{numstr}(\mathsf{i}, \mathsf{z}) \wedge \mathsf{numstr}(\mathsf{j}, \mathsf{z}0) \wedge \mathsf{numstr}(\mathsf{n}_a, 1\alpha) \wedge \mathsf{numstr}(\mathsf{n}_b, 1\beta)$$
$$\wedge \; (\mathsf{i} \leq \mathsf{n}_a) \wedge (\mathsf{n}_a + 1 \leq \mathsf{j}) \wedge (\mathsf{i} \leq \mathsf{n}_b) \wedge (\mathsf{n}_b + 1 \leq \mathsf{j}),$$

for integer variables $\mathsf{i}, \mathsf{j}, \mathsf{n}_a, \mathsf{n}_b$ and string variables z. Indeed, for a potential assignment $h \in \mathcal{H}_{A \cup \mathsf{Z}}$, we have $h(\mathsf{i}) = 2^{len(\mathsf{z})}$ and $h(\mathsf{j}) = 2^{len(\mathsf{z})+1}$. Then, we have $h(\mathsf{n}_a) = 2^{\text{len}(\alpha)} + A$ and $h(\mathsf{n}_b) = 2^{\text{len}(\beta)} + B$, where $\mathsf{numstr}(A, \alpha)$ and $\mathsf{numstr}(B, \beta)$ are true. Therefore, $2^{\text{len}(\mathsf{z})} \leq 2^{\text{len}(\alpha)} + A < 2^{\text{len}(\mathsf{z}+1)}$ and $2^{\text{len}(\mathsf{z})} \leq 2^{\text{len}(\beta)} + B < 2^{\text{len}(\mathsf{z})+1}$. It is immediate that $\text{len}(\alpha) = \text{len}(\beta) = \text{len}(\mathsf{z})$, so our claim holds.

We can also show that the theory of word equations with regular constraints and numstr predicate is equivalent to the theory $\mathcal{A}_{enc}$.

For one direction, we need to be able to express an equality predicate between string terms $eq \subseteq \text{Pat}_A \times \text{Pat}_A$. The regular constraints as well as those involving the numstr predicate are canonically encoded. This predicate is encoded as follows:

$$eq(\alpha, \beta) = eqLen(\alpha, \beta) \wedge \mathsf{numstr}(\mathsf{i}, 1\alpha 1\beta) \wedge \mathsf{numstr}(\mathsf{j}, 1\beta 1\alpha) \wedge (\mathsf{i} = \mathsf{j}),$$

for $\alpha, \beta \in \text{Pat}_A$. Indeed, this tests for a potential assignment $h \in \mathcal{H}_{A \cup \mathsf{Z}}$ that $\text{len}(\alpha) = \text{len}(\beta)$ and $h(1\alpha 1\beta) = h(1\beta 1\alpha)$. If these are true, it is immediate that $h(\alpha) = h(\beta)$.

For the converse, it is easy to see that each string constraint $\alpha \doteq R$ (respectively, $\neg(\alpha \doteq R)$), where $\alpha \in \text{Pat}_A$ and $R \in \text{RegExC}_A$, can be expressed as the word equation $\alpha \doteq \mathsf{x}_R$, where $\mathsf{x}_R \in \mathcal{X}$ is a fresh variable, which is constrained by the regular language defined by R (respectively, by the regular language defined by $\overline{R}$). This allows us to define a stronger length-comparison predicate. We will define a predicate $leqLen \subseteq \text{Pat}_A \times \text{Pat}_A$, whose semantics is defined by

$$leqLen(\alpha, \beta) \text{ iff } \text{len}(\alpha) \leq \text{len}(\beta),$$

for $\alpha, \beta \in \text{Pat}_A$. We can express $leqLen(\alpha, \beta)$ by $leqLen(\alpha, \beta) = (\mathsf{z} \in \{0,1\}^*) \wedge eqLen(\alpha \mathsf{z}, \beta)$.

Finally, we can now move on to $\mathcal{A}_{elnc}$ and show our statement. According to [14] the quantifier-free theory of word equations expanded with numstr predicate and length function (not only a length-comparison predicate) and linear arithmetic is undecidable. Thus, if we consider $\mathcal{A}_{elnc}$, this undecidability result immediately holds according to the above.

In conclusion, $\mathcal{A}_{enc}$ and $\mathcal{A}_{eln}$ are the only fragments of $\mathcal{A}_{elnc}$ where the decidability status of the satisfiability problem remains open.

5   Conclusion

Within this work we analysed 56993 string solving benchmarks containing regular expression membership queries and identified relevant sub-theories based around regular membership predicates. It turned out that the most frequently occurring sub-theory is decidable. Notably, the ideas of these proofs directly lead to a well-performing solver for regular expression membership predicates. This paper also shows that an interleaving between theory and practice potentially leads to new interesting solutions in both worlds. Our future work will continue on this trail to obtaining relevant sub-theories used in practice, always in the hope of finding decidable sub-theories which lead to the design of new decision procedures for solving practically relevant string constraints.

References

1. Abdulla, P.A., et al.: String constraints for verification. In: Biere, A., Bloem, R. (eds.) CAV 2014. LNCS, vol. 8559, pp. 150–166. Springer, Cham (2014). https://doi.org/10.1007/978-3-319-08867-9_10
2. Abdulla, P.A., et al.: Norn: an SMT solver for string constraints. In: Kroening, D., Păsăreanu, C.S. (eds.) CAV 2015. LNCS, vol. 9206, pp. 462–469. Springer, Cham (2015). https://doi.org/10.1007/978-3-319-21690-4_29
3. Balzarotti, D., et al.: Saner: composing static and dynamic analysis to validate sanitization in web applications. In: SP, pp. 387–401. IEEE (2008)
4. Barrett, C., et al.: CVC4. In: Gopalakrishnan, G., Qadeer, S. (eds.) CAV 2011. LNCS, vol. 6806, pp. 171–177. Springer, Heidelberg (2011). https://doi.org/10.1007/978-3-642-22110-1_14
5. Berzish, M., Ganesh, V., Zheng, Y.: Z3str3: a string solver with theory-aware heuristics. In: Proceedings of FMCAD, pp. 55–59. IEEE (2017)
6. Berzish, M., et al.: String theories involving regular membership predicates: from practice to theory and back (2021)
7. Berzish, M., et al.: An SMT solver for regular expressions and linear arithmetic over string length. In: Silva, A., Leino, K.R.M. (eds.) CAV 2021. LNCS, vol. 12760, pp. 289–312. Springer, Cham (2021). https://doi.org/10.1007/978-3-030-81688-9_14
8. Bjørner, N., Ganesh, V., Michel, R., Veanes, M.: An SMT-LIB format for sequences and regular expressions. SMT **12**, 76–86 (2012)
9. Blotsky, D., Mora, F., Berzish, M., Zheng, Y., Kabir, I., Ganesh, V.: StringFuzz: a fuzzer for string solvers. In: Chockler, H., Weissenbacher, G. (eds.) CAV 2018. LNCS, vol. 10982, pp. 45–51. Springer, Cham (2018). https://doi.org/10.1007/978-3-319-96142-2_6
10. Bultan, T., Yu, F., Alkhalaf, M., Aydin, A.: String Analysis for Software Verification and Security. Springer, Cham (2017). https://doi.org/10.1007/978-3-319-68670-7
11. Chrobak, M.: Finite automata and unary languages. Theor. Comput. Sci. **47**(3), 149–158 (1986)
12. CVE: Common vulnerabilities and exposures. http://www.cve.mitre.org. Accessed 3 Jan 2021

13. Day, J.D., Ehlers, T., Kulczynski, M., Manea, F., Nowotka, D., Poulsen, D.B.: On solving word equations using SAT. In: Filiot, E., Jungers, R., Potapov, I. (eds.) RP 2019. LNCS, vol. 11674, pp. 93–106. Springer, Cham (2019). https://doi.org/10.1007/978-3-030-30806-3_8

14. Day, J.D., Ganesh, V., He, P., Manea, F., Nowotka, D.: The satisfiability of word equations: decidable and undecidable theories. In: Potapov, I., Reynier, P.-A. (eds.) RP 2018. LNCS, vol. 11123, pp. 15–29. Springer, Cham (2018). https://doi.org/10.1007/978-3-030-00250-3_2

15. Diekert, V.: More than 1700 years of word equations. In: Maletti, A. (ed.) CAI 2015. LNCS, vol. 9270, pp. 22–28. Springer, Cham (2015). https://doi.org/10.1007/978-3-319-23021-4_2

16. Ebbinghaus, H.-D., Flum, J.: Finite Model Theory. SMM. Springer, Heidelberg (1995). https://doi.org/10.1007/3-540-28788-4

17. Ganesh, V., Berezin, S., Dill, D.L.: Deciding Presburger arithmetic by model checking and comparisons with other methods. In: Aagaard, M.D., O'Leary, J.W. (eds.) FMCAD 2002. LNCS, vol. 2517, pp. 171–186. Springer, Heidelberg (2002). https://doi.org/10.1007/3-540-36126-X_11

18. Gawrychowski, P.: Chrobak normal form revisited, with applications. In: Bouchou-Markhoff, B., Caron, P., Champarnaud, J.-M., Maurel, D. (eds.) CIAA 2011. LNCS, vol. 6807, pp. 142–153. Springer, Heidelberg (2011). https://doi.org/10.1007/978-3-642-22256-6_14

19. Holík, L., Janku, P., Lin, A.W., Rümmer, P., Vojnar, T.: String constraints with concatenation and transducers solved efficiently. PACMPL 2(POPL), 1–32 (2018)

20. Hospodár, M., Jirásková, G., Mlynárčik, P.: A survey on fooling sets as effective tools for lower bounds on nondeterministic complexity. In: Böckenhauer, H.-J., Komm, D., Unger, W. (eds.) Adventures Between Lower Bounds and Higher Altitudes. LNCS, vol. 11011, pp. 17–32. Springer, Cham (2018). https://doi.org/10.1007/978-3-319-98355-4_2

21. Jeż, A.: Word equations in nondeterministic linear space. In: Proceedings of ICALP. LIPIcs, vol. 80, pp. 95:1–95:13. Dagstuhl (2017)

22. Kulczynski, M., Manea, F., Nowotka, D., Poulsen, D.B.: The power of string solving: simplicity of comparison. In: Proceedings of AST, pp. 85–88. IEEE/ACM (2020)

23. Liang, T., Tsiskaridze, N., Reynolds, A., Tinelli, C., Barrett, C.: A decision procedure for regular membership and length constraints over unbounded strings. In: Lutz, C., Ranise, S. (eds.) FroCoS 2015. LNCS (LNAI), vol. 9322, pp. 135–150. Springer, Cham (2015). https://doi.org/10.1007/978-3-319-24246-0_9

24. Lothaire, M. (ed.): Algebraic Combinatorics on Words. Cambridge University Press, Cambridge (2002)

25. Makanin, G.S.: The problem of solvability of equations in a free semigroup. Sbornik: Math.32(2), 129–198 (1977)

26. Matiyasevich, Y.V.: A connection between systems of words-and-lengths equations and Hilbert's tenth problem. Zapiski Nauchnykh Seminarov POMI 8, 132–144 (1968)

27. Saxena, P., Akhawe, D., Hanna, S., Mao, F., McCamant, S., Song, D.: A symbolic execution framework for JavaScript. In: Proceedings of SP, pp. 513–528 (2010)

28. Stockmeyer, L.J.: The Complexity of Decision Problems in Automata Theory and Logic. Ph.D. Thesis, MIT (1974)

29. Thomé, J., Shar, L.K., Bianculli, D., Briand, L.: An integrated approach for effective injection vulnerability analysis of web applications through security slicing and hybrid constraint solving. IEEE TSE 46(2), 163–195 (2018)

30. Yu, F., Alkhalaf, M., Bultan, T.: STRANGER: an automata-based string analysis tool for PHP. In: Esparza, J., Majumdar, R. (eds.) TACAS 2010. LNCS, vol. 6015, pp. 154–157. Springer, Heidelberg (2010). https://doi.org/10.1007/978-3-642-12002-2_13

31. Zheng, Y., et al.: Z3str2: an efficient solver for strings, regular expressions, and length constraints. Formal Methods Syst. Des. **50**(2), 249–288 (2016). https://doi.org/10.1007/s10703-016-0263-6

Binary Cyclotomic Polynomials: Representation via Words and Algorithms

Antonio Cafure[1,2] and Eda Cesaratto[1,2(✉)]

[1] Universidad Nacional de General Sarmiento, Buenos Aires, Argentina
[2] Consejo Nacional de Investigaciones Científicas y Técnicas, Buenos Aires, Argentina
{acafure,ecesaratto}@campus.ungs.edu.ar

Abstract. Cyclotomic polynomials are basic objects in Number Theory. Their properties depend on the number of distinct primes that intervene in the factorization of their order, and the binary case is thus the first nontrivial case. This paper sees the vector of coefficients of the polynomial as a word on a ternary alphabet $\{-1, 0, +1\}$. It designs an efficient algorithm that computes a compact representation of this word. This algorithm is of linear time with respect to the size of the output, and, thus, optimal. This approach allows to recover known properties of coefficients of binary cyclotomic polynomials, and extends to the case of polynomials associated with numerical semi-groups of dimension 2.

Keywords: Binary cyclotomic polynomials · Words · Algorithms

1 Introduction

General Description. Cyclotomic polynomials are defined as the irreducible factors in $\mathbb{Q}[x]$ of the polynomial $x^n - 1$. The identity

$$x^n - 1 = \prod_{d \mid n} \Phi_d(x) \qquad (1)$$

holds between the d-th cyclotomic polynomials Φ_d. The polynomial $\Phi_d(x)$ equals the product $\prod(x - \alpha)$ where α ranges over the primitive d-th roots of unity. It has integer coefficients, is self-reciprocal, and irreducible over $\mathbb{Q}[x]$. Its degree is $\varphi(d)$, where φ stands for Euler's totient function. The following holds, for any prime p,

$$\begin{aligned}
\Phi_{pn}(x) &= \Phi_n(x^p) &&\text{if } p \text{ divides } n, \\
\Phi_n(x)\Phi_{pn}(x) &= \Phi_n(x^p) &&\text{if } p \text{ does not divide } n.
\end{aligned} \qquad (2)$$

It is thus enough to compute Φ_d when d is squarefree. Due to the equality

$$\Phi_p(x) = x^{p-1} + \cdots + x + 1,$$

Supported by Project UNGS 30/3307, Project Stic AmSud 20STIC-06 and LIA SIN-FIN.
Dedicated to Adrián Tato Álvarez.

T. Lecroq and S. Puzynina (Eds.): WORDS 2021, LNCS 12847, pp. 65–77, 2021.
https://doi.org/10.1007/978-3-030-85088-3_6

that holds when $d = p$ is prime, the first nontrivial case occurs when d is the product of two distinct primes p and q, and defines what is called a *binary cyclotomic polynomial*.

A cyclotomic polynomial Φ_d is usually given by its dense representation, described by the vector of its coefficients in $\mathbb{Q}^{\varphi(d)+1}$. Many properties of interest hold on the dense representation of a binary cyclotomic polynomial Φ_{pq} (see [3,4,6]), for instance:

(a) the polynomial Φ_{pq} has all its coefficients in the ternary alphabet $\{-1, 0, +1\}$;

(b) its nonzero coefficients alternate their signs: i.e. after a 1 follows a -1 .

(c) for $p < q$, the maximum number of consecutive zeros in the vector of coefficient equals $p - 2$.

Many algorithms are designed for computing these polynomials, most of them being based on identities of type (1) and (2). Around 2010, Arnold and Monagan wished to study the coefficients of cyclotomic polynomials in an experimental way. They thus designed efficient algorithms that compute cyclotomic polynomials of very large order. Their work [1] is the main reference on this subject.

Our Approach. We are concerned here with the representation and the computation of binary cyclotomic polynomials Φ_{pq}. We let $m = \varphi(pq) = (p-1)(q-1)$, and we consider the *cyclotomic vector* $\boldsymbol{a}_{pq}$, (that is moreover palindromic) formed with coefficients of Φ_{pq}, from two points of view:

(i) first, classically, as a vector of $\mathbb{Q}^{m+1}$,

(ii) but also as a word of length $m + 1$ on the ternary alphabet $\mathcal{A} = \{-1, 0, +1\}$.

With the first point of view, we use classical arithmetical operations on vectors. With the second point of view, we use operations on words as cyclic permutations, concatenations and fractional powers. In fact, we use *both* points of view and then all the operations of these *two* types. We have just to check that the sum of two words of $\mathcal{A}^{m+1}$ always belongs to $\mathcal{A}^{m+1}$. We will see in Lemma 1 that this is indeed the case in our context: the "bad" cases $(+1)+(+1)$ or $(-1)+(-1)$ never occur in the addition of two symbols of $\mathcal{A}$, and the sum remains "internal".

We design an algorithm, called BCW, that *only* uses simple operations on words (concatenation, shift, fractional power, internal addition). It takes as an input a pair (p, q) of two prime numbers p, q with $p < q$, together with the quotient and the remainder of the division of q by p; this input is thus of size $\Theta(\log q)$. The algorithm outputs the cyclotomic word $\boldsymbol{a}_{pq}$ of size $\Theta(pq)$; it performs a number $\Theta(pq)$ of operations on symbols of $\mathcal{A}$. The complexity of the algorithm is thus linear with respect to the size of the output, and thus optimal in this sense. The *proof* of the algorithm is based on arithmetical operations on polynomials, of type (1) or (2), that are further *transfered* into operations on words; however, the algorithm itself *does not perform* any polynomial multiplication or division.

To the best of our knowledge, our approach, based on combinatorics of words, is quite novel inside the domain of cyclotomic polynomials. The compact representation of binary cyclotomic polynomials described in Theorem 1 appears to be new and the algorithm BCW, whose complexity is $\Theta(pq)$, is more efficient that the already existing algorithms, whose complexity is $O(pq)E(p, q)$, where $E(p, q)$ is polynomial in $\log q$ (see Sect. 4).

2   Statement of the Main Results

We first define the main operations on words that will be used, then we state our two main results: the first one (Theorem 1) describes a compact representation of the cyclotomic word, whereas the second result (Theorem 2) describes the algorithm – so called the *binary cyclotomic word algorithm* (BCW algorithm for short) – that is used to obtain it. Theorem 1 will be proven in Sect. 3 and Theorem 2 in Sect. 4.

Operations on Words. We first define the operations on words that we will use along the paper: concatenation and fractional power, cyclic permutation, addition.

Concatenation and Fractional Power. Given two words $u = u_0 u_1 \ldots u_{k-1}$ and $v = v_0 v_2 \ldots v_{\ell-1}$ from the alphabet $\mathcal{A}$, its *concatenation* is denoted by $u \cdot v$. For any $s \in \mathbb{N}$, the *s-power* of a word is denoted by v^s, and $v^0 = \epsilon$ is the empty word.

Let k and p be positive integers. For any $v \in \mathcal{A}^p$, the *fractional power* $v^{k/p}$ of v is the element of $\mathcal{A}^k$ defined as follows:

(a) For $k < p$, then $v^{k/p}$ is the *truncation* of the word v to its prefix of length k.
(b) For $k \geq p$, then $v^{k/p} = v^{\lfloor k/p \rfloor} \cdot v^{(k \,(\mathrm{mod}\, p))/p}$.

Circular Permutation. The *left circular permutation* σ is the mapping $\sigma : \mathcal{A}^p \to \mathcal{A}^p$ defined as

$$\text{if } v = v_0\, v_1\, v_2 \cdots v_{p-2}\, v_{p-1}, \text{ then} \quad \sigma(v) = v_1\, v_2 \cdots v_{p-2}\, v_{p-1}\, v_0. \tag{3}$$

The mapping σ has order p: it satisfies for $s \in \mathbb{Z}$,

$$\sigma^s = \sigma^{s \,\mathrm{mod}\, p}, \qquad \sigma^p = \mathrm{Id}.$$

Addition. The paper is based on a transfer between algebra and combinatorics on words and leads to define addition between words; we begin with the usual sum between two vectors of $\mathbb{Z}^k$, component by component: the sum $u+v$ between two words $u = u_0 u_1 \ldots u_{k-1}$ and $v = v_0 v_2 \ldots v_{k-1}$ is

$$u + v = (u_0 + v_0)(u_1 + v_1) \cdots (u_{k-1} + v_{k-1}).$$

Then, the sum of two words from $\mathcal{A}$ may have symbols not in $\mathcal{A}$, due to the two bad cases $[+1+(+1)]$ and $[(-1)+(-1)]$. However, we will prove in Lemma 1 that this will never occur in our framework. Then, an addition between two symbols of $\mathcal{A}$ will be always an *internal* operation, with a sum that stays inside $\mathcal{A}$ and coincides with the sum in $\mathbb{Z}$:

$$0 + 0 = 0; \quad 1 + 0 = 1; \quad -1 + 0 = -1; \quad -1 + 1 = 0; \quad 1 + (-1) = 0.$$

A Compact Representation of Ninary Cyclotomic Polynomials.

Theorem 1. *Let $p < q$ be prime numbers. Consider the alphabet $\mathcal{A} = \{-1, 0, 1\}$ and the left circular permutation σ defined in (3). With the $(p-1)$ words $d_0, d_1, \ldots d_{p-2} \in \mathcal{A}^p$,*

$$d_0 = 1(-1)0 \cdots 0, \qquad d_i = \sigma^q(d_{i-1}) = \sigma^{iq}(d_0), \quad for \ i \geq 1,$$

define the $(p-1)$ words $\omega_0, \omega_1, \ldots, \omega_{p-2}$,

$$\omega_0 = d_0, \qquad \omega_i = \omega_{i-1} + d_i, \quad for \ i \geq 1. \tag{4}$$

Then, the following holds:

(a) The words $\omega_0, \omega_1, \ldots \omega_{p-2}$ belong to $\mathcal{A}^p$: each addition $\omega_{i-1} + d_i$ is internal in $\mathcal{A}$. These $(p-1)$ words only depend on the pair $(p, r = q \,(\mathrm{mod}\ p))$.

(b) Let $s = \lfloor q/p \rfloor$. The cyclotomic word a_{pq} coincides with the prefix of length $\varphi(pq) + 1$ of the word

$$b_{pq} = \omega_0^s \cdot \omega_0^{r/p} \cdot \omega_1^s \cdot \omega_1^{r/p} \cdot \ldots \cdot \omega_{p-2}^s \cdot \omega_{p-2}^{r/p} \in \mathcal{A}^{(p-1)q}, \tag{5}$$

and is written in fractional power notation as

$$a_{pq} = \omega_0^{q/p} \cdot \omega_1^{q/p} \cdots \omega_{p-3}^{q/p} \cdot \omega_{p-2}^{(q-p+2)/p}.$$

The Algorithm BCW. Given two positive primes p and q with $p < q$, together with the quotient $s = \lfloor q/p \rfloor$ and the remainder $r = q \ (\mathrm{mod}\ p)$ of q by p, the BCW algorithm proceeds in three main steps. The first two steps define the precomputation phase and only depend on the pair $(p, r = q \ (\mathrm{mod}\ p))$, whereas the third step performs the computation itself and depends on the pair (q, p).

Precomputation Phase. It has two steps and computes
(i) the words d_i for $i \in [\![0, p-2]\!]$ defined in (4),
(ii) the words ω_i defined in (4) and their prefixes ω_i^r for $i \in [\![0, p-2]\!]$.

Computation Phase. It computes the word b_{pq} defined in (5) with concatening powers of $\omega_0, \ldots, \omega_{p-2}$. It depends on s. The cyclotomic word a_{pq} is obtained from b_{pq} by deleting its suffix of length $p-2$.

The BCW algorithm only performs the operations described at the beginning of this section, on words of length at most p, each one having a cost $\Theta(p)$. The precomputation phase performs

- cyclic permutations of any order of a word of length p;
- truncations of a suffix of length $\ell < p$;
- internal additions between two words of length p.

The computation phase performs concatenations of a word of length at most p (at the end of an already existing word of any length);

Theorem 2 summarizes the analysis of the complexity of the BCW algorithm. It will be proven in Sect. 4.

Theorem 2. *Suppose that the two prime numbers $p < q$ are given together with the quotient $s = \lfloor q/p \rfloor$ and the remainder $r = q \pmod p$. Then, the BCW algorithm computes the cyclotomic word $\boldsymbol{a}_{pq}$ in $\Theta(pq)$ operations on words:*

(a) *The precomputation phase only depends on the pair $(p, r = q \pmod p)$ and its cost is $\Theta(p^2)$.*

(b) *The computation phase performs $s(p-2)$ concatenations of words of length p, and its cost is $\Theta(pq)$.*

3   A Compact Representation for Binary Cyclotomic Polynomials

This section is devoted to the proof of Theorem 1 and is organized into three main parts. We first consider the particular case $p = 2$, then we describe the general case $p > 2$, and we finally prove two auxiliary results (Proposition 1 and Lemma 1) that have been used in the proof of the general case.

Particular Case $p = 2$. The polynomial Φ_{2q} satisfies the relation $\Phi_{2q}(x) = \Phi_q(-x)$. Since $\Phi_q(x) = 1 + x + \cdots + x^{q-1}$, it turns out that Φ_{2q} is a polynomial of degree $q-1$ whose nonzero coefficients take the values 1 and -1 alternatively. The equality $m + 1 = \varphi(2q) + 1 = q$ holds and entails $\lfloor (m+1)/q \rfloor = 1$. There is only one word $\boldsymbol{\omega}_0 = 1(-1)$ which has to be concatenated with itself $\lfloor q/2 \rfloor$ times. The resulting word is

$$\boldsymbol{\omega}_0^{\lfloor q/2 \rfloor} = 1(-1)1(-1) \cdots 1(-1),$$

which coincides with the vector $\boldsymbol{a}_{2q}$ of coefficients of Φ_{2q}. This ends the proof for the case $p = 2$.

General Case $p > 2$. We let $m = \varphi(pq)$. We consider the set of words $\mathcal{A}^{m+1}$ as embedded in $\mathbb{Q}^{m+1}$ because we only deal with internal additions between words. This fact that is stated as the assertion (a) of Theorem 1 is proven in Lemma 1. We then do not distinguish between words or vectors, and the notation $\boldsymbol{v} = v_0 \cdots v_m$ denotes both a word $\boldsymbol{v} \in \mathcal{A}^{m+1}$ and/or a vector of $\mathbb{Q}^{m+1}$. We write $v_{i,j}$ to denote the j-th symbol (or coordinate) of a word $\boldsymbol{v}_i$.

The proof is described along three main steps. In Step 0, we deal with linear algebra over the vector space $\mathbb{Q}^{m+1}$, introduce the shift-matrix S and recall its main properties. Then Step 1 transfers polynomial identities into linear algebra equations on $\mathbb{Q}^{m+1}$. Finally, Step 2 adopts the point of view of words.

Step 0. Starting with Linear Algebra. We consider the vector space $\mathbb{Q}^{m+1}$, and the linear map

$$S : \mathbb{Q}^{m+1} \to \mathbb{Q}^{m+1} \qquad S\boldsymbol{v} = 0v_0 \ldots v_{m-1} \quad \text{for any } \boldsymbol{v} = v_0 \ldots v_m \in \mathbb{Q}^{m+1}.$$

We denote by $\boldsymbol{e}_0, \boldsymbol{e}_1, \ldots, \boldsymbol{e}_m$ the vectors of the canonical basis of $\mathbb{Q}^{m+1}$: the word $\boldsymbol{e}_i$ is the word whose symbols satisfy $e_{i,i} = 1$ and $e_{i,j} = 0$ for $j \neq i$. We deal with the matrix associated with the linear map S in the canonical basis that is also denoted by S and called the shift-matrix. This is a nilpotent matrix of size $(m + 1) \times (m + 1)$ having 1-s along the main lower subdiagonal and 0-s everywhere else, that satisfies

$$S^{m+1} = 0 \in \mathbb{Q}^{(m+1)\times(m+1)} \quad \text{and} \quad S^\ell e_0 = e_\ell, \text{ for any } \ell \in [\![0, m]\!].$$

With a polynomial $f(x) = a_0 + a_1 x + a_2 x^2 + \cdots + a_m x^m$, we associate the matrix $f(S) = a_0 \text{Id} + a_1 S + a_2 S^2 + \cdots + a_m S^m$ (here Id is the identity matrix) together with the vector of coefficients $\boldsymbol{a} = a_0 a_1 \ldots a_m \in \mathbb{Q}^{m+1}$. The following relation holds

$$\boldsymbol{a} = a_0 S^0 e_0 + a_1 S^1 e_0 + \cdots + a_m S^m e_0 = f(S)e_0. \tag{6}$$

Moreover, $f(S)$ is invertible if and only if $a_0 \neq 0$.

Step 1. From Polynomial Identities to Linear Algebra Equations. Our starting point is the polynomial identity (2). When applied to the case when p and q are distinct primes, it writes as

$$(1 - x^q)\Phi_{pq}(x) = (1 - x)\Phi_q(x^p). \tag{7}$$

We now specialize the previous identity in the shift-matrix S of size $(m + 1) \times (m + 1)$, apply it to the vector e_0, and obtain

$$(\text{Id} - S^q)\Phi_{pq}(S)e_0 = (\text{Id} - S)\Phi_q(S^p)e_0.$$

Relation (6) entails the equality

$$\boldsymbol{a}_{pq} = \Phi_{pq}(S)e_0 = (\text{Id} - S^q)^{-1}(\text{Id} - S)\Phi_q(S^p)e_0. \tag{8}$$

Since S is a nilpotent matrix, the inverse matrix $(\text{Id} - S^q)^{-1}$ equals

$$\left(\text{Id} - S^q\right)^{-1} = \sum_{i=0}^{\lfloor (m+1)/q \rfloor} S^{iq}. \tag{9}$$

Next, for $p > 2$, the following holds

$$m + 1 = q(p - 2) + q - (p - 1) + 1, \quad 0 \leq q - (p - 1) + 1 < q,$$

and entails the equality $\lfloor (m + 1)/q \rfloor = p - 2$. Following (8) and (9), the vector $\boldsymbol{a}_{pq}$ satisfies

$$\boldsymbol{a}_{pq} = \Phi_{pq}(S)e_0 = \sum_{i=0}^{p-2} S^{iq} \boldsymbol{c}_{pq} \quad \text{with} \quad \boldsymbol{c}_{pq} = (\text{Id} - S)\Phi_q(S^p)e_0. \tag{10}$$

The relation $S^\ell e_0 = e_\ell$ holds for integer $\ell \in [\![0, m]\!]$ and yields

$$\Phi_q(S^p)e_0 = \sum_{j=0}^{\lfloor \frac{m+1}{p} \rfloor} e_{jp}.$$

Then the vector $c_{pq} \in \mathbb{Q}^{m+1}$ defined in (10) is

$$c_{pq} = \sum_{j=0}^{\lfloor \frac{m+1}{p} \rfloor} (e_{jp} - e_{jp+1}) = \underbrace{1 - 10 \cdots 0}_{p} \cdots \underbrace{1 - 10 \cdots 0}_{p} \underbrace{1 - 10 \cdots 0}_{m+1 \pmod{p}}. \quad (11)$$

Step 2. From Vectors to Words. With (11), we now view the vector c_{pq} as a word of $\mathcal{A}^{m+1}$ that is written as a fractional power of the word $d_0 \in \mathcal{A}^p$ defined in the statement of Theorem 1:

$$c_{pq} = d_0^{(m+1)/p} \in \mathcal{A}^{m+1}. \quad (12)$$

We now use the next proposition (Proposition 1) that states that such a fractional power can be written in terms of the transforms d_i of d_0 via cyclic permutations. The proof of Proposition 1 will be found at the end of this section.

Proposition 1. *The fractional power $d_0^{(m+1)/p}$ of d_0 is written as a concatenation of fractional powers of the cyclic transforms $d_i = \sigma^{iq}(d_0)$:*

$$c_{pq} = d_0^{(m+1)/p} = d_0^{q/p} \cdot d_1^{q/p} \cdots d_{p-3}^{q/p} \cdot d_{p-2}^{(q-p+2)/p}.$$

We then return to the proof. In (10), we now view, for each $i \in [\![0, p-2]\!]$, $S^{iq} c_{pq}$ as a word in $\mathcal{A}^{m+1}$, written as

$$S^{iq} c_{pq} = 0^{iq} \cdot d_0^{q/p} \cdot d_1^{q/p} \cdots d_{p-2-i}^{(q-p+2)/p},$$

where 0^{iq} is the word obtained with the concatenation of iq consecutive zeros. From (10), the cyclotomic word a_{pq} is expressed in terms of c_{pq} as

$$a_{pq} = \sum_{i=0}^{p-2} S^{iq} c_{pq} = \sum_{i=0}^{p-2} 0^{iq} d_0^{q/p} \cdot d_1^{q/p} \cdots d_{p-2-i}^{(q-p+2)/p}. \quad (13)$$

We now explain how the sums ω_i that intervene in Theorem 1 appear. We consider the $(p-2)$ following vectors of $\mathbb{Q}^q$

$$f_0 = d_0^{q/p}, \quad f_1 = d_1^{q/p}, \quad f_2 = d_2^{q/p}, \ldots, \quad f_{p-3} = d_{p-3}^{q/p},$$

together with the $(p-2)$ following vectors of $\mathbb{Q}^{q-p+2}$:

$$g_0 = d_0^{(q-p+2)/p}, \, g_1 = d_1^{(q-p+2)/p}, \, \ldots, \, g_{p-2} = d_{p-2}^{(q-p+2)/p}. \quad (14)$$

We use the expression of the vector a_{pq} in (13) as a sum that uses $(p-1)$ lines, indiced from 0 to $p-2$, each line for each term of the sum,

$$\begin{aligned} a_{pq} = {} & f_0 \cdot f_1 \cdot f_2 \cdot f_3 \cdots f_{p-3} \cdot g_{p-2} \\ & + 0^q \cdot f_0 \cdot f_1 \cdot f_2 \cdots f_{p-4} \cdot g_{p-3} \\ & + 0^q \cdot 0^q \cdot f_0 \cdot f_1 \cdots f_{p-5} \cdot g_{p-4} \\ & \vdots \\ & + 0^q \cdot 0^q \cdot 0^q \cdot 0^q \cdots 0^q \quad \cdot \quad g_0. \end{aligned}$$

Each line exactly follows the same pattern: it is formed with $(p-2)$ blocks of length q, followed with a block of length $(q-p+2)$. The line of index i begins with i blocks of q zeroes, that perform a *shift* between blocks. Then, when we read the result of the previous sum by columns, indexed from 0 to $p-1$, the j-th column contains (for $0 \le j < p-2$) the element $f_0 + \cdots + f_j$, whereas the last column equals the sum of elements defined in (14). Provided that all the sums be internal, this implies the equality

$$a_{pq} = d_0^{q/p} \cdot (d_0 + d_1)^{q/p} \cdots (d_0 + \cdots + d_{p-3})^{q/p} \cdot (d_0 + \cdots + d_{p-2})^{(q-p+2)/p},$$

and introduces the words ω_i involved in Theorem 1.

The next lemma shows that the computation of the words $\omega_i = d_0 + \cdots + d_i$ defined in Theorem 1 indeed involves internal operations in $\mathcal{A}$. It will be proven at the end of the section. It provides the proof of Statement (a) of Theorem 1.

Lemma 1. *Under the assumptions and notations of Theorem 1, the words ω_i belong to $\mathcal{A}^p$ for each $i \in [\![0, p-2]\!]$.*

Now, with Lemma 1, the cyclotomic word a_{pq} is thus expressed as a concatenation of fractional powers,

$$a_{pq} = \omega_0^{q/p} \cdot \omega_1^{q/p} \cdots \omega_{p-3}^{q/p} \cdot \omega_{p-2}^{(q-p+2)/p},$$

it has length $m+1$ and coincides with the prefix of length $m+1$ of the word

$$b_{pq} = \omega_0^{q/p} \cdot \omega_1^{q/p} \cdots \omega_{p-3}^{q/p} \cdot \omega_{p-2}^{q/p}.$$

This ends the proof of Theorem 1.

We now provide the proof of the two results that we have used.

Proof of Proposition 1. The first step of the proof is based on a simple fact which holds for any positive integers m, p and q, with $m > p$ and $q \in [\![p+1, m]\!]$, and any word $d_0 \in \mathcal{A}^p$. The fact is the following: if $r = q \pmod{p}$, then

$$d_0^{(m+1)/p} = d_0^{q/p} \cdot d_1^{(m+1-q)/p} \quad \text{with} \quad d_1 = \sigma^q(d_0) = \sigma^r(d_0) \in \mathcal{A}^p.$$

Here we prove it. First, it is clear that

$$d_0^{(m+1)/p} = d_0^{q/p} \cdot v \quad \text{with} \quad v \in \mathcal{A}^{m+1-q}.$$

Let d_j and v_j denote the jth coordinates of d_0 and v, respectively. It is clear that v_j equals the $(j+r) \pmod{p}$ coordinate of d_0. This implies that $v = d_1^{(m+1-q)/p}$, with $d_1 = \sigma^q(d_0) = \sigma^r(d_0) = d_{r+1} \cdots d_p d_0 \cdots d_r$. This is what we wanted to prove.

For the particular choice $m = \varphi(pq)$, it turns out that $\lfloor (m+1)/q \rfloor = p-2$ and $m+1 = q-p+2 \pmod{q}$ (recall that $p > 2$).

Now, we apply an inductive argument. For $i = 0, \ldots, p-2$, set $d_i = \sigma^r(d_{i-1})$. Then

$$d_0^{(m+1)/p} = d_0^{q/p} \cdot d_1^{(m+1-q)/p} = d_0^{q/p} \cdot d_1^{q/p} \cdots d_{p-3}^{q/p} \cdot d_{p-2}^{(q-p+2)/p}.$$

Proof of Lemma 1. Fix $i \in [\![1, p-2]\!]$ and let j be in the interval $[\![0, p-1]\!]$. With $r = q \pmod{p}$, the relation

$$d_k = \sigma^r(d_{k-1}) = \sigma^{kr}(d_0)$$

shows that

$$d_{k,j} = \begin{cases} 1 & \text{if } j + kr = 0 \pmod{p}, \\ -1 & \text{if } j + kr = 1 \pmod{p}, \\ 0 & \text{otherwise}. \end{cases}$$

Now, we proceed with an inductive argument. The statement of the lemma is clear for $\omega_0 = d_0$. Suppose that $\omega_{i-1} \in \mathcal{A}^p$. From the definition of words ω_i, given in (4), it is clear that

$$\omega_{i-1} = d_0 + d_1 + \cdots + d_{i-1}.$$

Hence, if $\omega_{i-1,j} = 1$, then $j + kr = 0 \pmod{p}$ for some $k \in [\![0, i-1]\!]$; and similarly, if $\omega_{i-1,j} = -1$, then $j + kr = 1 \pmod{p}$ for some $k \in [\![0, i-1]\!]$. Remark that the converse might not be true because there might be cancellations between 1 and -1.

Since $\gcd(p, r) = 1$, we have $kr \neq ir \pmod{p}$ for any $k \in [\![0, i-1]\!]$. It follows that there is no $j \in [\![0, p-1]\!]$ such that $\omega_{i-1,j} = d_{i,j} = 1$ or $\omega_{i-1,j} = d_{i,j} = -1$. Then, the sum $\omega_i = \omega_{i-1} + d_i \in \mathcal{A}^p$, that is, the sums of two 1 or two -1 never occur.

4 Algorithms

This section considers algorithms, and analyses their complexities. It is organised into three main parts. It begins with the BCW algorithm, then explains how it may be used to compute tables of binary cyclotomic polynomials, and finally compares the BCW algorithms with other algorithms that have been previously proposed.

Analysis of the BCW Algorithm. We first analyse the BCW algorithm designed in Sect. 2. This will provide the proof of Theorem 2. The analysis of the BCW algorithm follows from the next two lemmas.

Lemma 2. [Precomputation step]. *The output $\mathcal{O}_{p,r}$ of the precomputation step on the input $(p, r = q \pmod{p})$ contains the words ω_i together with their prefixes $\omega_i^{r/p}$, for $i = 0, \ldots, p-2$. It is computed with $\Theta(p)$ operations on words of $\mathcal{A}^p$. The total cost for computing $\mathcal{O}_{p,r}$ is $\Theta(p^2)$.*

Proof. In Lemma 1, we already proved that all the additions between words $\boldsymbol{\omega}_i$ and words $\boldsymbol{d}_i$ are internal to the alphabet $\mathcal{A}$. Given $r = q \pmod{p}$, the $p - 2$ words $\boldsymbol{d}_i$ are obtained by applying $p - 2$ cyclic permutations of order r to a word of length p. The words $\boldsymbol{\omega}_i$ are obtained with $\Theta(p)$ internal additions of words in $\mathcal{A}^p$. The computation of the prefixes $\boldsymbol{\omega}_i^{r/p}$ only involves $p - 2$ truncations. All the words involved has length p. Thus, the total cost is $\Theta(p^2)$.

Lemma 3. [Concatenation step]. *The concatenation step computes $\boldsymbol{a}_{pq}$ from the output $\mathcal{O}_{p,r}$ of the precomputation step in $\Theta(q)$ words operations (mainly concatenations). The total cost of the concatenation step is $\Theta(pq)$.*

Proof. We consider the quotient s and the remainder r of the division of q by p. The word $\boldsymbol{b}_{pq}$ is obtained by concatenating the words $\boldsymbol{\omega}_i \in \mathcal{A}^p$ or $\boldsymbol{\omega}_i^{r/q} \in \mathcal{A}^r$ according to the order prescribed in Theorem 1. There are $\Theta(sp) = \Theta(q)$ such concatenations of words of length at most p. Finally, to obtain $\boldsymbol{a}_{pq}$ from $\boldsymbol{b}_{pq}$, it suffices to delete the last $p - 2$ symbols. The total cost of this step is $\Theta(pq)$.

Computing Tables of Binary Cyclotomic Polynomials with the BCW Algorithm. We denote by $\mathcal{O}_p$ the total precomputation output relative to a fixed prime p, that is the union of the precomputation outputs relative this fixed prime p and all possible values of $r \in [\![1, p - 1]\!]$. The cost of building the total output $\mathcal{O}_p$ is $\Theta(p^3)$.

We discuss the cost of computing a list of polynomials $\Phi_{p_0 q}$ for a fixed prime p_0 and a number $t = p_0^\beta$ of primes q that all satisfy $q \le p_0^\alpha$ (with $\alpha > 1$).

Case (a). Consider first the case when we wish to compute the polynomials $\Phi_{p_0 q}$ for a number $t = p_0^\beta$ of primes q that satisfy $q \le p_0^\alpha$ (with $\alpha \ge 1$) and $q \pmod{p} = r$. We then precompute only the output $\mathcal{O}_{p_0,r}$, with a cost $\Theta(p_0^2)$, then we perform t computations of cost $p_0^{1+\alpha}$. The total cost is

$$\Theta(p_0^2) + \Theta(p_0^{1+\alpha+\beta}) = \Theta(p_0^{1+\beta+\alpha}).$$

In this case, the cost of the precomputation is always smaller than the cost of the computation.

Case (b). Consider now the case when we wish to compute the polynomials $\Phi_{p_0 q}$ for a number $t = p_0^\beta$ of primes q that all satisfy $q \le p_0^\alpha$, with $\alpha \ge 1$ and various values of $q \pmod{p}$. We have to compute the total precomputation output $\mathcal{O}_{p_0}$ in $\Theta(p_0^3)$ steps, then we perform t computations of cost $p_0^{1+\alpha}$ The total cost is

$$\Theta(p_0^3) + \Theta(p_0^{1+\alpha+\beta}) = \Theta(p_0^{1+\max((\alpha+\beta),2)}).$$

The cost of the precomputation is larger than the cost of the computation, when $\alpha + \beta$ is smaller than 2.

Case (c). Consider finally the case when we wish to compute the polynomials $\Phi_{p_0 q}$ for any prime $q \le p_0^\alpha$, with $\alpha \ge 1$ with various values of $q \pmod{p}$. We assume that the list of primes $q \le p_0^\alpha$ is already built. There are $t = \Theta(1/\log p_0)p_0^\alpha$ such primes q. We have to compute the total output $\mathcal{O}_{p_0}$

in $\Theta(p_0^3)$ steps, then we perform $t = O(p_0^\alpha)$ computations of cost $p_0^{1+\alpha}$ The total cost is

$$\Theta(p_0^3) + \Theta(p_0^{1+2\alpha}) = \Theta(p_0^{1+2\alpha}).$$

The cost of the precomputation is always smaller than the cost of the computation.

Comparison with Other Algorithms. We now briefly describe algorithms that have been previously designed to compute cyclotomic polynomials. Even when they are designed in the case of a general cyclotomic polynomial, we only analyse their complexity in the binary case, and compare them to the complexity of the BCW algorithm. In the following, $M(\ell)$ denotes the cost of multiplying two integers of bit-length at most ℓ. With fast-multiplication algorithms, $M(\ell)$ is $O(\ell \log \ell \log \log \ell)$.

(i) The first algorithm is based on a formula due to Lenstra-Lam-Leung [4,5], and is only valid in the binary case. With this formula, the j-th coefficient of $\boldsymbol{a}_{pq}$ is computed via a solution (x,y) of an equation of the type $j = xp + yq$ with $|x| < q$ and $|y| < p$. The cost of computing one single coefficient is thus $O(M(\log q))$. The total cost for the whole vector $\boldsymbol{a}_{pq}$ is $\Theta(pq)M(\log q)$.

(ii) The other two algorithms are due to Arnold and Monagan and are described in [1]. They compute the cyclotomic vector in the case of a general order n, but we describe them for a binary order $n = pq$. They do not perform any polynomial divisions.

(ii)(a) The first algorithm is a recursive algorithm, called the sparse power series algorithm (SPS algorithm, for short) because it expresses cyclotomic polynomials as products of sparse power series. In the binary case, the recursion is made with $\Theta(pq)$ additions between integers less than (but close to) $\varphi(pq)$. Then, its complexity is $\Theta(pq) \log q$.

(ii)(b) The second algorithm is called the Big Prime algorithm (BP, for short). It is adapted to the case when the order n has a big prime as a factor. Here, in the binary case, the big prime is q. The algorithm is based on Identity (1) and performs a precomputation step which involves $\Theta(p)$ computations of residues modulo p of integers of size $O(\log q)$. The cost of this step is then $\Theta(p)M(\log q)$. The whole cyclotomic word $\boldsymbol{a}_{pq}$ is obtained via a recursive step which computes $\varphi(pq)$ residues modulo p of integers of size $O(\log q)$. Thus, the cost of computing the whole vector $\boldsymbol{a}_{pq}$ is $\Theta(pq)M(\log q)$.

The previous discussion shows that the existing algorithms are all of complexity $O(pq) \cdot E(p,q)$ where the factor $E(p,q)$ is polynomial in $\log q$. The complexity of the BCW algorithm does not contain such a factor $E(p,q)$.

5   Conclusions and Further Work

In this conclusion, we first explain how our results may be extended to another family of polynomials, related to semi-groups. We then discuss how the representation of Theorem 1 may entail results on the length of blocks of consecutive zeros in the cyclotomic words. The paper ends with a short discussion on the non-binary case.

Another Polynomial $F_{p,q}$. We consider two numbers (p, q) with $2 < p < q$, that are coprime but now *not necessarily primes*, and define the polynomial $F_{p,q} \in \mathbb{Z}[x]$ as follows:

$$F_{p,q}(x) = \frac{(x^{pq} - 1)(x - 1)}{(x^p - 1)(x^q - 1)}.$$

A folklore result relates the polynomial $F_{p,q}$ to numerical semi-groups (see, [6] for definitions and proofs). A numerical semi-group is the set

$$S(p, q) = \{ap + bq \mid a, b \in \mathbb{Z}_{\geq 0}\}.$$

The polynomial $F_{p,q}$ is called the semi-group polynomial associated with $S(p, q)$ due to the following identity

$$F_{p,q}(x) = 1 + (x - 1) \sum_{s \notin S(p,q)} x^s.$$

Clearly, $F_{p,q}$ coincides with the cyclotomic polynomial Φ_{pq} when p and q are primes. Also, if we define $F_q = 1 + x + \cdots + x^{q-1}$, the polynomial $F_{p,q}$ satisfies the identity (7) that is our starting point in the proof of Theorem 1, namely

$$(1 - x^q)F_{p,q}(x) = (1 - x)F_q(x^p).$$

This entails that all our results also hold for semi-groups polynomials $F_{p,q}$.

Proving Other Properties for Φ_{pq} and $F_{p,q}$. The representation provided in Theorem 1 implies that the coefficients of Φ_{pq} belong to $\mathcal{A}$. A parameter of interest is the maximum gap of these polynomials, that we now define: for a given polynomial $f(x) = b_1 x^{n_1} + b_2 x^{n_2} + \cdots + b_k x^{n_k}$, with b_i all nonzero and $n_1 < n_2 < \cdots < n_k$, the maximum gap is defined as

$$g(f) = \max_{1 \leq i < k} (n_{i+1} - n_i), \qquad g(f) = 0 \text{ if } k = 1.$$

The fact that $g(F_{p,q}) = p - 1$ was proven in [3] for binary cyclotomic polynomials Φ_{pq} and in [6] for semi-group polynomials. In [2,7], it is proven that the number of maximum gaps in binary cyclotomic polynomials is $2\lfloor q/p \rfloor$.

Our description in terms of words easily entails the inequality $g(F_{p,q}) \geq p - 1$ and proves that there are, at least, $2\lfloor q/p \rfloor$ maximum gaps. The upper bounds will be recovered from our Theorem 1 *provided that* the possible cancellations $1 + (-1)$ *be controlled* in the sums $\omega_{i-1} + d_i$. This is the aim of a further work.

Possible extensions to the non-binary case? Our results seem to be specific to the binary case where the order is $n = pq$. In the case when n is squarefree with at least three prime factors, the coefficients of the cyclotomic polynomial do not any longer belong to a finite alphabet $\mathcal{A}$. It seems thus difficult to easily deal with words on this alphabet $\mathcal{A}$.

Acknowledgments. The authors wish to thank Brigitte Vallée for many helpful conversations and suggestions.

References

1. Arnold, A., Monagan, M.: Calculating cyclotomic polynomials. Math. Comp. **80**(276), 2359–2379 (2011)
2. Camburu, O., Ciolan, E., Luca, F., Moree, P., Shparlinski, I.: Cyclotomic coefficients: gaps and jumps. J. Number Theor. **163**, 211–237 (2016)
3. Hong, H., Lee, E., Lee, H.S., Park, C.M.: Maximum gap in (inverse) cyclotomic polynomial. J. Number Theor. **132**(10), 2297–2315 (2012)
4. Lam, T., Leung, K.: On the cyclotomic polynomial $\phi_{pq}(x)$. Am. Math. Mon. **103**(7), 562–564 (1996)
5. Lenstra, H.W.: Vanishing sums of roots of unity. In: Proceedings, Bicentennial Congress Wiskundig Genootschap Vrije University, Amsterdam, 1978, Part II. Math. Centre Tracts, Mathematisch Centrum, Amsterdam, vol. 101, pp. 249–268 (1979)
6. Moree, P.: Numerical semigroups, cyclotomic polynomials, and Bernoulli numbers. Amer. Math. Mon. **121**(10), 890–902 (2014)
7. Zhang, B.: Remarks on the maximum gap in binary cyclotomic polynomials. Bull. Math. Soc. Sci. Math. Roumanie **59**(107), 109–115 (2016)

Computation of Critical Exponent
in Balanced Sequences

Francesco Dolce$^{(\boxtimes)}$, L'ubomíra Dvořáková, and Edita Pelantová

FNSPE, Czech Technical University, Prague, Czech Republic
{francesco.dolce,lubomira.dvorakova,edita.pelantova}@fjfi.cvut.cz

Abstract. We study balanced sequences over a d-letter alphabet. Each such sequence **v** is described by a Sturmian sequence and two constant gap sequences **y** and **y'**. We provide an algorithm which for a given **y**, **y'** and a quadratic slope of a Sturmian sequence computes the critical exponent of the balanced sequence **v**.

Keywords: Critical exponent · Balanced sequences · Return words · Bispecial factors

1 Introduction

An infinite sequence **v** over a finite alphabet is called balanced if for each pair u, v of its factors having the same length and for each letter a of the alphabet, the number of occurrences of a in u and v differs at most by one. Balanced aperiodic sequences over a binary alphabet were introduced already in 1940 by Hedlund and Morse under the name Sturmian sequences (see [7]). Balanced sequences over a d-letter alphabet were characterized by Hubert in [8]; in particular he showed that each aperiodic balanced sequence over a d-letter alphabet can be mapped by a letter-to-letter projection π to a Sturmian sequence. In this paper we focus on the critical exponent of a balanced sequence **v**. Roughly speaking, the critical exponent $E(\mathbf{v})$ describes the maximal repetition of factors in **v**. For Sturmian sequences, the formula to evaluate the critical exponent was provided by Carpi and de Luca in [3] (see also [4]). Recently, Rampersad, Shallit and Vandomme in [9] and Baranwal and Shallit in [1] and [2] started looking for balanced sequences over a d-letter alphabet having the least critical exponent. They used the automated theorem prover Walnut to show that the smallest possible critical exponent of a balanced sequence over d letters is $\frac{d-2}{d-3}$ for $d = 5, \ldots, 8$. For $d = 9, 10$ they showed that the least critical exponent can not be smaller than $\frac{d-2}{d-3}$ and conjectured that this value is attained by the sequences $\mathbf{x}_9$ and $\mathbf{x}_{10}$ (see Example 5).

In [5], we gave a general method to compute the critical exponent $E(\mathbf{v})$ and the asymptotic critical exponent $E^*(\mathbf{v})$ of any uniformly recurrent sequence **v**.

The research received funding from the Ministry of Education, Youth and Sports of the Czech Republic through the project CZ.02.1.01/0.0/0.0/16_019/0000778.

© Springer Nature Switzerland AG 2021
T. Lecroq and S. Puzynina (Eds.): WORDS 2021, LNCS 12847, pp. 78–90, 2021.
https://doi.org/10.1007/978-3-030-85088-3_7

Our method is based on looking for the shortest return words to bispecial factors in $\mathbf{v}$. The asymptotic critical exponent $E^*(\mathbf{v})$ reflects repetitions of factors of length growing to infinity. Since the letter-to-letter projection π maps every sufficiently long bispecial factor in a balanced sequence $\mathbf{v}$ to a bispecial factor in the underlying Sturmian sequence, we could apply our method to compute the asymptotic critical exponent of balanced sequences.

In this contribution we refine our approach to all bispecial factors, not only the long enough ones (Propositions 3, 5 and 6) and deduce an algorithm for computing the critical exponent of balanced sequences associated with Sturmian sequences with quadratic slopes (Sect. 6). In particular, we confirm the conjectured property of the sequences $\mathbf{x}_9$ and $\mathbf{x}_{10}$ (Example 7).

2 Preliminaries

An *alphabet* $\mathcal{A}$ is a finite set of symbols called *letters*. A (finite) *word* over $\mathcal{A}$ of *length* n is a string $u = u_0 u_1 \cdots u_{n-1}$, where $u_i \in \mathcal{A}$ for all $i = 0, 1, \ldots, n-1$. The length of u is denoted by $|u|$. The set of all finite words over $\mathcal{A}$ together with the operation of concatenation forms a monoid, denoted by $\mathcal{A}^*$. Its neutral element is the *empty word* ε and we denote $\mathcal{A}^+ = \mathcal{A}^* \setminus \{\varepsilon\}$. If $u = xyz$ for some $x, y, z \in \mathcal{A}^*$, then x is a *prefix* of u, z is a *suffix* of u and y is a *factor* of u. To any word u over $\mathcal{A}$ with cardinality $\#\mathcal{A} = d$, we assign its *Parikh vector* $\vec{V}(u) \in \mathbb{N}^d$ defined as $(\vec{V}(u))_a = |u|_a$ for all $a \in \mathcal{A}$, where $|u|_a$ is the number of letters a occurring in u. A *sequence* over $\mathcal{A}$ is an infinite string $\mathbf{u} = u_0 u_1 u_2 \cdots$, where $u_i \in \mathcal{A}$ for all $i \in \mathbb{N}$. In this paper we always denote sequences by bold letters. The shift of $\mathbf{u} = u_0 u_1 u_2 \cdots$ is the sequence $\sigma(\mathbf{u}) = u_1 u_2 u_3 \cdots$. A sequence $\mathbf{u}$ is *eventually periodic* if $\mathbf{u} = vwww \cdots = v(w)^\omega$ for some $v \in \mathcal{A}^*$ and $w \in \mathcal{A}^+$. It is *periodic* if $v = \varepsilon$. If $\mathbf{u}$ is not eventually periodic, then it is *aperiodic*. A *factor* of $\mathbf{u} = u_0 u_1 u_2 \cdots$ is a word u such that $u = u_i u_{i+1} u_{i+2} \cdots u_{j-1}$ for some $i, j \in \mathbb{N}$, $i \leq j$. The number i is called an *occurrence* of the factor u in $\mathbf{u}$. If each factor of $\mathbf{u}$ has infinitely many occurrences in $\mathbf{u}$, the sequence $\mathbf{u}$ is *recurrent*. Moreover, if for each factor the distances between its consecutive occurrences are bounded, $\mathbf{u}$ is said to be *uniformly recurrent*.

The *language* $\mathcal{L}(\mathbf{u})$ of a sequence $\mathbf{u}$ is the set of all its factors. A factor w of $\mathbf{u}$ is *right special* if wa, wb are in $\mathcal{L}(\mathbf{u})$ for at least two distinct letters $a, b \in \mathcal{A}$. Analogously, we define a *left special* factor. A factor is *bispecial* if it is both left and right special.

The central notion of our contribution is the *critical exponent* of an infinite sequence. Let $z \in \mathcal{A}^+$ be a prefix of a periodic sequence u^ω with $u \in \mathcal{A}^+$, and let us suppose that u is minimal in length with this property. We say that z has *fractional root* u and *exponent* $e = |z|/|u|$. We usually write $z = u^e$.

Definition 1. *Given a sequence $\mathbf{u}$, we define the* critical exponent *of $\mathbf{u}$ as*

$$E(\mathbf{u}) = \sup\{e \in \mathbb{Q} : \text{there exist } x, y \in \mathcal{L}(\mathbf{u}), \text{ with } |x| > 0 \text{ and } y = x^e\}.$$

If $E(\mathbf{u}) < +\infty$, we define the asymptotic critical exponent *of* $\mathbf{u}$ *as*

$$E^*(\mathbf{u}) = \lim_{n \to \infty} \sup\{e \in \mathbb{Q} : \text{there exist } x, y \in \mathcal{L}(\mathbf{u}), \text{ with } |x| > n \text{ and } y = x^e\}.$$

Otherwise $E^*(\mathbf{u}) = E(\mathbf{u}) = +\infty$.

In [5] we find a formula to compute $E(\mathbf{u})$ and $E^*(\mathbf{u})$ for a uniformly recurrent aperiodic sequence $\mathbf{u}$. This tool uses the notion of return words.

Let us consider a factor w of a recurrent sequence $\mathbf{u} = u_0 u_1 u_2 \cdots$. Let $i < j$ be two consecutive occurrences of w in $\mathbf{u}$. Then the word $u_i u_{i+1} \cdots u_{j-1}$ is a *return word* to w in $\mathbf{u}$. The set of all return words to w in $\mathbf{u}$ is denoted by $\mathcal{R}_{\mathbf{u}}(w)$. If $\mathbf{u}$ is uniformly recurrent, then the set $\mathcal{R}_{\mathbf{u}}(w)$ is finite for each prefix w. In this case $\mathbf{u}$ can be written as a concatenation $\mathbf{u} = r_{d_0} r_{d_1} r_{d_2} \cdots$ of return words to w. The *derived sequence* of $\mathbf{u}$ to w is the sequence $\mathbf{d}_{\mathbf{u}}(w) = d_0 d_1 d_2 \cdots$ over the alphabet of cardinality $\#\mathcal{R}_{\mathbf{u}}(w)$.

Proposition 1 ([5]). *Let* $\mathbf{u}$ *be a uniformly recurrent aperiodic sequence. Let* $(w_n)_{n \in \mathbb{N}}$ *be a sequence of all bispecial factors of* $\mathbf{u}$ *ordered by their length. For every* $n \in \mathbb{N}$, *let* v_n *be a shortest return word to* w_n *in* $\mathbf{u}$. *Then*

$$E(\mathbf{u}) = 1 + \sup_{n \in \mathbb{N}} \left\{ \frac{|w_n|}{|v_n|} \right\} \qquad \text{and} \qquad E^*(\mathbf{u}) = 1 + \limsup_{n \to \infty} \left\{ \frac{|w_n|}{|v_n|} \right\}.$$

3   Balanced Sequences

A sequence $\mathbf{u}$ over the alphabet $\mathcal{A}$ is *balanced* if for every letter $a \in \mathcal{A}$ and every pair of factors $u, v \in \mathcal{L}(\mathbf{u})$ with $|u| = |v|$, we have $||u|_a - |v|_a| \leq 1$. Aperiodic balanced sequences over binary alphabet, i.e., Sturmian sequences, can be characterized by many equivalent definitions. The definition we will need is based on return words. Vuillon in [10] shows that an infinite recurrent sequence $\mathbf{u}$ is Sturmian if and only if each of its factors has exactly two return words. Moreover, the derived sequence to a factor of a Sturmian sequence is Sturmian too. A Sturmian sequence $\mathbf{u}$ is called *standard* if each bispecial factor of $\mathbf{u}$ is a prefix of $\mathbf{u}$. To any Sturmian sequence $\mathbf{u}'$ there exists a standard Sturmian sequence $\mathbf{u}$ such that $\mathcal{L}(\mathbf{u}) = \mathcal{L}(\mathbf{u}')$. Balanced sequences over alphabets of higher cardinality can be constructed from Sturmian sequences. To describe the construction we need the following definition.

Definition 2. *A sequence* $\mathbf{y}$ *over an alphabet* $\mathcal{A}$ *is a* constant gap sequence *if for each letter* $a \in \mathcal{A}$ *appearing in* $\mathbf{y}$ *there is a positive integer, denoted* $\text{gap}_{\mathbf{y}}(a)$, *such that the distance between successive occurrences of* a *in* $\mathbf{y}$ *is always* $\text{gap}_{\mathbf{y}}(a)$.

Any constant gap sequence is periodic. We denote by $\text{Per}(\mathbf{y})$ the minimal period of $\mathbf{y}$. Note that $\text{gap}_{\mathbf{y}}(a)$ divides $\text{Per}(\mathbf{y})$ for each letter a appearing in $\mathbf{y}$. Given a constant gap sequence $\mathbf{y}$ and a word $y \in \mathcal{L}(\mathbf{y})$ we denote by $\text{gap}_{\mathbf{y}}(y)$ the length of the gap between two successive occurrences of y in $\mathbf{y}$. Note that $\text{gap}_{\mathbf{y}}(y) = \text{lcm}\{\text{gap}_{\mathbf{y}}(a) : a \in \mathcal{A} \text{ and } a \text{ occurs in } y\}$. Moreover $\text{gap}_{\mathbf{y}}(y)$ divides $\text{Per}(\mathbf{y})$ for every factor $y \in \mathcal{L}(\mathbf{y})$.

Example 1. In the sequel we will deal with the following constant gap sequences $\mathbf{y} = (01)^\omega$ and $\mathbf{y}' = (234567284365274863254768)^\omega$. The sequence $\mathbf{y}$ is evidently a constant gap sequence because $\mathrm{gap}_\mathbf{y}(0) = \mathrm{gap}_\mathbf{y}(1) = 2$. The sequence $\mathbf{y}'$ is also a constant gap sequence because $\mathrm{gap}_{\mathbf{y}'}(a) = 6$ for $a \in \{2,4,6\}$ and $\mathrm{gap}_{\mathbf{y}'}(a) = 8$ for $a \in \{3,5,7,8\}$. Moreover, for every $y \in \mathcal{L}(\mathbf{y}')$ with $|y| \geq 2$ we have $\mathrm{gap}_{\mathbf{y}'}(y) = 24$. The minimal periods are respectively $\mathrm{Per}(\mathbf{y}) = 2$ and $\mathrm{Per}(\mathbf{y}') = 24$.

Given a constant gap sequence $\mathbf{y}$ we define for every positive integer n the set $\mathrm{gap}(\mathbf{y}, n) = \{i \ : \ \exists y \in \mathcal{L}(\mathbf{y}), |y| = n, \ \mathrm{gap}_\mathbf{y}(y) = i\}$. It is clear that $\mathrm{gap}(\mathbf{y}, 0) = \{1\}$ for every constant gap sequence $\mathbf{y}$.

Example 2. Let $\mathbf{y}, \mathbf{y}'$ be the sequences in Example 1. One has $\mathrm{gap}(\mathbf{y}, n) = \{2\}$ for every $n \geq 1$, $\mathrm{gap}(\mathbf{y}', 1) = \{6, 8\}$ and $\mathrm{gap}(\mathbf{y}', n) = \{24\}$ for every $n \geq 2$.

Theorem 1 ([8]). *A recurrent aperiodic sequence $\mathbf{v}$ is balanced if and only if $\mathbf{v}$ is obtained from a Sturmian sequence $\mathbf{u}$ over $\{\mathsf{a}, \mathsf{b}\}$ by replacing the a's in $\mathbf{u}$ by a constant gap sequence $\mathbf{y}$ over some alphabet $\mathcal{A}$, and replacing the b's in $\mathbf{u}$ by a constant gap sequence $\mathbf{y}'$ over some alphabet $\mathcal{B}$ disjoint from $\mathcal{A}$.*

Let us recall that the frequencies of letters in any Sturmian sequence $\mathbf{u}$ are always well defined and irrational. We will assume here, without loss of generality, that $\rho_\mathsf{a} < \rho_\mathsf{b}$ and adopt the convention that the first component of the Parikh vector of a factor of $\mathbf{u}$ corresponds to the least frequent letter of $\mathbf{u}$ and the second component to the most frequent letter (even if we consider a Sturmian sequence over binary alphabets other than $\{\mathsf{a}, \mathsf{b}\}$).

Definition 3. *Let $\mathbf{u}$ be a Sturmian sequence over the alphabet $\{\mathsf{a}, \mathsf{b}\}$, and $\mathbf{y}, \mathbf{y}'$ be two constant gap sequences over two disjoint alphabets $\mathcal{A}$ and $\mathcal{B}$. The colouring of $\mathbf{u}$ by $\mathbf{y}$ and $\mathbf{y}'$, denoted $\mathbf{v} = \mathrm{colour}(\mathbf{u}, \mathbf{y}, \mathbf{y}')$, is the sequence over $\mathcal{A} \cup \mathcal{B}$ obtained by the procedure described in Theorem 1.*

For $\mathbf{v} = \mathrm{colour}(\mathbf{u}, \mathbf{y}, \mathbf{y}')$ we use the notation $\pi(\mathbf{v}) = \mathbf{u}$ and $\pi(v) = u$ for any $v \in \mathcal{L}(\mathbf{v})$ and the corresponding $u \in \mathcal{L}(\mathbf{u})$. Symmetrically, given a word $u \in \mathcal{L}(\mathbf{u})$, we denote by $\pi^{-1}(u) = \{v \in \mathcal{L}(\mathbf{v}) \ : \ \pi(v) = u\}$. We say that $\mathbf{u}$ (resp. u) is a *projection* of $\mathbf{v}$ (resp. v).

Example 3. Let us consider the sequence $\mathbf{x}_9$ (see Example 5 later for a more precise definition) obtained as colouring by the constant gap sequences $\mathbf{y}$ and $\mathbf{y}'$ given in Example 1 of a Sturmian sequence $\mathbf{u}$ starting as follows:

$$\mathbf{u} = \mathsf{bbabbabbabbbbabbabbabbbbabbabbabbabbbbabbabbabbbbabbabb} \cdots .$$

Thus $\mathbf{x}_9$ starts as follows:

$$\mathbf{x}_9 = 2304516702841360521748063125047168203415607281430 65 \cdots .$$

Such a sequence is balanced according to Theorem 1.

The language of balanced sequences has certain symmetries. In particular, the following result is proved in [5, Corollary 1].

Lemma 1 ([5]). *Let* $\mathbf{v} = \text{colour}(\mathbf{u}, \mathbf{y}, \mathbf{y}')$ *and* $v \in \mathcal{L}(\mathbf{v})$. *For any* $i, j \in \mathbb{N}$ *the word* v' *obtained from* $\pi(v)$ *by replacing the* a*'s by* $\sigma^i(\mathbf{y})$ *and the* b*'s by* $\sigma^j(\mathbf{y}')$ *is in* $\mathcal{L}(\mathbf{v})$.

Note that $\mathcal{L}(\mathbf{v})$ does not depend on the sequence $\mathbf{u}$ itself but only on $\mathcal{L}(\mathbf{u})$. Having in mind the formula for computing the critical exponent given in Proposition 1, we focus on return words to factors of balanced sequences.

In the sequel we will use the following notation:

$$\binom{a}{b} \mod \binom{n}{n'} := \binom{a \mod n}{b \mod n'}.$$

Proposition 2. *Let* $u, f \in \mathcal{L}(\mathbf{u})$ *such that* $fu \in \mathcal{L}(\mathbf{u})$ *and* u *is a prefix of* fu. *Then the two statements are equivalent:*

1. *there exist* w *and* v *such that* $vw \in \mathcal{L}(\mathbf{v})$, w *is a prefix of* vw, $|w| = |u|$ *and* $\pi(vw) = fu$;
2. $\vec{V}(f) = \binom{0}{0} \mod \binom{n}{n'}$ *for some* $n \in \text{gap}(\mathbf{y}, |u|_a)$ *and* $n' \in \text{gap}(\mathbf{y}', |u|_b)$.

Proof. Let v and w be as in Item 1. Then u is a prefix and a suffix of $\pi(vw)$ and $f = \pi(v)$. By Lemma 1, the factor w occurring as a prefix of vw is obtained from u by colouring the a's with $\sigma^s(\mathbf{y})$ and the b's with $\sigma^t(\mathbf{y}')$ for some $s, t \in \mathbb{N}$. Hence, the same factor w occurring as a suffix of vw is obtained from u by colouring the a's with $\sigma^S(\mathbf{y})$ and the b's with $\sigma^T(\mathbf{y}')$, where $S = s + |f|_a$ and $T = t + |f|_b$. Hence the prefixes of length $|u|_a$ of $\sigma^s(\mathbf{y})$ and $\sigma^S(\mathbf{y})$ coincide, and similarly the prefixes of length $|u|_b$ of $\sigma^t(\mathbf{y}')$ and $\sigma^T(\mathbf{y}')$ coincide. This implies that $|f|_a$ is divisible by some $n \in \text{gap}(\mathbf{y}, |u|_a)$ and that $|f|_b$ is divisible by some $n' \in \text{gap}(\mathbf{y}', |u|_b)$. In other words, $|f|_a = 0 \mod n$ and $|f|_b = 0 \mod n'$.

Let f, n and n' be as in Item 2. Let us consider $y \in \mathcal{L}(\mathbf{y})$ and $y' \in \mathcal{L}(\mathbf{y}')$ such that $\text{gap}_\mathbf{y}(y) = n$ with $|y| = |u|_a$ and $\text{gap}_\mathbf{y}(y') = n'$ with $|y'| = |u|_b$. Let $s, t \in \mathbb{N}$ be such that y is a prefix of $\sigma^s(\mathbf{y})$ and y' is a prefix of $\sigma^t(\mathbf{y}')$. Colouring the letters a's in fu with $\sigma^s(\mathbf{y})$ and the letters b's with $\sigma^t(\mathbf{y}')$, we get, by Lemma 1, a factor x of $\mathbf{v}$. Since $|f|_a$ is a multiple of $\text{gap}_\mathbf{y}(y)$ and $|f|_b$ is a multiple of $\text{gap}_\mathbf{y}(y')$, the prefix and the suffix of length $|u|$ of x coincide, i.e., $x = vw$, w is a prefix of vw, $|w| = |u|$ and $\pi(vw) = fu$.

As we have already mentioned, any factor of a Sturmian sequence has exactly two return words and thus any piece of $\mathbf{u}$ between occurrences of u is a concatenation of these two return words. This implies the following observation.

Observation 1. *Let* r *and* s *be respectively the most and the least frequent return words to* u *in* $\mathbf{u}$. *If* $fu \in \mathcal{L}(\mathbf{u})$ *and* u *is a prefix of* fu, *then* $\vec{V}(f) = k\vec{V}(r) + \ell\vec{V}(s)$, *where* $\binom{\ell}{k}$ *is the Parikh vector of a factor of the derived sequence* $\mathbf{d_u}(u)$.

4 Shortest Return Words to Factors in Balanced Sequences

The length of the return words to factors of a Sturmian sequence $\mathbf{u}$ is well-known. Our aim in this section is to find a formula for the length of the shortest return words to factors of a colouring of $\mathbf{u}$. As occurrences of a factor u in a Sturmian sequence $\mathbf{u}$ and occurrences of factors from $\pi^{-1}(u)$ in any colouring of $\mathbf{u}$ coincide, we can therefore be able to give a formula based on the knowledge of the length of return words in $\mathbf{u}$. Proposition 2 and Observation 1 justify the following definition.

Definition 4. *Let $u \in \mathcal{L}(\mathbf{u})$ and let r and s be respectively the most and the least frequent return words to u in $\mathbf{u}$. We denote $\mathcal{S}(u) = \mathcal{S}_1(u) \cap \mathcal{S}_2(u) \cap \mathcal{S}_3$, where*

$$\mathcal{S}_1(u) = \left\{ \binom{\ell}{k} : \binom{\ell}{k} \text{ is the Parikh vector of a factor of } \mathbf{d_u}(u) \right\};$$

$$\mathcal{S}_2(u) = \bigcup_{n \in \text{gap}(\mathbf{y}, |u|_\mathtt{a})} \bigcup_{n' \in \text{gap}(\mathbf{y}', |u|_\mathtt{b})} \left\{ \binom{\ell}{k} : k\vec{V}(r) + \ell\vec{V}(s) = \binom{0}{0} \bmod \binom{n}{n'} \right\};$$

$$\mathcal{S}_3 = \left\{ \binom{\ell}{k} : 1 \le k + \ell \le \text{Per}(\mathbf{y})\text{Per}(\mathbf{y}') \right\}.$$

Using the formula provided in Proposition 1, we can treat all bispecial factors of the same length simultaneously.

Proposition 3. *Let $\mathbf{v} = \text{colour}(\mathbf{u}, \mathbf{y}, \mathbf{y}')$ and $u \in \mathcal{L}(\mathbf{u})$. The shortest words in the set $\{v : v \in \mathcal{R}_\mathbf{v}(w) \text{ and } \pi(w) = u\}$ have length*

$$|v| = \min\{k|r| + \ell|s| : \binom{\ell}{k} \in \mathcal{S}(u)\}.$$

Proof. First, let us show that the length of any return word in $\mathbf{v}$ to a factor from $\pi^{-1}(u)$ is contained in the set $\{k|r| + \ell|s| : \binom{\ell}{k} \in \mathcal{S}_1(u) \cap \mathcal{S}_2(u)\}$. By Proposition 2 and Observation 1, a vector $\binom{\ell}{k}$ belongs to $\mathcal{S}_1(u) \cap \mathcal{S}_2(u)$ if and only if $k\vec{V}(r) + \ell\vec{V}(s)$ is the Parikh vector of $\pi(v)$, where v is a factor between two (possibly not consecutive) occurrences of a factor $w \in \pi^{-1}(u)$ in $\mathbf{v}$. Obviously, the length of v is $k|r| + \ell|s|$. It is evident that if we consider above $|v| = \min\{k|r| + \ell|s|\}$, where $\binom{\ell}{k} \in \mathcal{S}_1(u) \cap \mathcal{S}_2(u)$, then v is a return word to a factor $w \in \pi^{-1}(u)$.

To finish the proof, we have to show that the minimum value of $|v|$ is attained for k and ℓ satisfying $1 \le k + \ell \le \text{Per}(\mathbf{y})\text{Per}(\mathbf{y}')$. Let $\binom{\ell}{k} \in \mathcal{S}_1(u) \cap \mathcal{S}_2(u)$ and $k + \ell > \text{Per}(\mathbf{y})\text{Per}(\mathbf{y}')$. Thus $\vec{V}(d) = \binom{\ell}{k}$ for some $d = d_1 d_2 d_3 \cdots d_{k+\ell} \in \mathcal{L}(\mathbf{d_u}(u))$. For every $i = 1, 2, \ldots, k + \ell$, we denote $\binom{\ell_i}{k_i} = \vec{V}(d_1 d_2 \cdots d_i)$. We assign to each i the vector $X_i = k_i\vec{V}(r) + \ell_i\vec{V}(s)$. Since the number of equivalence classes mod $\binom{n}{n'}$ is $nn' \le \text{Per}(\mathbf{y})\text{Per}(\mathbf{y}')$, there exist i, j with $1 \le i < j \le k + \ell$ such that $X_i = X_j \bmod \binom{n}{n'}$. Denote $\binom{\ell'}{k'}$ the Parikh vector of $d_{i+1} d_{i+2} \cdots d_j$.

Obviously, $\left(\begin{smallmatrix}\ell'\\k'\end{smallmatrix}\right) \in \mathcal{S}_1(u)$, $1 \leq j - i = k' + \ell' < k + \ell$ and $k' \leq k$ and $\ell' \leq \ell$. Hence $k'|r| + \ell'|s| < k|r| + \ell|s|$. Since $k'\vec{V}(r) + \ell'\vec{V}(s) = X_j - X_i = \left(\begin{smallmatrix}0\\0\end{smallmatrix}\right)$ mod $\left(\begin{smallmatrix}n\\n'\end{smallmatrix}\right)$, the vector $\left(\begin{smallmatrix}\ell'\\k'\end{smallmatrix}\right) \in \mathcal{S}_2(u)$. Therefore, the minimum length can not be achieved for $k + \ell > \mathrm{Per}(\mathbf{y})\mathrm{Per}(\mathbf{y}')$.

Since a constant gap sequence is periodic, it is clear that any long enough factor in the sequence is neither right special nor left special. Let us define, for a given constant gap sequence $\mathbf{y}$, the number

$$\beta(\mathbf{y}) = \max\{|u| : u \text{ is a bispecial factor of } \mathbf{y}\}.$$

It immediately follows that for $n > \beta(\mathbf{y})$, we have $\mathrm{gap}(\mathbf{y}, n) = \{\mathrm{Per}(\mathbf{y})\}$.

Example 4. Let us consider the sequences $\mathbf{y}$ and $\mathbf{y}'$ from Example 1. One can easily check that $\beta(\mathbf{y}) = 0$ and $\beta(\mathbf{y}') = 1$.

The following result is analogous to [5, Lemma 3].

Lemma 2. *Let* $\mathbf{v} = \mathrm{colour}(\mathbf{u}, \mathbf{y}, \mathbf{y}')$ *and* $w \in \mathcal{L}(\mathbf{v})$.

1. *If* $\pi(w)$ *is bispecial in* $\mathbf{u}$, *then* w *is bispecial in* $\mathbf{v}$.
2. *If* w *is bispecial in* $\mathbf{v}$, $|\pi(w)|_{\mathsf{a}} > \beta(\mathbf{y})$ *and* $|\pi(w)|_{\mathsf{b}} > \beta(\mathbf{y}')$, *then* $\pi(w)$ *is bispecial in* $\mathbf{u}$. *Moreover, in this case* $\pi(\mathcal{R}_{\mathbf{v}}(w)) = \pi(\mathcal{R}_{\mathbf{v}}(w'))$ *for each* $w' \in \mathcal{L}(\mathbf{v})$ *with* $\pi(w') = \pi(w)$.

If a projection of a bispecial factor w in $\mathbf{v}$ is bispecial in $\mathcal{L}(\mathbf{u})$, we can deduce an explicit formula for $1 + \frac{|w|}{|v|}$, where $|v|$ is the length of a shortest return word to w in $\mathbf{v}$. These values are crucial for the computation of $E(\mathbf{v})$ and $E^*(\mathbf{v})$.

First, we list some important facts on Sturmian sequences. They are partially taken from [6]. Recall our convention for the frequencies of letters $\rho_{\mathsf{a}} < \rho_{\mathsf{b}}$. The language of the Sturmian sequence $\mathbf{u}$ is fully described by the coefficients of the continued fraction of the number θ associated with $\mathbf{u}$, that is

$$\theta = \theta(\mathbf{u}) := \frac{\rho_{\mathsf{a}}}{\rho_{\mathsf{b}}} = [0, a_1, a_2, a_3, \ldots].$$

The relation to the slope α of $\mathbf{u}$ is $\alpha = \frac{1}{1+\theta}$. The Parikh vectors of the bispecial factors in $\mathbf{u}$ and the corresponding return words can be easily expressed using the convergents $\frac{p_N}{q_N}$ to θ.

Proposition 4 ([6]). *Let* $\theta = [0, a_1, a_2, a_3, \ldots]$ *be the irrational number associated with a standard Sturmian sequence* $\mathbf{u}$ *and* b *a bispecial factor of* $\mathbf{u}$. *Then*

1. *there exists a unique pair* $(N, m) \in \mathbb{N}^2$ *with* $0 \leq m < a_{N+1}$ *such that the Parikh vectors of the most frequent return word* r *to* b, *of the least frequent return word* s *to* b *and of* b *itself are*

$$\vec{V}(r) = \begin{pmatrix} p_N \\ q_N \end{pmatrix}, \quad \vec{V}(s) = \begin{pmatrix} m\,p_N + p_{N-1} \\ m\,q_N + q_{N-1} \end{pmatrix} \quad and \quad \vec{V}(b) = \vec{V}(r) + \vec{V}(s) - \begin{pmatrix} 1 \\ 1 \end{pmatrix};$$

2. *the derived sequence* $\mathbf{d_u}(b)$ *to* b *in* $\mathbf{u}$ *is Sturmian and the irrational number associated with* $\mathbf{d_u}(b)$ *is* $\theta' = [0, a_{N+1} - m, a_{N+2}, a_{N+3}, \ldots]$.

Let us recall that the nominator p_N and the denominator q_N of the Nth convergent to θ satisfy for all $N \geq 1$ the recurrence relation $X_N = a_N X_{N-1} + X_{N-2}$, but that they differ in their initial values: $p_{-1} = 1, p_0 = 0; q_{-1} = 0, q_0 = 1$. The following statement is a direct consequence of Propositions 3 and 4.

Proposition 5. *Let* $\mathbf{v} = \text{colour}(\mathbf{u}, \mathbf{y}, \mathbf{y}')$ *and* $\left(\frac{p_N}{q_N}\right)_N$ *be the sequence of convergents to the irrational number* θ *associated with* $\mathbf{u}$. *Let* (N, m) *be the pair assigned in Proposition 4 to a bispecial factor* $b \in \mathcal{L}(\mathbf{u})$. *Then, a shortest return word* v *to a factor* $w \in \pi^{-1}(b)$ *satisfies*

$$I(N, m) := 1 + \frac{|w|}{|v|} = 1 + \max\left\{\frac{(1+m)Q_N + Q_{N-1} - 2}{(k+\ell m)Q_N + \ell Q_{N-1}} : \binom{\ell}{k} \in \mathcal{S}(b)\right\}, \quad (1)$$

where $Q_N := p_N + q_N$ *and* $Q_{N-1} := p_{N-1} + q_{N-1}$.

The following lemma helps us to recognize which vector is the Parikh vector of a factor of a given Sturmian sequence. This is important to decide whether $\binom{\ell}{k}$ belongs to $\mathcal{S}_1(b)$. The lemma can be shown using the facts that $\theta = \frac{\rho_a}{\rho_b}$ and that $\mathbf{u}$ is balanced.

Lemma 3. *Let* $\mathbf{u}$ *be a Sturmian sequence with associated irrational number* θ. *Then* $\mathbf{u}$ *contains a factor* u *such that* $|u|_b = k$ *and* $|u|_a = \ell$ *if and only if* $(k-1)\theta - 1 < \ell < (k+1)\theta + 1$ *and* $k, \ell \in \mathbb{N}$.

Example 5. In the sequel, we will illustrate our method for computing the critical exponent on the balanced sequences $\mathbf{x}_9$ and $\mathbf{x}_{10}$ introduced in [9] as candidates to be the balanced sequences having the minimal critical exponent over respectively a 9- and a 10-letter alphabet. Let us define $\mathbf{x}_9$ and $\mathbf{x}_{10}$.

- $\mathbf{x}_9 = \text{colour}(\mathbf{u}, \mathbf{y}, \mathbf{y}')$, where $\mathbf{u}$ is the standard Sturmian sequence associated with $\theta = [0, 2, 3, 2^\omega]$, and $\mathbf{y}, \mathbf{y}'$ are the constant gap sequences introduced in Example 1. Prefixes of $\mathbf{u}$ and $\mathbf{x}_9$ are displayed in Example 3.
- $\mathbf{x}_{10} = \text{colour}(\mathbf{u}', \mathbf{y}, \mathbf{y}'')$, where $\mathbf{u}'$ is the standard Sturmian sequence associated with $\theta = [0, 4, 2, 3^\omega]$, $\mathbf{y}$ is the constant gap sequence introduced in Example 1 and $\mathbf{y}'' = (2345672849632547682943652748 69)^\omega$.

5 Computation of the Asymptotic Critical Exponent

From now on we consider a standard Sturmian sequence $\mathbf{u}$ with associated irrational number θ having eventually periodic continued fraction expansion. The goal of this section is to compute the asymptotic critical exponent of a sequence $\mathbf{v}$ obtained by colouring of $\mathbf{u}$. By Proposition 1, to determine $E^*(\mathbf{v})$ we only need to consider long enough bispecial factors w.

For this purpose, we write the continued fraction expansion of θ as

$$\theta = [0, a_1, a_2, \ldots, a_h, (z_0, z_1, \ldots, z_{M-1})^\omega], \tag{2}$$

where the preperiod h is chosen so that each bispecial factor b associated with (N, m), $N \geq h$, satisfies $|b|_a > \beta(\mathbf{y})$ and $|b|_b > \beta(\mathbf{y}')$.

We then decompose the set $\mathcal{W}$ of all nonempty bispecial factors of $\mathbf{v} = \mathrm{colour}(\mathbf{u}, \mathbf{y}, \mathbf{y}')$ into two subsets:
$$\mathcal{W}^{long} := \{w \in \mathcal{W} : \pi(w) \text{ bispecial in } \mathbf{u} \text{ assigned to } (N, m) \text{ with } N \geq h\}.$$
$$\mathcal{W}^{short} := \mathcal{W} \setminus \mathcal{W}^{long}.$$

Using Proposition 5 and Lemma 2 in order to compute $E^*(\mathbf{v})$, we need to manipulate the numbers $I(N, m)$ defined in Eq. (1).

Our approach consists in partitioning the set of all possible pairs (N, m), $N \geq h$, into a finite number of subsets such that $\mathcal{S}(b)$ is the same for each Sturmian bispecial factor b assigned to a pair in the given subset. A suitable partition uses the following equivalence relation on the first component of the pair.

Definition 5. *Let $N_1, N_2 \in \mathbb{N}$ and $N_1, N_2 \geq h$. We say that N_1 is equivalent to N_2, and write $N_1 \sim N_2$, if the following three conditions are satisfied:*

1. $N_1 = N_2 \bmod M$,
2. $\begin{pmatrix} p_{N_1-1} \\ q_{N_1-1} \end{pmatrix} = \begin{pmatrix} p_{N_2-1} \\ q_{N_2-1} \end{pmatrix} \bmod \begin{pmatrix} \mathrm{Per}(\mathbf{y}) \\ \mathrm{Per}(\mathbf{y}') \end{pmatrix}$,
3. $\begin{pmatrix} p_{N_1} \\ q_{N_1} \end{pmatrix} = \begin{pmatrix} p_{N_2} \\ q_{N_2} \end{pmatrix} \bmod \begin{pmatrix} \mathrm{Per}(\mathbf{y}) \\ \mathrm{Per}(\mathbf{y}') \end{pmatrix}$.

The properties of the equivalence $\sim$ are summarized in the following lemma. They follow from the definition of convergents to θ and from the periodicity of the continued fraction expansion of θ.

Lemma 4. *Let $\sim$ be the equivalence on the set $\{N \in \mathbb{N} : N \geq h\}$ introduced in Definition 5 and let H denote the number of equivalence classes.*

1. *If $N_1 \sim N_2$, then $a_{N_1+1} = a_{N_2+1}$.*
2. *$N_1 \sim N_2$ if and only if $N_1 + 1 \sim N_2 + 1$.*
3. *$N_1 \sim N_2$ if and only if $N_2 = N_1 \bmod H$.*
4. *$H = \min\{i \in \mathbb{N}, i > 0 : h + i \sim h\} \leq M\mathrm{Per}(\mathbf{y})^2\mathrm{Per}(\mathbf{y}')^2$.*
5. *H is divisible by M.*

Definitions 4 and 5 together with Lemma 4 ensure the following property.

Corollary 1. *Let $b^{(1)}$ and $b^{(2)}$ be bispecial factors of $\mathbf{u}$ and (N_1, m_1) and (N_2, m_2), with $N_1 \geq h$ and $N_2 \geq h$, be the pairs assigned to $b^{(1)}$ and $b^{(2)}$ respectively.*
If $N_1 \sim N_2$ and $m_1 = m_2$, then $\mathcal{S}(b^{(1)}) = \mathcal{S}(b^{(2)})$.

Let us define a partition of the set $\mathcal{W}^{long}$ into subsets $C(i,m)$, where $0 \leq i < H$ and $0 \leq m < z_i \bmod M$, as follows: if $(h+i+NH, m)$ is the pair assigned to a bispecial factor $b = \pi(w)$ in $\mathbf{u}$, then we put w into the subset $C(i,m)$. Using Propositions 1 and 5, we have

$$E^*(\mathbf{v}) = \max\{E^*(i,m) : 0 \leq i < H, \ 0 \leq m < z_i \bmod M\}, \tag{3}$$

where $E^*(i,m) := \limsup\limits_{N \to \infty} I(h+i+NH, \ m)$.

To compute $E^*(i,m)$, we need, according to Eq. (1), to determine $\lim\limits_{N \to \infty} \frac{Q_{N-1}}{Q_N}$. A direct consequence of the Perron-Frobenius theorem serves this purpose.

Lemma 5. *Let $A \in \mathbb{N}^{2 \times 2}$ be a primitive matrix with $\det A = \pm 1$, and $(S_N)_N$, $(T_N)_N$ be two sequences of integers given by the recurrent relation $(S_{N+1}, T_{N+1}) = (S_N, T_N)A$ for each $N \in \mathbb{N}$, with $S_0, T_0 \in \mathbb{N}$ such that $S_0 + T_0 > 0$. Denote by $\binom{x}{y}$ an eigenvector of A to the non-dominant eigenvalue λ. Then*

1. $\lim\limits_{N \to \infty} \frac{S_N}{T_N} = -\frac{y}{x}$, *and*
2. $S_N + \frac{y}{x}T_N = \lambda^N(S_0 + \frac{y}{x}T_0)$ *for each $N \in \mathbb{N}$.*

Proof. As A is a primitive matrix with non-negative entries, the components x and y of an eigenvector to the non-dominant eigenvalue have opposite signs. In particular $x, y \neq 0$. Obviously, $(S_N, T_N) = (S_0, T_0)A^N$ for each $N \in \mathbb{N}$. Multiplying both sides of the equation by the eigenvector $\binom{x}{y}$, we obtain $xS_N + yT_N = \lambda^N(xS_0 + yT_0)$, i.e., Item 2 is proven.

As $|\lambda| < 1$, Item 2 implies that $\lim\limits_{N \to \infty} xT_N \left(\frac{S_N}{T_N} + \frac{y}{x}\right) = \lim\limits_{N \to \infty} (xS_N + yT_N) = 0$. Since $\lim\limits_{N \to \infty} T_N = +\infty$, necessarily $\lim\limits_{N \to \infty} \left(\frac{S_N}{T_N} + \frac{y}{x}\right) = 0$. This proves Item 1.

Periodicity of the continued fraction expansion of θ and the previous lemma ensure that the sequences $S_N := Q_{MN+h+i-1}$ and $T_N := Q_{MN+h+i}$ satisfy the recurrent relation $(S_{N+1}, T_{N+1}) = (S_N, T_N)A^{(i)}$ with

$$A^{(i)} = \begin{pmatrix} 0 & 1 \\ 1 & z_i \end{pmatrix}\begin{pmatrix} 0 & 1 \\ 1 & z_{i+1} \end{pmatrix} \cdots \begin{pmatrix} 0 & 1 \\ 1 & z_{M-1} \end{pmatrix}\begin{pmatrix} 0 & 1 \\ 1 & z_0 \end{pmatrix} \cdots \begin{pmatrix} 0 & 1 \\ 1 & z_{i-1} \end{pmatrix}, \tag{4}$$

and hence also the existence of the limit

$$L_i = \lim_{N \to \infty} \frac{S_N}{T_N} = \lim_{N \to \infty} \frac{Q_{HN+h+i-1}}{Q_{HN+h+i}} \quad \text{for } i = 0, 1, \ldots, H-1. \tag{5}$$

Moreover, the non-dominant eigenvalue λ of $A^{(i)}$ satisfies

$$S_N - L_iT_N = \lambda^N(S_0 - L_iT_0) \quad \text{for each } N \in \mathbb{N}. \tag{6}$$

By Corollary 1, for all bispecial factors w in $C(i,m)$ we obtain the same set $\mathcal{S}(\pi(w))$. Let us denote $\mathcal{S}(i,m) := \mathcal{S}(\pi(w))$. Formula (1) then immediately gives

$$E^*(i,m) = 1 + \max\left\{\frac{1+m+L_i}{k+\ell m + \ell L_i} : \binom{\ell}{k} \in \mathcal{S}(i,m)\right\}. \tag{7}$$

Example 6. Let us evaluate $E^*(\mathbf{x}_9)$, where $\mathbf{x}_9$ is the sequence defined in Example 5. It is easy to find that $H = 8$ and there are 16 distinct subsets $C(i,m)$ for $i \in \{0, 1, \ldots, 7\}$ and $m \in \{0, 1\}$. As $\theta = [0, 2, 3, 2^\omega]$ has period 1, the recurrence relation for $(Q_N)_N$ is $Q_{N+1} = 2Q_N + Q_{N-1}$ for $N \geq 2$. In particular $L_i = \lim_{N \to \infty} \frac{Q_{N-1}}{Q_N} = \sqrt{2} - 1$ for each i. Listing all elements of $\mathcal{S}(i,m)$ is more laborious (but possible to do by hand as well). Thanks to a program implemented by our student Daniela Opočenská we find that $E^*(\mathbf{x}_9) = E^*(2,1)$. Since $\mathcal{S}(2,1) = \{\binom{6}{10}\}$, we have $E^*(\mathbf{x}_9) = E^*(2,1) = 1 + \frac{2+L_0}{16+6L_0} = 1 + \frac{2\sqrt{2}-1}{14} \doteq 1,1306$.

Using the same program we also find that $E^*(\mathbf{x}_{10}) = 1 + \frac{\sqrt{13}}{26} \doteq 1,1387$.

6   Computation of the Critical Exponent

In order to evaluate the critical exponent of $\mathbf{v} = \mathrm{colour}(\mathbf{u}, \mathbf{y}, \mathbf{y}')$, we have to determine, by Proposition 1,

$$E(\mathbf{v}) = 1 + \sup \left\{ \frac{|w|}{|v|} : w \in \mathcal{L}(\mathbf{v}), w \text{ bispecial and } v \in \mathcal{R}_\mathbf{v}(w) \right\}.$$

To find the maximum value of $\frac{|w|}{|v|}$ among $w \in \mathcal{W}^{short}$ and $v \in \mathcal{R}_\mathbf{v}(w)$ we use Propositions 3 and 5. To determine $\sup \left\{ \frac{|w|}{|v|} : w \in \mathcal{W}^{long} \text{ and } v \in \mathcal{R}_\mathbf{v}(w) \right\}$ we use the partition of $\mathcal{W}^{long}$ into subsets $C(i,m)$ which have been introduced in the previous section to count the asymptotic critical exponent. For each $C(i,m)$ we have to find

$$E(i,m) := \sup \{I(h + i + NH, \ m) : N \in \mathbb{N}\} \geq E^*(i,m)$$

and then to determine the maximal value among $E(i,m)$. We show that $I(h + i + NH, m)$ may exceed $E^*(i,m)$ only for a finite number of indices $N \in \mathbb{N}$.

Proposition 6. *Let λ be the non-dominant eigenvalue of the matrix $A^{(i)}$ defined in Eq. (4) and L_i be the limit given in Eq. (5). Denote $\mu = |\lambda|^{H/M} < 1$. If $N_0 \in \mathbb{N}$ satisfies $\mu^{N_0} |Q_{h+i-1} - L_i Q_{h+i}| \leq 2L_i$, then $I(h + i + NH, m) \leq E^*(i,m)$ for all $N \geq N_0$ and $0 \leq m < z_{i \bmod M}$.*

Proof. Equation (6) gives $|Q_{h+i+NH-1} - L_i Q_{h+i+NH}| = \mu^N |Q_{h+i-1} - L_i Q_{h+i}|$. Thus, it is enough to show the implication:

If $I(h + i + NH, m) > E^*(i,m)$, then $|Q_{h+i+NH-1} - L_i Q_{h+i+NH}| > 2L_i$.

For this sake, we abbreviate notation by putting $S = Q_{h+i+NH-1}$, $T = Q_{h+i+NH}$ and $L = L_i$. Recall that $0 < L_i < 1$. Let $\binom{\ell}{k} \in \mathcal{S}(i,m)$ such that

$$I(h + i + NH, m) = 1 + \frac{(1+m)T + S - 2}{(k + \ell m)T + \ell S} > E^*(i,m) \geq 1 + \frac{1 + m + L}{k + \ell m + \ell L}.$$

Thus we have $(k - \ell)(S - LT) > 2(k + \ell m + \ell L) \geq 2L|k - \ell|$, hence $|S - LT| > 2L$.

Example 7. Let us show that $E(\mathbf{x}_9) = \frac{7}{6}$. To do that we have to consider the sets of short and long bispecial factors.

$\mathcal{W}^{short}$: It is easy to check that $\pi\left(\mathcal{W}^{short}\right) = \{\mathsf{a}, \mathsf{b}, \mathsf{ab}, \mathsf{ba}, \mathsf{b}^2, \mathsf{b}^3, \mathsf{b}^2\mathsf{ab}^2,$ $\mathsf{b}^2\mathsf{ab}^2\mathsf{ab}^2\}$. For each element w in $\mathcal{W}^{short}$ we have to prove that $1 + \frac{|w|}{|v|} \leq 1 + \frac{1}{6}$, i.e., that $\frac{|w|}{|v|} \leq \frac{1}{6}$, where v is a shortest return word to w.

Let $\pi(w) = \mathsf{a}$, which is not a bispecial factor in $\mathbf{u}$. We use Proposition 3. Looking into the prefix of $\mathbf{u}$ (as in Example 3) we see that the return words to a are $r = \mathsf{ab}^2$ and $s = \mathsf{ab}^3$. By Definition 4, each vector $\binom{\ell}{k} \in \mathcal{S}(\mathsf{a})$ satisfies $k + \ell \geq 1$ and $k\binom{1}{2} + \ell\binom{1}{3} = \binom{0}{0} \mod \binom{2}{1}$ since $\text{gap}(\mathbf{y}, 1) = \{2\}$ and $\text{gap}(\mathbf{y}', 0) = \{1\}$. This implies for each solution that $k + \ell \geq 2$. Since $3k + 4\ell \geq 3k + 3\ell \geq 6$, we have $\frac{|w|}{|v|} = $
$$\max\left\{\frac{|\mathsf{a}|}{|\mathsf{ab}^2|k + |\mathsf{ab}^3|\ell} : \binom{\ell}{k} \in \mathcal{S}(\mathsf{a})\right\} \leq \max\left\{\frac{1}{3k+4\ell} : k + \ell \geq 2\right\} \leq \frac{1}{6}.$$

On the other hand, $\frac{|w|}{|v|} \geq \frac{1}{3\cdot 2 + 4\cdot 0} = \frac{1}{6}$ as the solution $\binom{\ell}{k} = \binom{0}{2}$ is the Parikh vector of a factor of any Sturmian sequence, in particular of $\mathbf{d}_\mathbf{u}(\mathsf{a})$. Note that the value $1/6$ is attained for any w with $\pi(w) = \mathsf{a}$. For instance we can consider $w = 0$ and $v = 045167$.

Let $\pi(w) = \mathsf{ab}$. Again, $\mathcal{R}_\mathbf{u}(\mathsf{ab}) = \{\mathsf{ab}^2, \mathsf{ab}^3\}$. We have $\text{gap}(\mathbf{y}, 1) = \{2\}$ and $\text{gap}(\mathbf{y}', 1) = \{6, 8\}$. Thus $\binom{\ell}{k} \in \mathcal{S}(\mathsf{ab})$ satisfies
$$k\left(\begin{smallmatrix}1\\2\end{smallmatrix}\right) + \ell\left(\begin{smallmatrix}1\\3\end{smallmatrix}\right) = \left(\begin{smallmatrix}0\\0\end{smallmatrix}\right) \mod \left(\begin{smallmatrix}2\\6 \text{ or } 8\end{smallmatrix}\right). \tag{8}$$
If $k + \ell \geq 4$, then $\frac{|\mathsf{ab}|}{|\mathsf{ab}^2|k + |\mathsf{ab}^3|\ell} = \frac{2}{3k+4\ell} \leq \frac{2}{3k+3\ell} \leq \frac{2}{3\cdot 4} = \frac{1}{6}$. When $1 \leq k + \ell \leq 3$, the only vector $\binom{\ell}{k}$ satisfying Eq. (8) is $\binom{2}{0}$. However, this is never the Parikh vector of a Sturmian factor (cf. Lemma 3).

Let $\pi(w) = \mathsf{b}^2\mathsf{ab}^2$. Then $b = \mathsf{b}^2\mathsf{ab}^2$ is a bispecial factor of $\mathbf{u}$ associated with $(N, m) = (1, 1)$. We have $\text{gap}(\mathbf{y}, 1) = \{2\}$ and $\text{gap}(\mathbf{y}', 4) = \{24\}$. By Proposition 4 we know the Parikh vectors of r and s, thus $\binom{\ell}{k} \in \mathcal{S}(b)$ satisfies
$$k\left(\begin{smallmatrix}1\\2\end{smallmatrix}\right) + \ell\left(\begin{smallmatrix}1\\3\end{smallmatrix}\right) = \left(\begin{smallmatrix}0\\0\end{smallmatrix}\right) \mod \left(\begin{smallmatrix}2\\24\end{smallmatrix}\right). \tag{9}$$
It is not difficult to see that $\frac{5}{3k+4\ell} \leq \frac{1}{6}$.

Similar computations show that $\frac{|w|}{|v|} \leq \frac{1}{6}$ for each $w \in \mathcal{W}^{short}$ and $v \in \mathcal{R}_\mathbf{v}(w)$.

$\mathcal{W}^{long}$: From Example 6 it follows that $E^*(\mathbf{x}_9) \doteq 1,1306 < \frac{7}{6}$. Apply Proposition 6. We have $\mu = (\sqrt{2}-1)^8$. Since $\mu|Q_1 - L_0 Q_2| = \mu|3 - (\sqrt{2}-1)10| \leq 2(\sqrt{2}-1)$ and $|Q_2 - L_0 Q_3| = |10 - (\sqrt{2}-1)23| \leq 2(\sqrt{2}-1)$, we have $I(2 + i + NH, m) \leq E^*(i, m) \leq E^*(\mathbf{x}_9)$ for all i, m and N besides $i = 0, N = 0$, i.e., we have to consider separately the bispecial factors associated with the pairs $(2, 0)$ and $(2, 1)$. Again, both $I(2, 0)$ and $I(2, 1)$ are smaller than $\frac{7}{6}$.

A similar computation can be done for the sequence $\mathbf{x}_{10}$. In this case we can show that $E(\mathbf{x}_{10}) = 1 + \frac{1}{7}$ and that the value $\frac{8}{7}$ is attained for instance for $w = 2$ and $v = 2345067$.

References

1. Baranwal, A.R.: Decision algorithms for Ostrowski-automatic sequences, master thesis, University of Waterloo. http://hdl.handle.net/10012/15845 (2020)
2. Baranwal, A.R., Shallit, J.: Critical exponent of infinite balanced words via the pell number system. In: Mercaş, R., Reidenbach, D. (eds.) WORDS 2019. LNCS, vol. 11682, pp. 80–92. Springer, Cham (2019). https://doi.org/10.1007/978-3-030-28796-2_6
3. Carpi, A., de Luca, A.: Special factors, periodicity, and an application to Sturmian words. Acta Informatica **37**, 986–1006 (2000)
4. Damanik, D., Lenz, D.: The index of Sturmian sequences. Eur. J. Comb. **23**, 23–29 (2002)
5. Dolce, F., Dvořáková, L'., Pelantová, E.: On balanced sequences and their asymptotic critical exponent. In: Proceedings LATA 2021, LNCS vol. 12638, pp. 293–304 (2021)
6. Dvořáková, L'., Medková, K., Pelantová, E.: Complementary symmetric Rote sequences: the critical exponent and the recurrence function. Discrete Math. Theor. Comput. Sci. **20**(1) (2020)
7. Hedlund, G.A., Morse, M.: Symbolic dynamics II - Sturmian trajectories. Am. J. Math. **62**, 1–42 (1940)
8. Hubert, P.: Suites équilibrées. Theor. Comput. Sci. **242**, 91–108 (2000)
9. Rampersad, N., Shallit, J., Vandomme, É.: Critical exponents of infinite balanced words. Theor. Comput. Sci. **777**, 454–463 (2019)
10. Vuillon, L.: A characterization of Sturmian words by return words. Eur. J. Comb. **22**, 263–275 (2001)

The Range Automaton: An Efficient Approach to Text-Searching

Simone Faro$^{(\boxtimes)}$ and Stefano Scafiti

Department of Mathematics and Computer Science, University of Catania,
Catania, Italy
{simone.faro,stefano.scafiti}@unict.it

Abstract. *String matching* is one of the most extensively studied problems in computer science, mainly due to its direct applications to such diverse areas as text, image and signal processing, speech analysis and recognition, information retrieval, data compression, computational biology and chemistry. In the last few decades a myriad of alternative solutions have been proposed, based on very different techniques. However, automata have always played a very important role in the design of efficient string matching algorithms. In this paper we introduce the *Range Automaton*, a weak yet efficient variant of the non-deterministic suffix automaton of a string whose configuration can be encoded in a very simple form and which is particularly suitable to be used for solving text-searching problems. As a first example of its effectiveness we present an efficient string matching algorithm based on the Range Automaton, named Backward Range Automaton Matcher, which turns out to be very fast in many practical cases. Despite our algorithm has a quadratic worst-case time complexity, experimental results show that it obtains in most cases the best running times when compared against the most effective automata based algorithms. In the case of long patterns, the speed-up reaches 250%. This makes our proposed solution one of the most flexible algorithms in practical cases.

Keywords: String matching · Text processing · Automata · Experimental algorithms · Design and analysis of algorithms

1   Introduction

The *string matching* problem consists in finding all the occurrences of a pattern P of length m in a text T of length n, both strings defined over an alphabet Σ of size σ. In the last few decades a myriad of alternative solutions have been proposed, based on very different techniques [9].

The first linear-time solution to the problem was given by Knuth, Morris and Pratt (KMP [11]), whereas Boyer and Moore (BM) provided the first sub-linear solution on average [2]. The Backward-DAWG-Matching (BDM) algorithm [5] was instead the first solution to reach the optimal $\mathcal{O}(n \log_\sigma(m)/m)$ time complexity on the average. Both the KMP and the BDM algorithms are based on

© Springer Nature Switzerland AG 2021
T. Lecroq and S. Puzynina (Eds.): WORDS 2021, LNCS 12847, pp. 91–103, 2021.
https://doi.org/10.1007/978-3-030-85088-3_8

finite automata; in particular, they simulate a deterministic automaton for the language $\Sigma^{\star}P$ and a deterministic suffix automaton for the language of the suffixes of P, respectively. The subsequent solutions to the problem introduced in the literature (see for instance [3,6,8,14]) have amply demonstrated how the efficiency of such solutions is strictly affected by the encoding used for simulating the underlying automaton.

An efficient technique which has been extensively used for the simulation of non-deterministic automaton is *bit parallelism* [1]. It has been used, for instance, in the design of the Shift-Or (SO) algorithm [1] and the Backward-Non-deterministic-DAWG-Matching (BNDM) algorithm [13]. The first is based on the non-deterministic simulation of the KMP automaton, while the second is a very fast variant of the BDM algorithm, based on the bit-parallel simulation of the non-deterministic suffix automaton. Specifically, in the design of automata-based algorithms, bit-parallelism allows to take advantage of the intrinsic parallelism of the bitwise operations inside a computer word, potentially cutting down the number of transitions that an algorithm performs by a factor up to w, i.e. the number of bits in a computer word. However, one bit per pattern symbol is required for representing the states of the automaton, for a total of $\lceil m/w \rceil$ words. This implies that, as long as a pattern fits in a computer word, bit-parallel algorithms are extremely fast, otherwise their performances degrade considerably as $\lceil m/w \rceil$ grows. Although such limitation is intrinsic, several techniques have been developed which retain good performance also in the case of long patterns.

1.1   Previous Results

A common approach to overcome this problem consists in constructing an automaton for a substring of the pattern fitting in a single computer word, to filter possible candidate occurrences of the pattern. However, besides the costs of the additional verification phase, a drawback of this approach is that, in the case of the BNDM algorithm, the maximum possible shift length cannot exceed w, which could be much smaller than m.

The Long-BNDM [14] (LBNDM) and the BNDM with eXtended Shift [6] (BXS) algorithms are two efficient solutions specifically designed for simulating the suffix automaton using bit-parallelism in the case of long patterns. Specifically the LBNDM algorithm works by partitioning the pattern in $\lfloor m/k \rfloor$ consecutive substrings, each consisting in $k = \lfloor (m-1)/w \rfloor + 1$ characters. Similarly the BXS algorithm cuts the pattern into $\lceil m/w \rceil$ consecutive substrings of length w except for the rightmost piece which may be shorter. In both cases the substrings are superimposed getting a superimposed pattern of length w. The idea is to search using a filter approach: first the superimposed pattern is searched in the text, then an additional verification phase is run when a candidate occurrence of the pattern has been located.

Cantone *et al.* presented in [3] an alternative technique, still suitable for bit-parallelism, to encode the non-deterministic suffix automaton of a given string in a more compact way. Their encoding is based on factorization of strings in which no character occurs more than once in any factor. It turns out that the non-deterministic automaton can be encoded with k bits, where k is the size of the factorization. As a consequence, the resulting algorithm, called Factorized-BNDM (FBNDM) tends to be faster in the case of sufficiently long patterns.

Finally the Backward-SNR-DAWG-Matching (BSDM) algorithm, introduced by Faro and Lecroq in [8]. It is an efficient filtration algorithm based on a very simple encoding of the suffix automaton of a pattern x. The BSDM algorithm is based on the fact that a string where each character is repeated only once admits a deterministic suffix automaton which can be encoded with a simple integer.

1.2 Our Contribution

In this paper we introduce the *Range Automaton*, a weak yet efficient variant of the non-deterministic suffix automaton of a string whose configuration can be encoded in a very simple form and which is particularly suitable to be used within text-searching algorithms, naturally overcoming the intrinsic space limitations introduced by bit-parallelism.

The idea underlying the Range Automaton is to approximate the configuration of the non-deterministic Suffix Automaton by means of a simple pair of integers to represent the range within which the active states of the automaton are located. In order to prove the practical effectiveness of our approach, we also present an efficient string matching algorithm based on the Range Automaton and named Backward Range Automaton Matcher (BRAM), which turns out to be very effective in practical cases. From our experimental results it turns out that our algorithm, despite its quadratic worst-case running time, obtains in most cases the best searching speed when compared against the most effective automata based algorithms, especially in the case of long patterns.

The paper is organized as follows. In Sect. 2 we briefly introduce the basic notions which we use along the paper. Then in Sect. 3 we introduce the Range Automaton and present the new algorithm and some efficient implementation of it in Sect. 4. In Sect. 5 we compare the newly presented solution with the suffix automata based algorithms known in literature. We draw our conclusions in Sect. 6.

2   Basic Notions and Definitions

Given a finite alphabet Σ, we denote by Σ^m, with $m \geq 0$, the set of all strings of length m over Σ and put $\Sigma^* = \bigcup_{m \in \mathbb{N}} \Sigma^m$. We represent a string $P \in \Sigma^m$, also called an m-gram, as an array $P[0 \mathinner{.\,.} m - 1]$ of characters of Σ and write $|P| = m$ (in particular, for $m = 0$ we obtain the empty string ε). Thus, $P[i]$ is the $(i + 1)$-st character of P, for $0 \leq i < m$, and $P[i \mathinner{.\,.} j]$ is the substring of P contained between its $(i + 1)$-st and the $(j + 1)$-st characters, for $0 \leq i \leq j < m$. For any two strings P and P', we say that P' is a suffix of P if $P' = P[i \mathinner{.\,.} m - 1]$, for some $0 \leq i < m$, and write $Suff(P)$ for the set of all suffixes of P. Similarly, P' is a prefix of P if $P' = P[0 \mathinner{.\,.} i]$, for some $0 \leq i < m$. In addition, we write

$P \cdot P'$, or more simply PP', for the concatenation of P and P', and P^r for the reverse of the string P, i.e. $P^r = P[m-1]P[m-2]\cdots P[0]$.

Given a string $P \in \Sigma^m$, we indicate with $S(P) = (Q, \Sigma, \delta, I, F)$ the non-deterministic suffix automaton with ϵ-transitions for the language $\mathit{Suff}(P)$, where $Q = \{I, q_0, q_1, \ldots, q_m\}$ is the set of automaton states, I is the initial state, $F = \{q_m\}$ is the set of final states and the transition function $\delta : \mathscr{P}(Q) \times (\Sigma \cup \{\epsilon\}) \longrightarrow \mathscr{P}(Q)$, where $\mathscr{P}(Q)$ is the set of parts of Q. Specifically, for any $Q' \subseteq Q$ and $c \in \Sigma$, we have $q_{i+1} \in \delta(Q', c)$ if $q_i \in Q'$ and $c = P[i]$, for $0 \le i < m$. In addition we have $\delta(I, \varepsilon) = Q$. In all other cases we agree that $\delta(Q', c) = \emptyset$. For simplicity, in what follows, we will use the notation $\delta(q, c)$ instead of $\delta(\{q\}, c)$.

The valid configurations $\delta^*(I, W)$ which are reachable by the automaton $S(P)$ on input $W \in \Sigma^*$ and starting from the initial state I are defined recursively as:

$$\delta^*(I, W) := \begin{cases} \{q_0, q_1, \ldots, q_m\} & \text{if } W = \epsilon, \\ \bigcup_{q' \in \delta^*(I, W')} \delta(q', c) & \text{if } W = W'c, \text{for some } c \in \Sigma, \text{and } W' \in \Sigma^*. \end{cases}$$

3   The Range Automaton

Let P be a string of length m over the alphabet Σ. The *Range Automaton* of a pattern P is a *weaker* version of the non-deterministic Suffix Automaton of P in the sense that, while using an encoding that can allow to keep track of the set of all active states of the automaton, it adopts a *weak* transition approach, meaning that also transitions not tagged with the current character may be activated.

Despite this weak transition approach, the Range Automaton has the interesting feature of operating as an Oracle: the recognized language contains all the factors of P and (possibly) other strings as well. This is the price to pay for an automaton that can allow a simpler encoding and a more efficient simulation.

Before entering into the details of the description of the Range Automaton it is advisable that some useful notions are introduced, some definitions are provided and some properties are proved.

For each character $c \in \Sigma$, we define the *position function*, $\rho : \Sigma \longrightarrow \mathscr{P}(\{0, 1, \ldots, m-1\})$, as the function which maps each character $c \in \Sigma$ to the set of positions where c occurs in P. If c doesn't occur in P we agree to set $\rho(c) = \emptyset$. More formally, $\rho(c) := \{i \mid P[i] = c, 0 \le i < m\}$, for each $c \in \Sigma$. Particularly important for our discussion is the following definition of a range-set.

Definition 1 (Range-Set). *Given a string P of length m and a termination symbol $\$ \notin \Sigma$, a range-set of P is a set of contiguous positions in the string $P\$$. We use the notation $[i : j]$ to denote the range-set of positions in P from i to j, extremes included. Formally $[i : j] = \{i, i+1, \ldots, j\}$, where $0 \le i \le j \le m$.*

The symbol $\$$ is concatenated at the end of P in order to extend its length of one character and allow the value m to be included in any range-set.

We denote by $\mathcal{R}^m$ the set of all possible range-sets associated to a given string on length m. Formally $\mathcal{R}^m = \{[i : j] \mid 0 \leq i \leq j \leq m\}$.

We also define the *range function*, denoted by $r : \Sigma \longrightarrow \mathcal{R}^m$, as the function which maps each character c to the tightest set-range where the character c occurs in the pattern. More formally, for $c \in \Sigma$, $r(c)$ is defined as follows.

$$r(c) = \begin{cases} [\min \rho(c) : \max \rho(c)] & \text{if } \rho(c) \neq \emptyset \\ \emptyset & \text{otherwise.} \end{cases}$$

Example 1. Given the pattern $P =$ banana, we have that $r(\mathsf{a}) = [1 : 5] = \{1, 2, 3, 4, 5\}$, $r(\mathsf{b}) = [0 : 0] = \{0\}$ and $r(\mathsf{n}) = [2 : 4] = \{2, 3, 4\}$, while $r(c) = \emptyset$ for any other character c not appearing in P.

Given a range-set R, we denote by $R \ll k$ the *left shift operation* on R by k positions. The result of such shift operation is a new range-set obtained by decreasing each element of R by k. More formally, if $R = [i : j]$, we have:

$$R \ll k := \begin{cases} \emptyset & \text{if } R = \emptyset \text{ or } j < k, \\ \{0, 1, .., j - k\} & \text{if } i < k \text{ and } j \geq k. \\ \{i - k, i, .., j - k\} & \text{if } i \geq k. \end{cases}$$

Example 2. Given a range-set $R = [2 : 5] = \{2, 3, 4, 5\}$ of size 4, we have that $R \ll 1 = [1 : 4] = \{1, 2, 3, 4\}$ and $R \ll 2 = [0 : 3] = \{0, 1, 2, 3\}$. In addition we have also $R \ll 4 = [0 : 1] = \{0, 1\}$ and $R \ll 6 = \emptyset$.

We notice that a one-to-one correspondence can be defined between the states of the suffix automaton $S(P)$ and the positions within the string $P\$$. Consequently it is possible to map any range-set in $\mathcal{R}$ to a set of states in the suffix automaton. Formally we can map any position i with the state q_i, for $0 \leq i \leq m$, and any range-set $[i : j]$ with the set of states $\{q_i, q_{i+1}, .., q_j\}$, for $0 \leq i \leq j \leq m$.

We are now ready to define the Range Automaton used in our approach. Using the correspondence between any range-set of the pattern P and the set of states in the suffix automaton of P, in the following definition we will deal with the sets of states as *range-sets*. In this context a configuration of the Range Automaton of P is maintained as a single range-set, which identifies the set of all active states of the automaton. In other words if $[i : j]$ is the range-set which represents the configuration of the Range Automaton, each state q_k, with $k \in [i : j]$, is an active state.

Definition 2 (The Range Automaton). *Given a string $P \in \Sigma^m$, we indicate with $A(P) = (Q, \Sigma, \gamma, I_r, F)$ the non-deterministic range suffix automaton of P. It is defined as follows:*

- $Q = [0 : m] = \{0, 1, \ldots, m\}$ *is the set of states of the automaton;*
- $I_r = [0 : m] = Q$ *is the set of initial states;*
- $\gamma : \mathcal{R}^m \times \Sigma \longrightarrow \mathcal{R}^m$ *is the transition function, where $\gamma(R, c)$ is defined as $\gamma(R, c) = (R \ll 1) \cap r(c)$, for any $R \in \mathcal{R}^m$ and $c \in \Sigma$;*
- $F = [0 : 0] = \{0\}$ *is the set of final states.*

The valid configurations $\gamma^*(I_r, W)$ which are reachable by the Range Automaton $A(P)$ on input $W \in \Sigma^*$, with $|W| = n$, are defined recursively as follows

$$\gamma^*(I_r, W) = \begin{cases} [0:m] & \text{if } n = 0 \\ \gamma(\gamma^*(I_r, W[0..n-2]), W[n-1]) & \text{if } n > 0 \end{cases}$$

The following technical lemma allows to characterize the Range Automaton as an oracle, proving that it recognizes (at least) all the factors of the pattern.

Lemma 1. *Let P be a string of length m and let $S(P)$ be the non-deterministic suffix automaton with ϵ-transitions for the language $Suff(P)$. In addition let $A(P)$ be the Range Automaton of P. We have that if $q_i \in \delta^*(I, W)$, for a string $W \in \Sigma^*$, then $i \in \gamma^*(I_r, W)$.*

Proof. Let W be a string of length n. We proceed by induction on n.

For the base case, we have $n = 0$, i.e. $W = \epsilon$. In this case $\delta^*(I, W) = \{q_0, q_1, .., q_m\}$ and $\gamma^*(I_r, W) = [0:m]$, so the lemma trivially holds.

Let now $n > 0$ and let us suppose that the lemma holds for every string of length $l \leq n - 1$. Since $|W| = n > 0$ we can write $W = W'c$, with $W' \in \Sigma^{n-1}$. By inductive hypothesis, if $q_i \in \delta^*(I, W')$, then $i \in \gamma^*(I_r, W')$. Since $\gamma^*(I_r, W')$ is a range, then $[i' : j'] \subseteq \gamma^*(I_r, W')$, where i' and j' are, respectively, the minimum and the maximum of the set $\{i \mid q_i \in \delta^*(I, W')\}$. Remembering that $\delta^*(I, W) = \bigcup_{q' \in \delta^*(q_0, W')} \delta(q', c)$, we have that if $q_k \in \delta^*(I, W)$ then, by the definition of δ the following inequalities hold:

- $\max(0, i' - 1) \leq k \leq \max(0, j' - 1)$,
- $f \leq k \leq l$, where $[f : l] = r(c)$.

By the first inequality it follows that $k \in ([i' : j'] \ll 1)$. Since for the second inequality $k \in r(c)$, then $k \in ([i' : j'] \ll 1) \cap r(c)$. Moreover, we observe that, since $[i' : j'] \subseteq \gamma^*(I_r, W')$, then $([i' : j'] \ll 1) \cap r(c) \subseteq \gamma^*(I_r, W)$ holds. Thus, we can conclude that $k \in \gamma^*(I_r, W)$ too.

The following Corollary allows to characterize the range automaton as a useful tool to search for a pattern in a text. It trivially follows from Lemma 1.

Corollary 1. *Let P be a pattern of length m and let T be a text of length n. In addition let $A(P)$ be the Range Automaton of P. If the prefix $P[0..i]$ occurs in T at position j, i.e. $P[0..i] = T[j..j + i]$, then $0 \in \gamma^*(I, (T[j..j + i])^r)$.*

4   The Backward Range Automaton Matcher

In this section we describe the Backward Range Automaton Matcher (BRAM) and discuss its time and space complexity. In our presentation we will refer to the pseudo-code of the BRAM algorithm depicted in Fig. 1.

As before, let P be a pattern of length m and let T be a text of length n, both strings defined over an alphabet Σ of size σ.

The preprocessing phase of the BRAM algorithm consists in the computations of the function $r(c)$, for each $c \in \Sigma$ (lines 1–6) by means of two simple for loops, taking time $\mathcal{O}(\sigma)$ and $\mathcal{O}(m)$, respectively. Thus, the preprocessing phase achieves an overall $\mathcal{O}(m + \sigma)$-time and $\mathcal{O}(\sigma)$-space complexity.

The searching phase of the algorithm proceeds along the same line of the BNDM algorithm, where the configuration of the Range Automaton is maintained as a range set R. The algorithm works by sliding a window W of length m along the text starting form the left end of the text and proceeding from left to right. At the end of each attempt the window is shifted to the right by a given amount $s > 0$. This process continues until the right end of the text is reached.

Suppose we are in any of the attempts of the search phase, assuming that $W = T[j..j + m - 1]$. At the beginning of the attempt the configuration of the automaton is initialized to the set of initial states I_r. This is done by setting $R = [0 : m] = \{0, 1, ..., m\}$ (line 11). Thus all automaton states are active.

While proceeding in the backward scan of the window the configuration of the automaton is updated accordingly. Specifically, after reading a character c, the configuration of the automaton is updated by the following operation:

$$R \leftarrow (R \ll 1) \cap r(c),$$

which is always performed at the beginning of each iteration of the while cycle at line 12.

The algorithm keeps track of the length of the prefixes recognized during the backward scan by maintaining a variable p which is initialized to 0 at the beginning of each attempt. By Lemma 1, a prefix of P is recognized whenever $0 \in R$. When this condition occurs, the algorithm updates the length of the prefix just identified (line 17). This information will later be used to carry out the correct advancement of the window along the text (line 22).

The backward scan proceeds until R becomes empty, a condition which occurs when the substring $T[j + m - i..j + m - 1]$ is not recognized by the automaton and no state in the automaton is active. In this case the window is advanced in order to align the first character of P with the starting position of the last recognized prefix (line 22).

However, observe that $R = \emptyset$ can occur also when exactly m characters have been scanned. If such condition occurs (line 18) then a candidate occurrence of the pattern has been located and a naive check if performed to verify the occurrence of the whole pattern starting from position j of the text.

Regarding the space and time complexity of the resulting algorithm it is straightforward to observe that the searching phase of the BRAM algorithm runs in $\mathcal{O}(mn)$-time and $\mathcal{O}(\sigma)$-space.

Example 3. Let $P = \mathsf{banana}$ be a pattern of length 6 and assume $W = \mathsf{anaban}$ is the current window of the text. Plainly we have $r(\mathsf{a}) = [1 : 5]$, $r(\mathsf{b}) = [0 : 0]$ and $r(\mathsf{n}) = [2 : 4]$. The following table shows the configurations of the Range Automaton obtained during the backward scan of the string W.

Iteration	Operation	Range-set computation	Configuration
Iteration 0	Initial state	$R_0 = [0 : 5]$	[b a n a n a]
Iteration 1	Read n	$R_1 = [0 : 4] \cap [2 : 4] = [2 : 4]$	b a[n a n]a
Iteration 2	Read a	$R_2 = [1 : 3] \cap [1 : 5] = [1 : 3]$	b[a n a]n a
Iteration 3	Read b	$R_3 = [0 : 2] \cap [0 : 0] = [0 : 0]$	[b]a n a n a
Iteration 4		$R_4 = \emptyset$	

```
BRAM (P, m, T, n)
  1. for each c ∈ Σ do
  2.       r(c) ← ∅
  3. for c ∈ P do
  4.       i ← min ρ(c)
  5.       j ← max ρ(c)
  6.       r(c) ← [i : j]
  7. j ← 0
  8. while j ≤ n − m do
  9.       p ← 0
 10.       i ← m
 11.       R ← [0 : m]
 12.       do
 13.             R ← (R ≪ 1) ∩ r(T[j + i − 1])
 14.             i ← i − 1
 15.             if (0 ∈ R) then
 16.                   if (i > 0) then
 17.                         p ← m − i
 18.                   else
 19.                         if P = T[j ... j + m − 1] then
 20.                               Output(j)
 21.       while (R ≠ ∅)
 22.       j ← j + m − p
```

Fig. 1. The pseudocode of the BRAM algorithm and its auxiliary procedures.

4.1   Speeding-Up Searching by Condensed Alphabets

In order to further improve the efficiency of the BRAM algorithm it is possible to adopt a well-known strategy based on the use of condensed alphabet, whose characters are obtained by combining groups of q characters, for a fixed value q.

As before let P be a pattern of length m over the alphabet Σ. An efficient method for computing a condensed alphabet was presented in [16], and has been then adopted extensively (see for instance [4,8,12]). It makes use of a hash function $hash : \Sigma^q \longrightarrow \{0, ..., \text{MAX} - 1\}$ to combine groups of q characters, for a fixed constant value MAX. Thus a new condensed pattern P_q of length $m - q + 1$, over the alphabet $\{0, \dots, \text{MAX} - 1\}$, is obtained from P. Specifically we have $P_q[i .. j] = hash(P[i] \cdots P[i + q - 1]) \cdots hash(P[j] \cdots P[j + q - 1])$, for $0 \leq i, j \leq m - q$, where $P_q = P_q[0 .. m - q]$.

For instance if $q = 3$ the pattern $P = \mathtt{banana}$ of length 6 is condensed in a new pattern P_3 of length 4, and specifically $P_3 = hash(\mathtt{ban}) \cdot hash(\mathtt{ana}) \cdot hash(\mathtt{nan}) \cdot hash(\mathtt{ana})$.

The hash function is implemented as a **shift-and-addition** procedure, defined as $hash(c_1, c_2, ..., c_q) = (\sum_{i=1}^{q}(c_i \ll (sh \cdot (q - i)))) \mod \text{MAX}$.

The choice of the value sh is related to MAX and q. In our experiments, we set $\text{MAX} = 2^{16}$ while we set $sh = 2$ when $1 \leq q \leq 4$, and $sh = 1$ otherwise.

Figure 2 shows experimental evaluations to compare the performances of the BRAM algorithm under various conditions and for different values of the parameter q (a description of the experimental settings can be found in Sect. 5).

It turns out from experimental evaluations shown in Fig. 2 that the performances of the algorithm strongly depend on the values of m, q and σ. When the size of the alphabet is small then larger values of the parameter q are more effective. Such difference is less sensible when the size of the alphabet gets larger. However it turns out that the smaller is the length of the pattern the lower is the performance of the algorithm. This behavior is more evident for larger values of the parameter q. Thus, the choice of the parameter q should be directed to larger values when the size of alphabet decreases or when the length of the pattern increases. Conversely the values of q should get smaller.

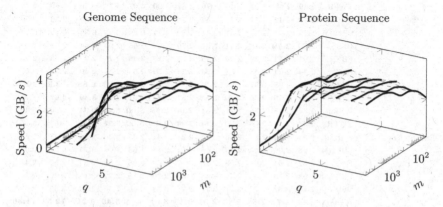

Fig. 2. Running times of the BRAM algorithm extended with condensed alphabets using groups of q characters. We report searching speed of the algorithms for different values of q. Experimental test have been conducted on a genome sequence and a protein sequence. Speed is reported in GB/s.

5   Experimental Results

We report in this section the results of an extensive experimentation of the BRAM algorithm against the most efficient solutions known in the literature for the online exact string matching problem, mostly focusing on those algorithms which make use of the suffix automaton. Specifically, the following 10 algorithms (implemented in 33 variants, depending on the values of their parameters) have been compared: the $BNDM_q$ algorithm [13] implemented with q-grams, for $2 \leq q \leq 6$; the LBNDM algorithm [14]; the BSX_q algorithm [6] implemented using q-grams, with $2 \leq q \leq 4$; the FBNDM algorithm [3] of the BNDM algorithm [13]; the $BSDM_q$ algorithm [8] using q-grams characters, with $3 \leq q \leq 7$; the $BRAM_q$ algorithm, implemented using q-grams characters, with $3 \leq q \leq 7$.

Table 1. Experimental results obtained for searching on a genome sequence (at the top), a protein sequence (in the center) and an English text (in the bottom). Searching speed is reported in GB/s. Best results have been bold faced.

	m	32	64	128	256	512	1024	2048	4096	8192	16384	32768	65536
GENOME SEQUENCE	$BNDM_q$	**3.05**	3.07	3.01	2.96	3.05	3.01	3.03	2.96	2.94	3.00	2.94	2.94
	LBNDM	1.56	1.78	2.00	0.94	0.19	0.20	0.23	0.25	0.25	0.25	0.25	0.26
	BXS_q	2.96	2.86	2.98	3.07	2.96	3.03	2.96	3.09	2.98	2.98	2.76	2.89
	FBNDM	1.76	2.25	2.42	2.26	2.26	2.47	2.34	2.48	2.41	2.37	2.29	2.38
	$BSDM_q$	2.48	2.53	2.58	2.65	2.65	2.79	2.74	2.82	2.74	2.76	2.73	2.73
	$BRAM_q$	2.81	**3.11**	**3.32**	**3.41**	**3.76**	**3.97**	**3.91**	**4.21**	**6.98**	**11.36**	**13.56**	**11.10**
	MAS	0.96	1.18	1.40	1.64	-	-	-	-	-	-	-	-
	MAS_4	2.19	2.82	3.26	3.37	-	-	-	-	-	-	-	-
	TMAS	1.18	1.54	1.74	1.74	-	-	-	-	-	-	-	-
	EPSM	**3.37**	**3.41**	**3.62**	**3.59**	3.76	3.94	3.67	3.97	4.00	4.98	4.65	4.88
	WFR_q	3.05	3.26	3.39	**3.59**	**3.81**	3.84	3.94	4.10	6.78	9.57	6.60	2.38
	$TWFR_q$	2.49	2.63	3.21	3.30	3.76	3.81	4.00	4.07	6.78	9.39	6.69	2.48
PROTEIN SEQUENCE	$BNDM_q$	2.42	2.42	2.41	2.43	2.44	2.38	2.41	2.39	2.41	2.39	2.56	2.56
	LBNDM	1.76	2.01	2.14	2.31	2.41	2.20	1.11	0.49	0.39	0.39	0.40	0.39
	BXS_q	**2.67**	**2.70**	**2.67**	**2.70**	2.67	2.70	2.65	2.68	2.63	2.64	2.63	2.61
	FBNDM	1.97	2.13	2.29	2.29	2.29	2.28	2.26	2.26	2.25	2.27	2.45	2.25
	$BSDM_q$	2.33	2.44	2.48	2.52	2.54	2.54	2.52	2.56	2.54	2.50	2.79	2.57
	$BRAM_q$	2.21	2.38	2.52	2.61	**2.82**	**2.77**	**2.96**	**2.98**	**5.37**	**8.88**	**12.21**	**10.85**
	EPSM	2.48	2.56	2.65	**2.81**	**2.86**	2.87	2.87	2.91	3.01	3.49	4.07	3.79
	WFR_q	2.34	2.48	2.58	2.71	2.81	**2.96**	**2.98**	**3.00**	5.43	9.04	10.61	5.49
	$TWFR_q$	2.37	2.49	2.56	2.70	**2.86**	2.94	**2.98**	**3.00**	5.37	**8.88**	10.39	5.55
ENGLISH TEXT	$BNDM_q$	2.50	2.57	2.58	2.56	2.61	2.57	2.41	2.35	2.36	2.34	2.35	2.3
	LBNDM	1.68	2.06	2.35	2.53	2.58	2.61	2.23	1.63	1.05	0.74	0.61	0.52
	BXS_q	2.57	2.65	2.6	2.58	2.58	2.57	2.6	2.58	2.58	2.53	2.50	2.44
	FBNDM	1.93	2.16	2.42	2.44	2.43	2.43	2.20	2.18	2.21	2.15	2.20	2.17
	$BSDM_q$	**2.60**	2.67	2.74	2.73	2.77	2.77	2.54	2.60	2.58	2.60	2.65	2.65
	$BRAM_q$	2.45	**2.70**	**2.81**	**2.94**	3.05	**3.15**	**3.01**	**3.05**	4.99	**9.21**	**12.52**	**11.36**
	EPSM	**2.65**	**2.71**	**2.84**	**3.00**	**3.07**	3.07	2.94	2.98	3.26	3.62	3.81	4.14
	WFR_q	2.50	2.65	2.74	2.82	3.00	2.89	3.00	3.03	5.25	8.00	7.40	3.81
	$TWFR_q$	2.64	**2.71**	2.76	2.91	3.03	3.11	2.94	3.00	5.25	8.00	7.40	3.81

For completeness, we also evaluated some among the most effective algorithms in practice and specifically: the Maximal Average Shift algorithm and its variants [15] (MAS, MAS_4 and TMAS), specifically designed for genome sequences and short patterns;[1] the Weak Factors Recognition (WFR) algorithm [4], implemented using q-grams, with $3 \leq q \leq 7$ and its variant (TWFR); the Exact Packed String Matching (EPSM) algorithm [7] based on SIMD instructions.[2]

---

[1] Search speed of MAS and its variants, MAS_4 and TMAS, has been omitted starting from $m = 256$, since the preprocessing time of such solutions become prohibitive as the length of the pattern increases.

[2] We notice that the EPSM algorithm is designed for simply counting the number of matching occurrences without reporting the corresponding positions.

All algorithms have been implemented in the C programming language[3] and have been tested using the SMART tool [10]. All experiments have been executed locally on a computer running Linux Ubuntu 20.04.1 with an Intel Core i5 3.40 GHz processor and 8 GB RAM.

Our tests have been run on a genome sequence, a protein sequence, and an English text (each of size 6 MB). Such sequences are provided by the SMART research tool and are available online for download.[4] In the experimental evaluation, patterns of length m were randomly extracted from the sequences, with m ranging over the set of values $\{2^i \mid 5 \leq i \leq 16\}$. In all cases, the mean over the search speed (expressed in Gigabytes per seconds) of 1000 runs has been reported. Table 1 summarises our evaluations. Each table is divided into two blocks. The first block presents results of the most effective automata based algorithms while the second block concerns the speed search obtained by other algorithms. Best results have been boldfaced both among automata-based algorithms and among the entire set of algorithms.

Among the automata-based algorithms the new algorithm turns out to be the best in many cases, obtaining increasingly higher performances as the length of the pattern increases, showing considerable speed ups, especially in the case of long patterns. In particular, other algorithms are superior only for $m = 32$, and, in the case of protein sequences, up to $m = 256$. However, as m grows beyond 1024, the BRAM algorithm becomes by far faster than the previous solutions, reaching a search speed up to 4.6 times higher than the second best solution.

Extending the comparison also to non-automata-based solutions, it is interesting to note how the BRAM algorithm scales better as the size of the pattern increases, outperforming all the remaining algorithms starting from $m = 1024$, both in the case of genome sequences and for texts in natural language. In the case of protein sequences, both WFR_q and TWFR_q turn out to be competitive up to $m = 8192$, but fail to scale-up with respect to the new approach for larger values of m. Moreover, we also notice how the BRAM algorithm is still very competitive also for patterns of medium size, since the search speed never deviates too much from the best results.

6   Conclusions

In this paper we introduced the Range Automaton, a weak version of the nondeterministic suffix automaton of a string whose configuration can be encoded as a simple pair of integers. Such encoding turns out to be effective in order to overcome the intrinsic space limitation of bit-parallel simulations of the suffix automaton. We then introduced a new efficient string matching algorithm, named Backward Range Automaton Matcher (BRAM), based on the Range Automaton of the pattern and conducted an extensive experimental evaluation

---

[3] The source code of the new BRAM algorithm is available at the following link: https://github.com/ostafen/range-automaton.

[4] Additional details on the sequences can be found in Faro et al. [10].

from which it turns out that our newly presented algorithm is very competitive when compared with the most efficient algorithms known in literature.

The good performances obtained by the BRAM algorithm suggest that the encoding of the automaton proposed in this work is simple and flexible and allows us to imagine that it can be adapted to other relevant text-processing problems. Among these, the application to multiple string matching is one of the most promising ways, as is its application to non-standard text processing problems such as approximate string matching.

References

1. Baeza-Yates, R.A., Gonnet, G.H.: A new approach to text searching. Commun. ACM **35**(10), 74–82 (1992). https://doi.org/10.1145/135239.135243
2. Boyer, R.S., Strother Moore, J.: A fast string searching algorithm. Commun. ACM **20**(10), 762–772 (1977). https://doi.org/10.1145/359842.359859
3. Cantone, D., Faro, S., Giaquinta, F.: A compact representation of nondeterministic (suffix) automata for the bit-parallel approach. Inf. Comput. **213**, 3–12 (2012). https://doi.org/10.1016/j.ic.2011.03.006
4. Cantone, D., Faro, S., Pavone, A.: Linear and efficient string matching algorithms based on weak factor recognition. ACM J. Exp. Algorithmics **24**(1), 1..8:1-1.8:20 (2019). https://doi.org/10.1145/3301295
5. Crochemore, M.: Text Algorithms. Oxford University Press, Oxford (1994).http://www-igm.univ-mlv.fr/%7Emac/REC/B1.html
6. Durian, B., Peltola, H., Salmela, L., Salmela, J.: Bit-parallel search algorithms for long patterns. In: Festa, P. (ed.) SEA 2010. LNCS, vol. 6049, pp. 129–140. Springer, Heidelberg (2010). https://doi.org/10.1007/978-3-642-13193-6_12
7. Faro, S., Oguzhan Külekci, M.: Fast and flexible packed string matching. J. Discrete Algorithms **28**, 61–72 (2014)
8. Faro, S., Lecroq, T.: A fast suffix automata based algorithm for exact online string Matching. In: Moreira, N., Reis, R. (eds.) CIAA 2012. LNCS, vol. 7381, pp. 149–158. Springer, Heidelberg (2012). https://doi.org/10.1007/978-3-642-31606-7_13
9. Faro, S., Lecroq, T.: The exact online string matching problem: a review of the most recent results. ACM Comput. Surv **45**(2), 13:1-13:42 (2013). https://doi.org/10.1145/2431211.2431212
10. Faro, S., Lecroq, T., Borzi, S., Di Mauro, S., Maggio, A.: The string matching algorithms research tool. In: 2016 Proceedings of the Prague Stringology Conference, pp. 99–111 (2016). http://www.stringology.org/event/2016/p09.html
11. Knuth, D.E., Morris, J.H., Jr., Pratt, V.R.: Fast pattern matching in strings. SIAM J. Comput., **6**(2), 323–350 (1977). https://doi.org/10.1137/0206024
12. Lecroq, T.: Fast exact string matching algorithms. Inf. Process. Lett **1012**(6), 229–235 (2007). https://doi.org/10.1016/j.ipl.2007.01.002
13. Navarro, G., Raffinot, M.: A bit-parallel approach to suffix automata: fast extended string matching. In: Farach-Colton, M. (ed.) CPM 1998. LNCS, vol. 1448, pp. 14–33. Springer, Heidelberg (1998). https://doi.org/10.1007/BFb0030778
14. Peltola, H., Tarhio, J.: Alternative algorithms for bit-parallel string matching. In: Nascimento, M.A., de Moura, E.S., Oliveira, A.L. (eds.) SPIRE 2003. LNCS, vol. 2857, pp. 80–93. Springer, Heidelberg (2003). https://doi.org/10.1007/978-3-540-39984-1_7

15. Ryu, C., Lecroq, T., Park, K.: Fast string matching for DNA sequences. Theor. Comput. Sci. **812**, 137–148 (2020). https://doi.org/10.1016/j.tcs.2019.09.031
16. Uratani, N., Takeda, M.: A fast string-searching algorithm for multiple patterns. Inf. Process. Manag **29**(6), 775–792 (1993). https://doi.org/10.1016/0306-4573(93)90106-N

A Numeration System for Fibonacci-Like Wang Shifts

Sébastien Labbé[1] and Jana Lepšová[1,2(✉)]

[1] Univ. Bordeaux, CNRS, Bordeaux INP, LaBRI, UMR 5800,
33400 Talence, France
{sebastien.labbe,jana.lepsova}@labri.fr
[2] FNSPE, CTU in Prague, Trojanova 13, 120 00 Praha, Czech Republic

Abstract. Motivated by the study of Fibonacci-like Wang shifts, we define a numeration system for $\mathbb{Z}$ and $\mathbb{Z}^2$ based on the binary alphabet $\{0,1\}$. We introduce a set of 16 Wang tiles that admits a valid tiling of the plane described by a deterministic finite automaton taking as input the representation of a position $(m,n) \in \mathbb{Z}^2$ and outputting a Wang tile.

1 Introduction

A theorem of Cobham (1972) says that a sequence $u = (u_n)_{n \geq 0}$ is k-automatic with $k \geq 2$ if and only if it is the image, under a coding, of a fixed point of a k-uniform morphism [AS03, Theorem 6.3.2]. This result was extended to non-uniform morphisms [RM02], see also [BR10, Theorem 3.4.1], by replacing the usual base-k expansion of nonnegative integers by an abstract numeration system and a regular language [LR01]. It was later extended to configurations $x : \mathbb{N}^d \to \Sigma$ in dimension $d \geq 1$ based on the notion of shape-symmetric morphic words [CKR10], see also [BR10, Theorem 3.4.26] and [AA20, Sect. 5].

In this article, we explore an extension of Cobham's result beyond the non-negative octant $\mathbb{N}^d$ to include configurations $x : \mathbb{Z}^d \to \Sigma$ defined on the whole lattice $\mathbb{Z}^d$. We concentrate on one example in dimension $d = 2$. The example is motivated by the study of Wang tilings of the plane. Given an alphabet $\mathcal{C}$ of colors, a Wang tile is a 4-tuple $(a,b,c,d) \in \mathcal{C}^4$ that represents the labeling of the edges of a unit square, by convention, in the order corresponding to a positive rotation on the complex plane, i.e., a is the east edge label, b is the north edge label, c is the west edge label and d is the south edge label.

We introduce a set $\mathcal{Z} = \{z_0, \ldots, z_{15}\}$ of 16 Wang tiles shown in Fig. 1. The set $\mathcal{Z}$ is a simplification of an existing aperiodic set of 19 Wang tiles [Lab19] after identification of few colors, which was shown to be related [Lab21] to the smallest set of aperiodic Wang tiles found by Jeandel and Rao [JR21].

A valid configuration over the set of Wang tiles $\mathcal{Z}$ is a function $f : \mathbb{Z}^2 \to \{0, \ldots, 15\}$ such that adjacent tiles have the same label on their common edge, i.e., for every $n \in \mathbb{Z}^2$, the east label of the tile $z_{f(n)}$ is equal to the west label of the tile $z_{f(n+e_1)}$ and the north label of the tile $z_{f(n)}$ is equal to the south label of the tile $z_{f(n+e_2)}$. A partial valid configuration is shown in Fig. 2. It is this

© Springer Nature Switzerland AG 2021
T. Lecroq and S. Puzynina (Eds.): WORDS 2021, LNCS 12847, pp. 104–116, 2021.
https://doi.org/10.1007/978-3-030-85088-3_9

O J 0 D O	O H 1 D L	M D 2 J P	M D 3 D K	P J 4 H P	P H 5 H N	K D 6 H P	O I 7 B O
L E 8 I O	L C 9 I L	L I 10A O	P I 11E P	P I 12I K	K B 13 I M	K A 14 I K	N I 15C P

Fig. 1. The set $\mathcal{Z} = \{z_0, \ldots, z_{15}\}$ of 16 Wang tiles. The index i of the tile z_i is written in the center of each tile.

particular configuration that is linked with a numeration system in Theorem 1. The set $\Omega_{\mathcal{Z}}$ of valid configurations $f : \mathbb{Z}^2 \to \{0, \ldots, 15\}$ is called the Wang shift associated to the set of Wang tiles $\mathcal{Z}$.

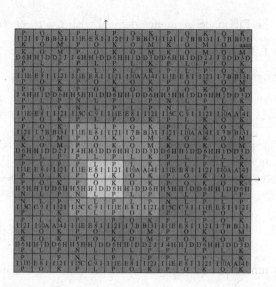

Fig. 2. A partial valid configuration $[-5, 8[^2 \to \{0, \ldots, 15\}$ with the set $\mathcal{Z}$ of Wang tiles. (Color figure online)

We show the following result which states a link between a specific configuration over the set $\mathcal{Z}$ of Wang tiles shown in Fig. 1, a numeration system for $\mathbb{Z}^2$ and a deterministic finite automaton with output (DFAO). Definition of DFAO is recalled in Sect. 3 and corresponds to the classic definition [AS03].

Theorem 1. *Let $\mathcal{Z}$ be the set of 16 Wang tiles shown in Fig. 1. There exist a valid Wang configuration $x \in \Omega_{\mathcal{Z}}$ and a DFAO $\mathcal{A}$ and a numeration system $\mathcal{F}$ for $\mathbb{Z}^2$ with a representation function $\mathrm{rep}_{\mathcal{F}} : \mathbb{Z}^2 \to \{\binom{0}{0}, \binom{0}{1}, \binom{1}{0}, \binom{1}{1}\}^*$ such that the tile at position $n \in \mathbb{Z}^2$ in x is $x_n = \mathcal{A}(\mathrm{rep}_{\mathcal{F}}(n))$.*

In fact, the Wang shift $\Omega_{\mathcal{Z}}$ is self-similar, minimal and aperiodic. Moreover, it is topologically conjugate to the Wang shift $\Omega_{\mathcal{U}}$ generated by the set $\mathcal{U}$ of 19

Wang tiles introduced in [Lab19]. These results are not shown here due to lack of space and will be part of an extended version of this article.

The article is structured as follows. In Sect. 2, we introduce a Fibonacci numeration system for $\mathbb{Z}$ and $\mathbb{Z}^2$. In Sect. 3, we illustrate how to change the usual automaton in Cobham's theorem for $\mathbb{N}$ so that it can read the representation of integers including the negative ones. In Sect. 4, we recall the definitions and notations for two-dimensional words, languages and morphisms. The self-similarity of the Wang shift $\Omega_{\mathcal{Z}}$ is stated in Sect. 5 (the proof is available in the appendix of the preprint version) from which the automaton of Theorem 1 is deduced, see Fig. 5. The proof of Theorem 1 is done in Sect. 6.

2   A Fibonacci Numeration System for $\mathbb{Z}$ and $\mathbb{Z}^2$

Let $(F_n)_{n \geq 1}$ be the Fibonacci sequence defined with the reccurent relation

$$F_0 = 1, F_1 = 1, F_2 = 2, F_{n+2} = F_{n+1} + F_n \quad \text{for all } n \geq 1.$$

By the Zeckendorf theorem [Zec72] every nonnegative integer n can be represented as a unique sum of nonconsecutive Fibonacci numbers $n = \sum_{i=1}^{\ell} w_i F_i$, where $\ell = \max \{i \in \mathbb{N}_0 : F_i \leq n\}$, $w_i \in \{0,1\}$ and $w_i w_{i+1} = 0$, for all $i \in \{1, 2, ..., \ell - 1\}$.

Inspired by the Two's complement, "*the most common method of representing signed integers on computers*",[1] we introduce a numeration system $\mathcal{F}$ which extends the Fibonacci numeration system to all $n \in \mathbb{Z}$ as follows. For each binary word $w = w_{2k+1} w_{2k} \cdots w_1 \in \Sigma^{2k+1}$ of odd length over the alphabet $\Sigma = \{0,1\}$, we define

$$\mathrm{val}_{\mathcal{F}}(w) = \sum_{i=1}^{2k} w_i F_i - w_{2k+1} F_{2k}.$$

The following lemma is an exercise based on the Fibonacci recurrence.

Lemma 2. *Let $k \in \mathbb{N}$ and $w \in \Sigma^{2k} \setminus \Sigma^* 11 \Sigma^*$. We have*

1. $\mathrm{val}_{\mathcal{F}}(0w) = \mathrm{val}_{\mathcal{F}}(000w) = \mathrm{val}_{\mathcal{F}}(110w)$,
2. $\mathrm{val}_{\mathcal{F}}(1w) = \mathrm{val}_{\mathcal{F}}(101w)$,
3. $\mathrm{val}_{\mathcal{F}}(100w) = \mathrm{val}_{\mathcal{F}}(000w) - F_{2k+2}$,
4. $0 \leq \mathrm{val}_{\mathcal{F}}(0w) < F_{2k+1}$,
5. $-F_{2k+2} \leq \mathrm{val}_{\mathcal{F}}(100w) < 0$. $\qquad\qquad\square$

Thus, the first digit of $w \in \Sigma^{2k+1} \setminus \Sigma^* 11 \Sigma^*$ gives the sign (nonnegative or negative) of the value $\mathrm{val}_{\mathcal{F}}(w)$. We can show the following.

Proposition 3. *For every $n \in \mathbb{Z}$ there exists a unique odd-length word*

$$w \in \Sigma(\Sigma\Sigma)^* \setminus (\Sigma^* 11 \Sigma^* \cup 000 \Sigma^* \cup 101 \Sigma^*)$$

such that $n = \mathrm{val}_{\mathcal{F}}(w)$.

---

[1] https://en.wikipedia.org/wiki/Two's_complement.

Proof. (Unicity). Let $w, w' \in \Sigma(\Sigma\Sigma)^* \setminus (\Sigma^*11\Sigma^* \cup 000\Sigma^* \cup 101\Sigma^*)$ of minimal length such that $\text{val}_{\mathcal{F}}(w) = \text{val}_{\mathcal{F}}(w')$. If $w \in 1\Sigma^*$, then $\text{val}_{\mathcal{F}}(w) = \text{val}_{\mathcal{F}}(w') < 0$ and $w' \in 1\Sigma^*$ as well. In fact, we must have $w, w' \in 100\Sigma^*$. Thus $w = 10u$ and $w = 10u'$ for some words u, u' such that $\text{val}_{\mathcal{F}}(u) = \text{val}_{\mathcal{F}}(u')$. This contradicts the minimality of the lengths of w and w'. If $w \in 0\Sigma^*$, then $\text{val}_{\mathcal{F}}(w) = \text{val}_{\mathcal{F}}(w') \geq 0$ and $w' \in 0\Sigma^*$ as well. But $w, w' \notin 000\Sigma^*$, thus $w \in \{01, 001\}u$ and $w' \in \{01, 001\}u'$ for some $u, u' \in \Sigma^*$. From Zeckendorf's theorem applied to $1u$ and $1u'$, we conclude that $u = u'$.

(Existence). If $n = 0$, then $n = 0 = \text{rep}_{\mathcal{F}}(0)$. Assume that $n > 0$. From Zeckendorf's theorem, there exists a unique $u \in 1\Sigma^* \setminus \Sigma^*11\Sigma^*$ such that $n = \text{val}_F(u)$. If u has odd-length, then $n = \text{val}_{\mathcal{F}}(00u)$. If u has even-length, then $n = \text{val}_{\mathcal{F}}(0u)$. Now assume that $n < 0$. Let $k \in \mathbb{N}$ be such that $-F_{2k} \leq n < -F_{2k-2}$. We have $0 \leq n + F_{2k} < F_{2k} - F_{2k-2} = F_{2k-1}$. Let $w \in \Sigma^{2k-2} \setminus \Sigma^*11\Sigma^*$ such that $\text{val}_{\mathcal{F}}(000w) = n + F_{2k}$. We thus have $n = \text{val}_{\mathcal{F}}(100w)$. $\qquad\square$

Definition 4. (Numeration system $\mathcal{F}$ for $\mathbb{Z}$). *For each $n \in \mathbb{Z}$, we denote by* $\text{rep}_{\mathcal{F}}(n)$ *the unique word satisfying the proposition.*

The numeration system $\mathcal{F}$ is illustrated in Fig. 3.

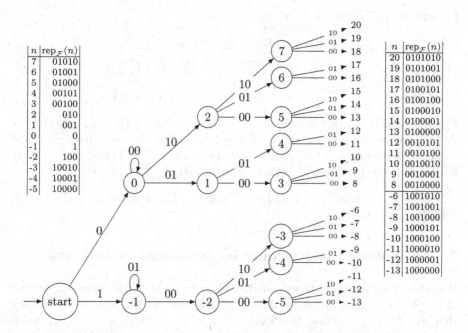

n	$\text{rep}_{\mathcal{F}}(n)$
7	01010
6	01001
5	01000
4	00101
3	00100
2	010
1	001
0	0
-1	1
-2	100
-3	10010
-4	10001
-5	10000

n	$\text{rep}_{\mathcal{F}}(n)$
20	0101010
19	0101001
18	0101000
17	0100101
16	0100100
15	0100010
14	0100001
13	0100000
12	0010101
11	0010100
10	0010010
9	0010001
8	0010000
-6	1001010
-7	1001001
-8	1001000
-9	1000101
-10	1000100
-11	1000010
-12	1000001
-13	1000000

Fig. 3. Representations in the numeration system $\mathcal{F}$ of $n \in [-13, 21[$.

We now extend that numeration system to $\mathbb{Z}^2$. If $\boldsymbol{n} = (n_1, n_2) \in \mathbb{Z}^2$ is such that $|n_1| \gg |n_2|$, then the word representing n_2 is smaller than the word representing n_1. We handle this issue by padding the smaller word so that it becomes of the same length as the longer one. The padding is done differently according to the sign of the number involved: nonnegative numbers are padded with 00, while negative numbers are padded with 10. This is consistent with the numeration system $\mathcal{F}$, refer also to the two loops in Fig. 3.

Definition 5. (Numeration system $\mathcal{F}$ for $\mathbb{Z}^2$). *Let $n = (n_1, n_2) \in \mathbb{Z}^2$. We define*

$$\mathrm{rep}_{\mathcal{F}}(n) = \begin{pmatrix} \mathrm{pad}_t(\mathrm{rep}_{\mathcal{F}}(n_1)) \\ \mathrm{pad}_t(\mathrm{rep}_{\mathcal{F}}(n_2)) \end{pmatrix}$$

where $t = \max\{|\mathrm{rep}_{\mathcal{F}}(n_1)|, |\mathrm{rep}_{\mathcal{F}}(n_2)|\}$ and

$$\mathrm{pad}_t(w) = \begin{cases} (00)^{\frac{1}{2}(t-|w|)}w & \text{if } w \in 0\Sigma^*, \\ (10)^{\frac{1}{2}(t-|w|)}w & \text{if } w \in 1\Sigma^*. \end{cases}$$

The set of all canonical representations $\mathrm{rep}_{\mathcal{F}}(n)$ for $n \in \mathbb{Z}^2$ are words in $\{\binom{0}{0}, \binom{0}{1}, \binom{1}{0}, \binom{1}{1}\}^*$ of odd length such that there are no consecutive one's in each row. E.g. $\mathrm{rep}_{\mathcal{F}}(-2, 9) = \binom{1010100}{0010001}, \mathrm{rep}_{\mathcal{F}}(14, 2) = \binom{0100001}{0000010}$. The length of the representation splits $\mathbb{Z}$ and $\mathbb{Z}^2$ into levels.

Lemma 6. *For every $k \in \mathbb{N}$, we have*

$$\{n \in \mathbb{Z} : |\mathrm{rep}_{\mathcal{F}}(n)| = 2k + 1\} = I_k \setminus I_{k-1},$$
$$\{n \in \mathbb{Z}^2 : |\mathrm{rep}_{\mathcal{F}}(n)| = 2k + 1\} = I_k^2 \setminus I_{k-1}^2$$

where $I_k = \{i \in \mathbb{Z} \mid -F_{2k} \leq i < F_{2k+1}\}$ for $k \geq 0$ and $I_{-1} = \varnothing$.

Proof. The first equality follows from the fact that $I_k = \{n \in \mathbb{Z} : |\mathrm{rep}_{\mathcal{F}}(n)| \leq 2k + 1\}$ where the minimal value $-F_{2k}$ is attained by the word $1(00)^k$ and the maximal value $F_{2k+1} - 1$ is attained by the word $0(10)^k$. The second equality follows from the fact that $I_k^2 = \{n \in \mathbb{Z}^2 : |\mathrm{rep}_{\mathcal{F}}(n)| \leq 2k + 1\}$. $\square$

In Fig. 2, the levels $I_0^2 \setminus I_{-1}^2$, $I_1^2 \setminus I_0^2$ and $I_2^2 \setminus I_1^2$ are shown in yellow, green and blue respectively.

3   An Automaton Not only for Nonnegative Integers

We introduce the terms based on [BR10] to be used in this section. Let $\sigma : A \mapsto A^*$ be a non-erasing morphism prolongable on the letter $a \in A$ such that $x = (x_n)_{n \geq 0} = \sigma^\omega(a)$ is infinite. Let $C = \{0, ..., \max_{b \in A} |\sigma(b)| - 1\}$ be an alphabet. The deterministic finite automaton with output (DFAO) associated to the morphism σ and letter a is the 5-tuple,[2] $\mathcal{A}_{\sigma,a} = (A, C, \delta, a, A)$, where $\delta : A \times C \to A$ is a partial function such that $\delta(b, i) = c$ if and only if $c = u_i$ and $\sigma(b) = u_0 \ldots u_{|\sigma(b)|-1}$. Let L be the language accepted by $\mathcal{A}_{\sigma,a}$. Then the

---

[2] In contrast to [BR10] we omit the coding as it is the identity map.

triple $\mathcal{S} = (L \setminus 0C^*, C, <)$ is an abstract numeration system, where $(C, <)$ is the totally ordered alphabet with the natural order on $\mathbb{N}$ and the language $L \setminus 0C^*$ is radix ordered. Radix order $<_{rad}$ is defined for words $u, v \in L \setminus 0C^*$ as follows: $u <_{rad} v$ if and only if $|u| < |v|$ or $|u| = |v|$ and $u <_{lex} v$. The map $\text{rep}_{\mathcal{S}} : \mathbb{N} \to L \setminus 0C^*$ maps $n \in \mathbb{N}$ to the $(n+1)^{th}$ word in the language $L \setminus 0C^*$ and the map $\text{val}_{\mathcal{S}} : L \to \mathbb{N}$ maps a word w to the number n such that $\text{rep}_{\mathcal{S}}(n) = w'$, where $w = 0^p w'$ for a $p \geq 0$. For some state $r \in A$ and word $w \in C^*$, we denote by $\mathcal{A}_{\sigma,a}(r, w)$ the state reached by the automaton after following the path labeled by w from the state r. We denote it by $\mathcal{A}_{\sigma,a}(w)$ when r is the initial state. The following is essentially a reformulation of [BR10, Corollary 3.4.14].

Proposition 7. *For every integer $n > 0$, there exist integers $m \in \mathbb{N}$ and $\ell \in C$ such that $x_n = \sigma(x_m)[\ell]$ and $\text{rep}_{\mathcal{S}}(n) = \text{rep}_{\mathcal{S}}(m) \cdot \ell$. Moreover, for any $i \in \mathbb{N}$, $|\text{rep}_{\mathcal{S}}(n)| = i$ if and only if $|\sigma^{i-1}(a)| \leq n < |\sigma^i(a)|$.*

Proof. Let $n \in \mathbb{N}$. Let $u \in L \setminus 0C^*$ such that $\text{rep}_{\mathcal{S}}(n) = u$. Let $w \in C^*$ and $\ell \in C$ such that $u = w\ell$. We have $\text{val}_{\mathcal{S}}(w\ell) = n$. Let $m = \text{val}_{\mathcal{S}}(w)$. Since $w \in L \setminus 0C^*$, then $\text{rep}_{\mathcal{S}}(m) = w$. From [BR10, Corollary 3.4.14], we have

$$\sigma(x_{\text{val}_{\mathcal{S}}(w)}) = x_{\text{val}_{\mathcal{S}}(w0)} x_{\text{val}_{\mathcal{S}}(w1)} \cdots x_{\text{val}_{\mathcal{S}}(w\ell)} \cdots x_{\text{val}_{\mathcal{S}}(w \cdot (K-1))}$$

where $K = |\sigma(x_{\text{val}_{\mathcal{S}}(w)})|$. Thus $x_n = x_{\text{val}_{\mathcal{S}}(w\ell)} = \sigma(x_{\text{val}_{\mathcal{S}}(w)})[\ell] = \sigma(x_m)[\ell]$. The other statement follows from the equation [BR10, (3.12)]. □

Let φ be the morphism $\varphi : a \mapsto ab, b \mapsto a$. The automaton $\mathcal{A}_{\varphi^2, a}$ associated to the right-infinite fixed point of φ^2 starting with letter a is shown in Fig. 4. We construct another automaton $\mathcal{A}_{\varphi^2, s}$ associated to the bi-infinite fixed point of φ^2 defined from the seed $s = b.a$, see Fig. 4. The bi-infinite Fibonacci word is $x = \lim_{k \to +\infty} \varphi^{2k}(b.a)$, where the dot represents the origin between positions -1 and 0. When referring to $\varphi^{2k}(b.)$ we mean the finite word $\varphi^{2k}(b) = x_{-|\varphi^{2k}(b)|} \cdots x_{-2} x_{-1}$.

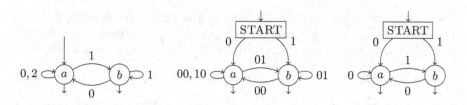

Fig. 4. Automata $\mathcal{A}_{\varphi^2, a}$, $\mathcal{A}_{\varphi^2, s}$ and $\mathcal{A}_{\varphi, s}$ with seed $s = b.a$.

Lemma 8. *Let $x = \lim_{k \to +\infty} \varphi^{2k}(b.a)$ and let $n \in \mathbb{Z} \setminus I_0$. Then there exist integers $m \in \mathbb{Z}$ and $0 \leq \ell < 3$ such that*

$$x[n] = \varphi^2(x[m])[\ell], \quad \text{with } \text{rep}_{\mathcal{F}}(n) = \text{rep}_{\mathcal{F}}(m)h(\ell),$$

where h is the morphism $h : \{0, 1, 2\}^* \to \{0, 1\}^*$ defined as $h : 0 \mapsto 00, 1 \mapsto 01, 2 \mapsto 10$. Moreover, if $n \in I_i \setminus I_{i-1}$ for some $1 \leq i \leq k$, then $m \in I_{i-1} \setminus I_{i-2}$.

Proof. First assume that $n > 0$. Let $\mathcal{A}_{\varphi^2,a}$ and $\mathcal{G} = (L \setminus 0C^*, C, <)$. Then $x = \lim_{k\to+\infty} \varphi^{2k}(a)$ is $\mathcal{G}$-automatic, i.e., for any $n \in \mathbb{N}$, we have $x_n = \mathcal{A}_{\varphi^2,a}(\mathrm{rep}_\mathcal{G}(n))$, where $\mathrm{rep}_\mathcal{G}(n) \in \{\varepsilon, 1, 2, 10, 11, 20, 21, 22, 100, 101, \dots \}$. From the definition of $\mathcal{A}_{\varphi^2,s}$ in Fig. 4, we have $x_n = \mathcal{A}_{\varphi^2,s}(0h(\mathrm{rep}_\mathcal{G}(n)))$. Moreover, from Proposition 7, for $n > 0$ there exist integers $m \in \mathbb{N}$ and ℓ such that $h(\mathrm{rep}_\mathcal{G}(n)) = h(\mathrm{rep}_\mathcal{G}(m))h(l)$. On the other hand, $0h(\mathrm{rep}_\mathcal{G}(n)) = \mathrm{rep}_\mathcal{F}(n)$ for all $n \in \mathbb{N}$. This follows from the following reasons. Applying Proposition 7, we obtain that $|0h(\mathrm{rep}_\mathcal{G}(n))| = 2i+1$ if and only if $n \in (I_i \setminus I_{i-1}) \cap \mathbb{N}$. The alphabet $h(C)$ has the same ordering as the alphabet C. Finally, a word 12 is not accepted by the automaton $\mathcal{A}_{\varphi^2,a}$ and therefore a word 11 is forbidden in $0h(L)$. Moreover, as $|h(\ell)| = 2$, we observe $|\mathrm{rep}_\mathcal{F}(m)| = |\mathrm{rep}_\mathcal{F}(n)| - 2$, thus $m \in I_{i-1} \setminus I_{i-2}$.

If $n = -2$, then we denote $\ell = 0, m = -1$. Let $n < -2$ and let $k \geq 2$ be such that $n \in I_k \setminus I_{k-1}$. We have

$$\varphi^{2k}(b.)[n] = \varphi^{2k-1}(a)[n + |\varphi^{2k}(b)|], \tag{1}$$

where $|\varphi^{2k}(b)| = F_{2k}$. As $\varphi^{2k-1}(a)$ is a prefix of $\varphi^{2k}(a)$, we can write

$$\varphi^{2k-1}(a)[n + F_{2k}] = \varphi^{2k}(a)[n + F_{2k}].$$

Then $0 \leq n + F_{2k} < F_{2k-1}$ and we denote $0 \leq i \leq k - 1$ such that $n + F_{2k} \in I_i \setminus I_{i-1}$. In case that $i > 0$, we use the previous paragraph for positive $n > 0$ on a long enough prefix z of x ($z = \varphi^{2k-2}(a)$, therefore $n + F_{2k} < |z| = F_{2k-1}$) and we find $m_P \in I_{i-1} \setminus I_{i-2}$ and ℓ such that $\mathrm{rep}_\mathcal{F}(n + F_{2k}) = \mathrm{rep}_\mathcal{F}(m_P)h(\ell)$ and

$$\varphi^2(\varphi^{2k-2}(a))[n + F_{2k}] = \varphi^2(\varphi^{2k-2}(a)[m_P])[\ell].$$

As $0 \leq m_P < F_{2k-3} < F_{2k-1} = |\varphi^{2k-2}(a)|$, we restrict the last relation just to a prefix $\varphi^{2k-3}(a)$ and use the relation (1) again to get $m = m_P - F_{2k-2} < 0$

$$\varphi^{2k}(b.)[n] = \varphi^2(\varphi^{2k-3}(a)[m_P])[\ell] = \varphi^2(\varphi^{2k-2}(b.)[m_P - F_{2k-2}])[\ell].$$

The representation $\mathrm{rep}_\mathcal{F}(n) = 10w$ has the first digit corresponding to $-F_{2k}$. Then $(00)^{k-i} \mathrm{rep}_\mathcal{F}(n + F_{2k}) = 00w$. As $m_P \in I_{i-1} \setminus I_{i-2}$, then $m_P - F_{2k-2} \in I_{k-1} \setminus I_{k-2}$ and $\mathrm{rep}_\mathcal{F}(m_P - F_{2k-2}) = 10(00)^{k-i-1} \mathrm{rep}_\mathcal{F}(m_P)$. As a whole,

$$\mathrm{rep}_\mathcal{F}(n) = 10(00)^{k-i-1} \mathrm{rep}_\mathcal{F}(m_P)h(\ell) = \mathrm{rep}_\mathcal{F}(m_P - F_{2k-2})h(\ell) = \mathrm{rep}_\mathcal{F}(m)h(\ell).$$

If $i = 0$, then we denote $\ell = 0, m = -F_{2k-2}$ and the statement holds true.   □

We show the following result.

Proposition 9. *The DFAO $\mathcal{A}_{\varphi,s}$ associated to the seed $s = b.a$ satisfies*

$$x_n = \mathcal{A}_{\varphi,s}(\mathrm{rep}_\mathcal{F}(n)) \text{ for all } n \in \mathbb{Z}.$$

Proof. Let $\mathcal{A}_{\varphi,s}$ be the automaton shown in Fig. 4, i.e., the DFAO $\mathcal{A}_{\varphi,s} = (\{a, b\} \cup \{\mathrm{START}\}, \{0, 1\}, \delta, \mathrm{START}, \{a, b\})$ with the partial function δ such that

- $\delta(\mathrm{START}, \mathrm{rep}_\mathcal{F}(n)) = s_n$ for every $n \in I_0 = \{-1, 0\}$, where $s_{-1} = b, s_0 = a$,

$-\ \delta(c,i) = d$ for any $c, d \in \{a, b\}$ if and only if $\varphi(c) = u$ and $u_i = d$.

Assume $n \in I_0$. If $n = 0$, then we have $x_0 = a = \mathcal{A}_{\varphi,s}(0) = \mathcal{A}_{\varphi,s}(\text{rep}_{\mathcal{F}}(0))$. If $n = -1$, then $x_{-1} = b = \mathcal{A}_{\varphi,s}(1) = \mathcal{A}_{\varphi,s}(\text{rep}_{\mathcal{F}}(-1))$. Induction hypothesis: we assume for some $k \in \mathbb{N}$ that $x_m = \mathcal{A}_{\varphi,s}(\text{rep}_{\mathcal{F}}(m))$ for all $m \in I_k \setminus I_{k-1}$. Let $n \in I_{k+1} \setminus I_k$. Then from Lemma 8 there exist $m \in I_k \setminus I_{k-1}$ and $\ell \in \{0, 1, 2\}$ such that $x_n = \varphi^2(x_m)[\ell]$ and $\text{rep}_{\mathcal{F}}(n) = \text{rep}_{\mathcal{F}}(m) \cdot h(\ell)$, where $h(\ell) = \ell_0 \cdot \ell_1$ for some $\ell_0, \ell_1 \in \{0, 1\}$. From the induction hypothesis, we have

$$x_n = \varphi^2(x_m)[\ell] = \varphi^2(\mathcal{A}_{\varphi,s}(\text{rep}_{\mathcal{F}}(m)))[\ell] = \varphi(\varphi(\mathcal{A}_{\varphi,s}(\text{rep}_{\mathcal{F}}(m)))[\ell_0])[\ell_1]$$
$$= \mathcal{A}_{\varphi,s}(\text{rep}_{\mathcal{F}}(m) \cdot \ell_0 \cdot \ell_1) = \mathcal{A}_{\varphi,s}(\text{rep}_{\mathcal{F}}(m)h(\ell)) = \mathcal{A}_{\varphi,s}(\text{rep}_{\mathcal{F}}(n)).$$

$\square$

4   Two-Dimensional Words, Languages and Morphisms

In this section, we introduce 2-dimensional words, languages and morphisms following the notations of [CKR10,Lab21]. Let $k \in \mathbb{N}$ and $\mathcal{A} = \{0, 1, \ldots, k\}$ be a finite alphabet and let $u : \{0, \ldots, n_1 - 1\} \times \{0, \ldots, n_2 - 1\} \to \mathcal{A}$ be a 2-dimensional word of shape $\boldsymbol{n} = (n_1, n_2) \in \mathbb{N}^2$. Let $\mathcal{A}^{\boldsymbol{n}}$ denote the set of all 2-dimensional words of shape $\boldsymbol{n}$. We refer to the words of shape $(1,2)$, $(2,1)$ as to the vertical, horizontal dominoes, respectively. We represent a 2-dimensional word u of shape $(n_1, n_2) \in \mathbb{N}^2$ as a matrix with Cartesian coordinates:

$$u = \begin{pmatrix} u_{0,n_2-1} & \cdots & u_{n_1-1,n_2-1} \\ \cdots & \cdots & \cdots \\ u_{0,0} & \cdots & u_{n_1-1,0} \end{pmatrix}.$$

Let $\mathcal{A}^{*2} = \bigcup_{\boldsymbol{n} \in \mathbb{N}^2} \mathcal{A}^{\boldsymbol{n}}$ the set of all 2-dimensional words. Let $u, v \in \mathcal{A}^{*2}$ be of shape (n_1, n_2), $(\tilde{n}_1, \tilde{n}_2)$, respectively. If $n_2 = \tilde{n}_2$, the concatenation in direction e_1 is defined as a 2-dimensional word $u \odot^1 v$ of shape $(n_1 + \tilde{n}_1, n_2)$ given as

$$u \odot^1 v = \begin{pmatrix} u_{0,n_2-1} & \cdots & u_{n_1-1,n_2-1} & v_{0,n_2-1} & \cdots\cdots & v_{\tilde{n}_1-1,n_2-1} \\ \cdots & \cdots & \cdots & \cdots & \cdots & \cdots \\ u_{0,0} & \cdots & u_{n_1-1,0} & v_{0,0} & \cdots\cdots & v_{\tilde{n}_1-1,0} \end{pmatrix}.$$

If $n_1 = \tilde{n}_1$, the concatenation in direction e_2 is defined analogically. A word $v \in \mathcal{A}^{*2}$ is a *subword* of a word $u \in \mathcal{A}^{*2}$ if there exist words $u_1, u_2, u_3, u_4 \in \mathcal{A}^{*2}$ such that $u = u_3 \odot^2 (u_1 \odot^1 v \odot^1 u_2) \odot^2 u_4$.

A subset $L \subseteq \mathcal{A}^{*2}$ is called a 2-dimensional *factorial* language if $u \in L$ implies that $v \in L$ for all 2-dimensional subwords v of u.

Let $\mathcal{A}$ and $\mathcal{B}$ be two alphabets. Let $L \subseteq \mathcal{A}^{*2}$ be a factorial language. A function $\omega : L \to \mathcal{B}^{*2}$ is a 2-*dimensional morphism* if for every i with $1 \leq i \leq 2$, and every $u, v \in L$ such that $u \odot^i v$ is defined and $u \odot^i v \in L$, we have that the concatenation $\omega(u) \odot^i \omega(v)$ in direction e_i is defined and

$$\omega(u \odot^i v) = \omega(u) \odot^i \omega(v).$$

A 2-dimensional morphism $L \to \mathcal{B}^{*^2}$ is thus completely defined from the image of the letters in $\mathcal{A}$ and can be denoted as a rule $\mathcal{A} \to \mathcal{B}^{*^2}$.

A subset $X \subseteq \mathcal{A}^{\mathbb{Z}^2}$ is called a *subshift* if it is closed under the shift[3] σ and closed with respect to compact product topology. Let $L \subseteq \mathcal{A}^{*^2}$ be a factorial language and $\mathcal{L}(x)$ be the factorial language containing all subwords of the configuration $x \in \mathcal{A}^{\mathbb{Z}^2}$. Then, $\mathcal{X}_L = \{x \in \mathcal{A}^{\mathbb{Z}^2} \mid \mathcal{L}(x) \subset L\}$ is a subshift generated by L. A 2-dimensional morphism $\omega : L \to \mathcal{B}^{*^2}$ can be extended to a continuous map $\omega : \mathcal{X}_L \to \mathcal{B}^{\mathbb{Z}^2}$ in such a way that the origin of $\omega(x)$ is at zero position in the word $\omega(x_{(0,0)})$ for all $x \in \mathcal{X}_L$.

In general, the closure under the shift of the image of a subshift $X \subseteq \mathcal{A}^{\mathbb{Z}^2}$ under ω is the subshift $\overline{\omega(X)}^\sigma = \{\sigma^k \omega(x) \in \mathcal{B}^{\mathbb{Z}^2} \mid k \in \mathbb{Z}^2, x \in X\} \subseteq \mathcal{B}^{\mathbb{Z}^2}$.

A 2-dimensional morphism $\omega : \mathcal{A} \to \mathcal{A}^{*^2}$ is said *expansive* if the width and height of $\omega^k(a)$ goes to ∞ for all letters $a \in \mathcal{A}$. A subshift $X \subset \mathcal{A}^{\mathbb{Z}^2}$ is *self-similar* if there exists an expansive 2-dimensional morphism $\mathcal{A} \to \mathcal{A}^{*^2}$ such that $X = \overline{\omega(X)}^\sigma$.

5   Self-similarity of the Wang Shift $\Omega_{\mathcal{Z}}$

Proposition 10. *The Wang shift $\Omega_{\mathcal{Z}}$ is self-similar satisfying $\overline{\phi(\Omega_{\mathcal{Z}})}^\sigma = \Omega_{\mathcal{Z}}$ where ϕ is the 2-dimensional morphism over the alphabet $\mathcal{H} = \{0, \ldots, 15\}$*

$$\phi : \mathcal{H} \to \mathcal{H}^{*^2}$$

$$
\begin{cases}
0 \mapsto (14), & 1 \mapsto (13), & 2 \mapsto (12, 10), & 3 \mapsto (11, 8), \\
4 \mapsto (14, 7), & 5 \mapsto (13, 7), & 6 \mapsto (12, 7), & 7 \mapsto \binom{6}{12}, \\
8 \mapsto \binom{3}{14}, & 9 \mapsto \binom{3}{13}, & 10 \mapsto \binom{2}{12}, & 11 \mapsto \binom{6\ \ 1}{12\ 10}, \\
12 \mapsto \binom{6\ 1}{11\ 8}, & 13 \mapsto \binom{5\ 1}{15\ 9}, & 14 \mapsto \binom{4\ 1}{11\ 8}, & 15 \mapsto \binom{2\ 0}{12\ 7}.
\end{cases}
\tag{2}
$$

The proof of Proposition 10 is done in the appendix. It is using algorithms to desubstitute Wang shifts based on the notion of marker tiles [Lab21].

Similarly to the 1-dimensional case, we can build an automaton associated to a fixed point of the 2-dimensional morphism ϕ defined in Eq. (2). Let $s = \left(\begin{smallmatrix} 8 & 12 \\ 1 & 6 \end{smallmatrix}\right) \in \mathcal{H}^{(2,2)}$ be the seed associating one letter to each quadrant. We observe that $\phi^2(s)$ prolongates s at the origin. Therefore, $\lim_{k \to \infty} \phi^{2k}(s)$ defines a configuration in $\mathcal{H}^{\mathbb{Z}^2}$ which is a fixed point of ϕ^2.

Associated to the morphism ϕ and to the seed $s = \left(\begin{smallmatrix} s_{(-1,0)} & s_{(0,0)} \\ s_{(-1,-1)} & s_{(0,-1)} \end{smallmatrix}\right) \in \mathcal{H}^{(2,2)}$, we construct a DFAO $\mathcal{A}_{\phi,s} = (\mathcal{H} \cup \{\text{START}\}, \Sigma, \delta, I, \mathcal{H})$ such that $\Sigma = \{\left(\begin{smallmatrix}0\\0\end{smallmatrix}\right), \left(\begin{smallmatrix}0\\1\end{smallmatrix}\right), \left(\begin{smallmatrix}1\\0\end{smallmatrix}\right), \left(\begin{smallmatrix}1\\1\end{smallmatrix}\right)\}$, $I = \{\text{START}\}$ and $\delta : Q \times \Sigma \to Q$ is a partial function such that

- $\delta(\text{START}, \text{rep}_{\mathcal{F}}(n)) = s_n$ for every $n \in I_0^2 = \{(0,0), (-1,0), (0,-1), (-1,-1)\}$,
- $\delta(a, e) = b$ for any $a, b \in \mathcal{H}$ and $e \in \Sigma$ if and only if b is in $\phi(a)$ at position e.

The automaton $\mathcal{A}_{\phi,s}$ associated to the morphism ϕ and seed $s = \left(\begin{smallmatrix} 8 & 12 \\ 1 & 6 \end{smallmatrix}\right)$ is shown in Fig. 5.

---

[3] Note that from now on, σ denotes the shift action and not a morphism.

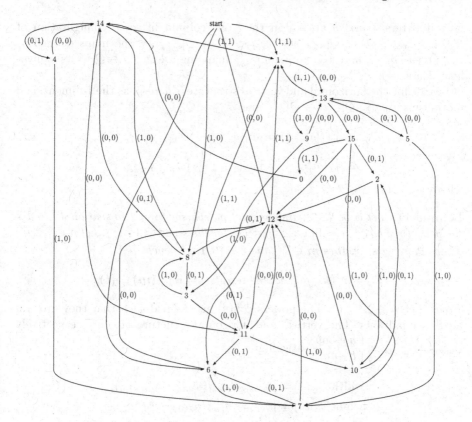

Fig. 5. The automaton $\mathcal{A}_{\phi,s}$ associated to the 2-dimensional morphism ϕ and seed $s = \left(\begin{smallmatrix} 8 & 12 \\ 1 & 6 \end{smallmatrix}\right)$.

6   Proof of Main Results

In this section, we prove Theorem 1. The strategy is to extract the horizontal and vertical structure (expressed as 1-dimensional morphisms) of a 2-dimensional morphism and exploit results proved for the 1-dimensional case in Sect. 3, in particular Lemma 8.

Let ω be a 2-dimensional morphism on the alphabet $\mathcal{Q}$ and $\mathcal{X}_\omega$ be the associated substitutive subshift. Since $\omega : \mathcal{X}_\omega \to \mathcal{X}_\omega$ is well-defined, it imposes that the horizontal width of $\omega(a)$ equals the horizontal width of $\omega(b)$ for every pair of letters $a, b \in \mathcal{Q}$ appearing in the same column. This holds also for the height of the images of letters appearing in the same row. However, more can be said.

We define $\sim_{\mathrm{col}}$ the equivalence relation as the reflexive, symmetric and transitive closure of the relation $\{(a,b) \mid \left(\begin{smallmatrix} b \\ a \end{smallmatrix}\right) \in \mathcal{L}_\omega\}$ made of the vertical dominoes in the language. We define $\sim_{\mathrm{row}}$ the equivalence relation as the reflexive, symmetric and transitive closure of the relation $\{(a,b) \mid (\, a\, b\,) \in \mathcal{L}_\omega\}$ made of the horizontal dominoes in the language. We have that $a \sim_{\mathrm{col}} b$ if and only if letters a and b appear in the same column in some configuration of $\mathcal{X}_\omega$. Let $\pi_{\text{col}} : \mathcal{Q} \to \mathcal{Q}|_{\sim_{\text{col}}}$ and $\pi_{\text{row}} : \mathcal{Q} \to \mathcal{Q}|_{\sim_{\text{row}}}$ be the maps defined as $\pi_{\text{col}} : a \mapsto [a]_{\sim_{\text{col}}}$, and $\pi_{\text{row}} : a \mapsto [a]_{\sim_{\text{row}}}$ mapping a letter $a \in \mathcal{Q}$ to its equivalence class.

We define the horizontal and vertical structure of ω resp. as the 1-dimensional morphisms $\omega_{\text{HORIZ}} : \mathcal{Q}|^*_{\sim_{\text{col}}} \to \mathcal{Q}|^*_{\sim_{\text{col}}}$ and $\omega_{\text{VERT}} : \mathcal{Q}|^*_{\sim_{\text{row}}} \to \mathcal{Q}|^*_{\sim_{\text{row}}}$ by

$$\omega_{\text{HORIZ}}([a]_{\sim_{\text{col}}}) = [w_{0,0}]_{\sim_{\text{col}}} \cdots [w_{m-1,0}]_{\sim_{\text{col}}},$$

$$\omega_{\text{VERT}}([a]_{\sim_{\text{row}}}) = [w_{0,0}]_{\sim_{\text{row}}} \cdots [w_{0,n-1}]_{\sim_{\text{row}}}$$

where $w = \omega(a) \in \mathcal{Q}^{(m,n)}$.

Lemma 11. *Let ϕ be the 2-dimensional morphism from Proposition 10 and x the point fixed by ϕ^2, $x = \lim_{k \to +\infty} \phi^{2k}(s)$ for the seed $s = \left(\begin{smallmatrix} 8 & 12 \\ 1 & 6 \end{smallmatrix}\right)$. Let $\boldsymbol{n} \in \mathbb{Z}^2 \backslash I_0^2$. Then, there exist vectors $\boldsymbol{m} \in \mathbb{Z}^2$ and $\boldsymbol{\ell} \in \{0,1,2\}^2$ such that*

$$x_{\boldsymbol{n}} = \phi^2(x_{\boldsymbol{m}})[\boldsymbol{\ell}], \text{ where } \text{rep}_{\mathcal{F}}(\boldsymbol{n}) = \text{rep}_{\mathcal{F}}(\boldsymbol{m}) \cdot h(\boldsymbol{\ell}).$$

Proof. Let $\boldsymbol{n} \in \mathbb{Z}^2 \setminus I_0^2$. The powers $\phi^{2k}(s)$ are defined such that they grow in all four quadrants. The vertical and horizontal structure of ϕ are respectively $\phi_{\text{VERT}} = \phi_{\text{HORIZ}} = \begin{cases} a \mapsto ab \\ b \mapsto a \end{cases}$ satisfying

$$\text{width}(\phi(w)) = |\phi_{\text{HORIZ}} \circ \pi_{col}(w_{0,0} \cdots w_{m-1,0})|$$
$$\text{height}(\phi(w)) = |\phi_{\text{VERT}} \circ \pi_{row}(w_{0,0} \cdots w_{0,n-1})|$$

for all 2-dimensional words $w \in \mathcal{L}_\phi$ of shape (m,n). Therefore, the vectors $\boldsymbol{m}$ and $\boldsymbol{\ell}$ we are searching for can be found coordinate by coordinate.

Let $y = \lim_{k \to \infty} \phi^{2k}_{\text{HORIZ}}(b.a) = \lim_{k \to \infty} \phi^{2k}_{\text{VERT}}(b.a)$. From Lemma 8, there exist integers $m_1, m_2 \in \mathbb{Z}$ and $0 \le \ell_1, \ell_2 < 3$ such that

$$y[n_1] = \phi^2_{\text{HORIZ}}(y[m_1])[\ell_1] \qquad \text{and} \qquad y[n_2] = \phi^2_{\text{VERT}}(y[m_2])[\ell_2],$$

where $\text{rep}_{\mathcal{F}}(n_1) = \text{rep}_{\mathcal{F}}(m_1) \cdot h(\ell_1)$ and $\text{rep}_{\mathcal{F}}(n_2) = \text{rep}_{\mathcal{F}}(m_2) \cdot h(\ell_2)$. Moreover, it satisfies $x_{(n_1,n_2)} = \phi^2(x_{(m_1,m_2)})[(\ell_1, \ell_2)]$. We conclude

$$\text{rep}_{\mathcal{G}}(\boldsymbol{n}) = \begin{pmatrix} \text{pad}_t(\text{rep}_{\mathcal{F}}(n_1)) \\ \text{pad}_t(\text{rep}_{\mathcal{F}}(n_2)) \end{pmatrix} = \begin{pmatrix} \text{pad}_{t-2}(\text{rep}_{\mathcal{F}}(m_1) \cdot h(\ell_1)) \\ \text{pad}_{t-2}(\text{rep}_{\mathcal{F}}(m_2) \cdot h(\ell_2)) \end{pmatrix} = \text{rep}_{\mathcal{F}}(\boldsymbol{m}) \cdot h(\boldsymbol{\ell}).$$

where $t = \max\{|\text{rep}_{\mathcal{F}}(n_1)|, |\text{rep}_{\mathcal{F}}(n_2)|\}$. $\qquad \square$

Lemma 12. *For any state $r \in Q \setminus \{\text{START}\}$ in the automaton $\mathcal{A}_{\phi,s}$ and any $\boldsymbol{\ell} \in \{0, ..., \text{width}(\phi^2(r)) - 1\} \times \{0, ..., \text{height}(\phi^2(r)) - 1\}$ we have*

$$\mathcal{A}_{\phi,s}(r, h(\boldsymbol{\ell})) = \phi^2(r)[\boldsymbol{\ell}].$$

Proof. Let r and ℓ be according to the assumptions. Therefore, $\ell \in \{0,1,2\}^2$. Then, there exist unique vectors $\ell_0, \ell_1 \in \{0,1\}^2$ such that $\phi^2(s)[\ell] = \phi(\phi(s)[\ell_0])[\ell_1]$. As $\ell_0, \ell_1 \in \{0,1\}^2$, they belong to the set of edges Σ of the automaton $\mathcal{A}_{\phi,s}$. Then, using the general properties of an automaton we have

$$\phi^2(r)[\ell] = \phi(\phi(r)[\ell_0])[\ell_1] = \phi(\mathcal{A}_{\phi,s}(r, \ell_0))[\ell_1]$$
$$= \mathcal{A}_{\phi,s}(\mathcal{A}_{\phi,s}(r, \ell_0), \ell_1) = \mathcal{A}_{\phi,s}(r, \ell_0\ell_1) = \mathcal{A}_{\phi,s}(r, h(\ell)),$$

where the last equation holds for the following reasons.

I) If $\ell \in \{0,1\}^2$, $\binom{i}{j} = \ell$, then $h(\ell) = \binom{0\ i}{0\ j}$. Also, $\ell_0 = \binom{0}{0}$ and $\ell_1 = \binom{i}{j}$.

II) If $\ell \in \{2\} \times \{0,1\}$, $\binom{2}{j} = \ell$, then $h(\ell) = \binom{1\ 0}{0\ j}$. On the other hand, $\ell_0 = \binom{1}{0}$ and $\ell_1 = \binom{0}{j}$. The case $\ell \in \{0,1\} \times \{2\}$ is analogical.

III) If $\ell \in \{2\}^2$, then $h(\ell) = \binom{1\ 0}{1\ 0}$. Also, $\ell_0 = \binom{1}{1}$ and $\ell_1 = \binom{0}{0}$. $\qquad\square$

Theorem 13. *Let ϕ be a 2-dimensional morphism and x the point fixed by ϕ^2, $x = \lim_{k \to +\infty} \phi^{2k}(s)$ for a seed $s = \binom{8\ 12}{1\ 6}$. Then, there exists an automaton $\mathcal{A}$ such that $x_n = \mathcal{A}(\mathrm{rep}_{\mathcal{F}}(n))$.*

Proof. Let $\mathcal{A} = \mathcal{A}_{\phi,s}$ the automaton associated to the morphism ϕ and seed $s = \binom{8\ 12}{1\ 6}$. If $n \in I_0^2 = \{(0,0), (-1,0), (-1,-1), (0,-1)\}$, then $x_n = \mathcal{A}(\mathrm{rep}_{\mathcal{F}}(n))$. Induction hypothesis: we assume for some $k \in \mathbb{N}$ that $x_m = \mathcal{A}(\mathrm{rep}_{\mathcal{F}}(m))$ for all $m \in I_k^2 \setminus I_{k-1}^2$. Let $n \in I_{k+1}^2 \setminus I_k^2$. Then, from Lemma 11 there exist $m \in \mathbb{Z}^2$ and $\ell \in \{0,1,2\}^2$ such that $x_n = \phi^2(x_m)[\ell]$ where $\mathrm{rep}_{\mathcal{F}}(n) = \mathrm{rep}_{\mathcal{F}}(m) \cdot h(\ell)$.

This implies $|\mathrm{rep}_{\mathcal{F}}(m)| = |\mathrm{rep}_{\mathcal{F}}(n)| - 2$, and therefore by Lemma 6, $m \in I_k^2 \setminus I_{k-1}^2$. From the induction hypothesis, $x_m = \mathcal{A}(\mathrm{rep}_{\mathcal{F}}(m))$. Then, from the induction hypothesis and Lemma 12, we have

$$x_n = \phi^2(x_m)[\ell] = \phi^2(\mathcal{A}(\mathrm{rep}_{\mathcal{F}}(m)))[\ell]$$
$$= \mathcal{A}(\mathrm{rep}_{\mathcal{F}}(m)h(\ell)) = \mathcal{A}(\mathrm{rep}_{\mathcal{F}}(n)).$$

$\qquad\square$

Proof (of Theorem 1). Let ϕ be the 2-dimensional morphism from Proposition 10 and let $x = \phi^2(x)$ be the point fixed by ϕ^2, where $x = \lim_{k \to +\infty} \phi^{2k}(s)$ of the seed $s = \binom{8\ 12}{1\ 6}$. Let $\mathcal{A} = \mathcal{A}_{\phi,s}$ (see Fig. 5). The conclusion follows from Theorem 13. $\square$

Example 14. Let $n = (-1,6) \in \mathbb{Z}^2$. Then, $\mathrm{rep}_{\mathcal{F}}(n) = \binom{10101}{01001}$ and $\mathcal{A}_{\phi,s}$ gives

$$\text{START} \xrightarrow{(1,0)} 8 \xrightarrow{(0,1)} 3 \xrightarrow{(1,0)} 8 \xrightarrow{(0,0)} 14 \xrightarrow{(1,1)} 1.$$

The tile at position n in the tiling x in Fig. 2 is indeed $x_n = 1$.

Since $\Omega_{\mathcal{Z}}$ is minimal, we believe that Theorem 1 can be extended to all configurations in $\Omega_{\mathcal{Z}}$ provided an additional input is given. Moreover, we believe that Theorem 1 holds for a large family of self-similar subshifts and not only for Fibonacci-like examples, thus extending Cobham's theorem to $\mathbb{Z}^2$ and $\mathbb{Z}^d$. This asks for further research and is part of an ongoing work.

Acknowledgements. This work was supported by the Agence Nationale de la Recherche through the project Codys (ANR-18-CE40-0007).

References

[AA20]   Akiyama, S., Arnoux, P. (eds.): Substitution and Tiling Dynamics: Introduction to Self-inducing Structures. LNM, vol. 2273. Springer, Cham (2020). https://doi.org/10.1007/978-3-030-57666-0

[AS03]   Allouche, J.-P., Shallit, J.: Automatic Sequences: Theory, Applications, Generalizatio. Cambridge University Press, Cambridge (2003)

[BR10]   Berthé, V., Rigo, M. (eds.): Combinatorics, automata, and number theory, vol. 135. Cambridge University Press, Cambridge (2010)

[CKR10] Charlier, E., Kärki, T., Rigo, M.: Multidimensional generalized automatic sequences and shape-symmetric morphic words. Discrete Math. **310**(6–7), 1238–1252 (2010)

[JR21]   Jeandel, E., Rao, M.: An aperiodic set of 11 Wang tiles. Advances in Combinatorics, January 2021

[Lab19] Labbé, S.: A self-similar aperiodic set of 19 Wang tiles. Geom. Dedicata **201**, 81–109 (2019)

[Lab21] Labbé, S.: Substitutive structure of Jeandel-Rao aperiodic Tilings. Discrete Comput. Geom. **65**(3), 800–855 (2021). https://doi.org/10.1007/s00454-019-00153-3

[LR01]   Lecomte, P.B.A., Rigo, M.: Numeration systems on a regular language. Theory Comput. Syst. **34**(1), 27–44 (2001). https://doi.org/10.1007/s002240010014

[RM02]   Rigo, M., Maes, A.: More on generalized automatic sequences. J. Autom. Lang. Comb. **7**(3), 351–376 (2002)

[Zec72] Zeckendorf, E.: Représentation des nombres naturels par une somme de nombres de Fibonacci ou de nombres de Lucas. Bull. Soc. Roy. Sci. Liège **41**, 179–182 (1972)

Perfectly Clustering Words are Primitive Positive Elements of the Free Group

Mélodie Lapointe[(✉)]

Université de Paris, IRIF, CNRS, 75006 Paris, France
lapointe@irif.fr

Abstract. A word over a totally ordered alphabet is perfectly clustering if its Burrows-Wheeler transform is a non-increasing word. A famous example of a family of perfectly clustering words are Christoffel words and their conjugates. In this paper, we show another similarity between perfectly clustering words and Christoffel words: Both are positive primitive elements of the free group.

Keywords: Perfectly clustering words · Free group · Bases

1 Introduction

Perfectly clustering words are a natural generalization of Christoffel words. Which properties of Christoffel words can be extended to perfectly clustering words? Our main focus is to show that as Christoffel words, perfectly clustering words are positive primitive elements of the free group. An element of the free group is called primitive if it is an element of a basis of the free group.

A word over a totally ordered alphabet is called perfectly clustering if its Burrows-Wheeler transform is a non-increasing word. They were introduced in [13]. The Burrows-Wheeler transform [4] is an invertible function introduced in data compression algorithms. On a binary alphabet the Burrows-Wheeler transform of a word is a non-increasing word if and only if it is a power of a conjugate of a Christoffel word [10]. This characterization of Christoffel words leads to perfectly clustering words as a way to generalize Christoffel words on a larger alphabet.

In the last decade, several properties of perfectly clustering words have been studied. Puglisi and Simpson [13] have used two morphisms and a function defined on the factors of length two of a word to construct the set of perfectly clustering words on a ternary alphabet, as well as proven that these words are the product of two palindromes on any alphabet. Moreover, the square of a word is a factor of an infinite word describing a minimal symmetric discrete interval exchange transformation if and only if the word is a clustering word as shown by Ferenczi and Zamboni [5]. Using the link between interval exchange transformations and perfectly clustering words lead to a family of free group morphisms that can generate perfectly clustering words [8].

© Springer Nature Switzerland AG 2021
T. Lecroq and S. Puzynina (Eds.): WORDS 2021, LNCS 12847, pp. 117–128, 2021.
https://doi.org/10.1007/978-3-030-85088-3_10

Words are elements of the free monoid, but also of the free group on that alphabet. In that sense, we aim to describe the positive bases of the free group using a set of words on the free monoid. For the free group on two generators, it is fully done: conjugates of Christoffel words are exactly the positive primitive elements of the free group with two generators [6,11]. Some sets of words on a larger alphabet are already known to appear in a basis of the free group such as return words of dendric words [3]. Our main result is that perfectly clustering words are positive primitive elements of the free group (see Theorem 2).

This paper is organized as follows. In Sect. 2, we recall properties of the free group and its bases as well as defining perfectly clustering words. In Sect. 3, we describe a set of morphisms such that any perfectly clustering word can be obtained by composing a sequence of those morphisms applied to the word a. In Sect. 4, we prove some properties of primitive elements of the free group and our main result.

2   Definitions

Let $A = \{a_1, \ldots, a_r\}$ be a totally ordered alphabet, where $a_1 < a_2 < \cdots < a_r$. A word w on the free monoid A^* is a sequence of letters, i.e., $w = w_1 \ldots w_n$. The length of $w = w_1 \ldots w_n$, denoted by $|w|$, is n. The number of occurrences of a letter a in w is denoted by $|w|_a$. The Parikh vector of w is the integer vector $(d_1, \ldots, d_r)$ where $d_i = |w|_{a_i}$. The function $Alph$ is defined by $Alph(w) = \{x \in A \mid |w|_a \geq 1\}$. If $w \in A^*$, then $Alph(w) \subseteq A$. A word w is called *complete* if $Alph(w) = A$.

A word w is called *primitive* if it is not the power of another word, that is, if there exists a word z such that $w = z^n$, then $n = 1$. The *conjugates* of a word w of length n are the words $w_i \ldots w_n w_1 \ldots w_{i-1}$. If a word w is primitive, then it has exactly n distinct conjugates.

The lexicographic order is an extension of the total order on A defined by the following: if $u, v \in A^*$, we have $u < v$ if either u is a proper prefix of v, or $u = rxs$ and $v = ryt$ such that $x < y$ and $x, y \in A$ and $r, s, t \in A^*$. A word w is called a *Lyndon word* if it is a primitive word, and it is the minimal word in lexicographic order among its conjugates.

A word u is a factor of w if there exist two words $x, y \in A^*$ such that $w = xuy$. The set of factors of a word is denoted by $Fact(w)$ and $Fact_n(w)$ denotes the set of factors of length n of w.

2.1   Perfectly Clustering Words

To define perfectly clustering words, we first need to introduce the *Burrows-Wheeler transform* of a word. Let w be a primitive word of length n on a totally ordered alphabet. Let $w_1 < w_2 < \cdots < w_n$ be its conjugates, lexicographically ordered. Let l_i be the last letter of the word w_i for $1 \leq i \leq n$. The *Burrows-Wheeler transform* of the word w, denoted by $bwt(w)$, is the word $l_1 \ldots l_n$. For example, the Burrows-Wheeler transform of the word *apartment* is bwt(*apartment*) = *tpmteaanr* (see Fig. 1). One can check that two words u and v are conjugates if and only if bwt(u) = bwt(v) (see [10]).

$$
\begin{array}{ccccccc|c}
a & p & a & r & t & m & e & n & t \\
a & r & t & m & e & n & t & a & p \\
e & n & t & a & p & a & r & t & m \\
m & e & n & t & a & p & a & r & t \\
n & t & a & p & a & r & t & m & e \\
p & a & r & t & m & e & n & t & a \\
r & t & m & e & n & t & a & p & a \\
t & a & p & a & r & t & m & e & n \\
t & m & e & n & t & a & p & a & r \\
\end{array}
$$

Fig. 1. The conjugates of the word *apartment* sorted in lexicographic order. The last column is its Burrows-Wheeler transform

A primitive word w is π-*clustering* if bwt(w) = $a_{\pi(1)}^{|w|_{a_{\pi(1)}}} \ldots a_{\pi(r)}^{|w|_{a_{\pi(r)}}}$, where π is a permutation on $\{1, \ldots, r\}$. Note that π cannot be the identity permutation: Suppose that $\pi = Id$, then $w_1 = ua$ where a is the smallest letter, then $au < w_1$ contradicting the fact that w_1 is the smallest word among its conjugates. For example, the word *aluminium* = $a_1 a_3 a_6 a_4 a_2 a_5 a_2 a_6 a_4$ is 451623-clustering, since bwt(*aluminium*) = *mmnauuiil*. A word is *perfectly clustering* if the permutation π is the symmetric permutation, i.e., $\pi(i) = r - i + 1$, for all $i \in \{1, \ldots, r\}$. A well-known family of examples of perfectly clustering words are Christoffel words and their conjugates (see [10,12]).

2.2 Christoffel Word

Let us recall the definition of Christoffel words. Let p and q be two relatively prime integers, and $A = (q, p)$ a point in the discrete plane $\mathbb{Z} \times \mathbb{Z}$. The *Christoffel path* of slope p/q is the discrete path, formed by elementary steps right and upward, from the origin O to A, lying under the segment OA, and such that the polygon delimited by this segment and the path does not contain any integral point, except those lying on the path (see Fig. 2). The *Christoffel word* of slope p/q is the word on the free monoid $\{a, b\}^*$, which represents the Christoffel path of the slope p/q, where a represents a right step and b represents an upward step. For example, the Christoffel word of slope 4/7 is the word *aabaabaabab* (see Fig. 2). A lot is known about Christoffel words, but let us only recall that Christoffel words are Lyndon words [2] and perfectly clustering words [10].

2.3 Free Group

Let F and G be groups (resp. monoids). A group (monoid) *homomorphism* f is a map from F into G such that $f(uv) = f(u)f(v)$ for any $u, v \in F$ (and

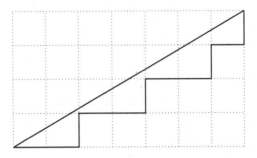

Fig. 2. The Christoffel path of the slope 4/7.

$f(\varepsilon) = \varepsilon$). An *isomorphism* is a bijective homomorphism, and an *automorphism* is an isomorphism from F to itself. An *antimorsphism* f is a map from F into G such that $f(uv) = f(v)f(u)$ for any $u, v \in F$.

Let us denote by F_A the free group generated by the alphabet A. Each element of the free group F_A may be represented by a *reduced word*, which is a word on the alphabet $A \cup A^{-1}$ without the factors aa^{-1} or $a^{-1}a$ for any letter $a \in A$. The monoid $A \cup A^{-1}$ has an involutory antimorphism called *inversion*, $w \mapsto w^{-1}$, which exchange a for a^{-1} for any letter $a \in A$.

For any word w on $A \cup A^{-1}$, there is a unique reduced word equivalent to w modulo the relations $aa^{-1} \equiv a^{-1}a \equiv \varepsilon$ for any $a \in A$. If u is the reduced word equivalent to w, we say that w *reduces* to u and we denote it as $w \equiv u$. The *length* of an element of F_A is the length of its unique reduced word. The *product* of two elements u, v of F_A is the reduced word w equivalent to uv.

The free monoid A^* is a submonoid of F_A, since each word in A^* is a reduced word. An element of the free group F_A that is also an element of A^* is called a *positive* element.

Recall that $A = \{a_1, \ldots, a_r\}$. A *basis* of F_A, denoted by B, is a subset of F_A such that for any function f from the set B to a group H, there exists a unique extension of f to a homomorphism f^* from F_A into H. The alphabet A is a basis of F_A by definition. Any basis B has cardinality equal to $|A|$, since all basis of a free group has the same cardinality. An element g of the free group F_A is *primitive* if there exist elements $h_1, \ldots, h_{|A|-1}$ such that $\{g, h_1, \ldots, h_{|A|-1}\}$ is a basis of F_A.

Remark 1. Some properties of words on the free monoid and elements of the free group have the same name although they are not equivalent, such as primitive words and primitive elements. To avoid confusion we call w a word if $w \in A^*$, and an element, if $w \in F_A$. Therefore, a primitive element is an element of a basis of F_A and a primitive word is not a power of another word.

In fact, a primitive element is also a primitive word, but a primitive word is not necessarily a primitive element. The word $aabb$ is a primitive word, but it is not a primitive element of $F_{\{a,b\}}$; only Christoffel words and their conjugates are primitive elements of $F_{\{a,b\}}$ (see [12]).

If F_A is a free group, then any automorphism of F_A carries A onto another basis. Conversely, every one-to-one map from A onto any basis of F_A uniquely extends to an automorphism. A more complete introduction to free groups can be found in [9].

3 Generation of Perfectly Clustering Words Using Free Group Morphisms

Even if perfectly clustering words are on the free monoid A^*, they can be generated by group morphisms which are not monoid morphisms. Since the definition of those group morphisms on an arbitrary alphabet is technical, we first begin by an example on the ternary alphabet $\{a, b, c\}$.

3.1 Example on the Ternary Alphabet

Let $A = \{a, b, c\}$ be an alphabet such that $a < b < c$. Let λ_a be the group morphism defined by

$$\lambda_a(a) = a, \; \lambda_a(b) = ab, \; \lambda_a(c) = ac.$$

Let λ_b be the group morphism defined by

$$\lambda_b(a) = ab^{-1}, \; \lambda_b(b) = b, \; \lambda_b(c) = bc.$$

Let λ_c be the group morphism defined by

$$\lambda_c(a) = ac^{-1}, \; \lambda_c(b) = bc^{-1}, \; \lambda_c(c) = c.$$

Let ρ_a, ρ_b and ρ_c be group morphisms satisfying $\rho_a = \lambda_a^{-1}$, $\rho_b = \lambda_b^{-1}$, and $\rho_c = \lambda_c^{-1}$. Let f_a and f_b be morphisms from $\{a, b\}^*$ to $\{a, b, c\}^*$ defined by $f_a = Id$ and $f_b(a) = b$, and $f_b(b) = c$. If w is a perfectly clustering word on the ternary alphabet $\{a, b, c\}$, then there exists a sequence of group morphisms, namely $g_1, \ldots, g_n \in \{\lambda_a, \lambda_b, \lambda_c, \rho_a, \rho_b, \rho_c\}$ and $f \in \{f_a, f_b\}$ such that

$$(g_1 \circ \cdots \circ g_n \circ f)(m_w) = w \tag{1}$$

where m_w is a Christoffel word (this is a rewriting of [8, Theorem 4.29] on the ternary alphabet). For example, the word $acbcacc$ is perfectly clustering and Lyndon; It is given by $\rho_c \circ \lambda_a \circ f_a(ab) = acbcacc$.

Remark 2. We denote the Christoffel word above by m_w to highlight that any Christoffel word can be used to generate perfectly clustering words. However, we cannot take a fixed Christoffel word over the alphabet $\{a, b\}$ or $\{b, c\}$ to obtain all perfectly clustering words on the ternary alphabet. If g is a sequence of morphisms in $\{\lambda_a, \lambda_b, \lambda_c, \rho_a, \rho_b, \rho_c\}$, then $g(m_w)$ is not necessarily a perfectly clustering word. It might even be a non-positive element of F_A, e.g. $\lambda_c \circ \lambda_a(ab) = ac^{-1}ac^{-1}bc^{-1}$ which is not an element of A^*.

3.2   General Case

Now, we can present the morphisms needed to generate perfectly clustering words over an arbitrary ordered alphabet in full generality. Let $A_r = \{a_1, a_2, \ldots, a_r\}$ be a totally ordered alphabet of cardinality r. For each letter ℓ in A_r, we define the group morphisms λ_ℓ and ρ_ℓ by

$$\lambda_\ell(a) = \begin{cases} a\ell^{-1}, & \text{if } a < \ell; \\ a, & \text{if } a = \ell; \\ \ell a, & \text{if } a > \ell; \end{cases} \quad \text{and} \quad \rho_\ell(a) = \begin{cases} a\ell, & \text{if } a < \ell; \\ a, & \text{if } a = \ell; \\ \ell^{-1}a, & \text{if } a > \ell. \end{cases}$$

In our example of Sect. 3.1, we define ρ_ℓ as the inverse of λ_ℓ. Now, we need to prove that this holds for the definition we give of ρ_ℓ and λ_ℓ. This will be done in Proposition 3.

Furthermore, we use all Christoffel words (on alphabets $\{a, b\}$ and $\{b, c\}$) to generate the perfectly clustering words on the ternary alphabet. On an arbitrary alphabet, we generate any perfectly clustering word starting with the binary word a. One can easily check that λ_a and ρ_b when defined on the alphabet $\{a, b\}$ coincide with G and $\widetilde{D}$ the usual sturmian morphisms used to construct Christoffel words (see [12] for example). Then, any Christoffel word w is equal to $g(a)$ with $g = g_1 \circ \cdots \circ g_k$ and $g_i \in \{\lambda_a, \rho_b\}$. Hence, m_w can be replaced by $g(a) = m_w$ in Eq. (1). Note that λ_a, defined on $\{a, b\}$ or $\{a, b, c\}$, sends a to the same word and b to the same word. So we simply use λ_ℓ and ρ_ℓ on any finite alphabet.

We generate perfectly clustering on the totally ordered alphabet A_r by starting from the set of perfectly clustering words on the totally ordered alphabet A_{r-1}. Hence, we also need monoid morphisms analogue to f. Let $A_r = \{a_1, \ldots, a_r\}$ be an alphabet of cardinality r such that $a_1 < a_2 < \cdots < a_r$. We define a larger totally ordered alphabet A_{r+1} of cardinality $r + 1$ by adding a new letter to A_r, namely $a_{r+1} \notin A$ such that for all $a \in A_r$, $a < a_{r+1}$, i.e., $A_{r+1} = A_r \cup \{a_{r+1}\}$. Let f_{ℓ, A_r} be a monoid morphism A_r^* to A_{r+1}^* defined by

$$f_{\ell, A_r}(a_i) = \begin{cases} a_i & \text{if } a_i < \ell, \\ a_{i+1} & \text{otherwise,} \end{cases}$$

where $a_i \in A_r$. For example, $f_{a_3, A_4}(a_1) = a_1$, $f_{a_3, A_4}(a_2) = a_2$, $f_{a_3, A_4}(a_3) = a_4$ and $f_{a_3, A_4}(a_4) = a_5$. In other words, if w is a word on the alphabet A_r, then $f_{a_i, A_r}(w)$ is a word on the alphabet $\{a_1, \ldots, a_{i-1}, a_{i+1}, \ldots, a_r\} \subset A_{r+1}$. The natural extension of f_{ℓ, A_r} on the free group is $f_{\ell, A_r}(a^{-1}) = f_{\ell, A_r}(a)^{-1}$ for all $a \in A_r$. If w is a perfectly clustering word, there is a composition of those morphisms applied to a which is equal to w as shown in [8, Theorem 4.29].

Theorem 1. *Let w be a Lyndon complete perfectly clustering word on the totally ordered alphabet A. There exists a sequence of free group morphisms, namely $g = g_1 \circ \cdots \circ g_k$, such that*

$$g(a) = w$$

and $g_i \in \{\lambda_{\ell_j}, \rho_{\ell_j}, \lambda_{\ell_j} \circ f_{\ell_j,B}, \rho_{\ell_{i+1}} \circ f_{\ell_{j+1},B} \mid \ell_j \in A$ *and* $B \subset A\}$ *for* $i \in \{1,\ldots,k\}$.

For example, the Lyndon complete perfectly clustering word $adbcbdadbd$ is given by the sequence of morphisms $\lambda_b \circ f_{b,\{a,b,c\}} \circ \rho_c \circ \lambda_a \circ f_{a,\{a,b\}} \circ \rho_b$.

Remark 3. We choose to construct only Lyndon perfectly clustering, since the Burrows-Wheeler transform of two words is equal if and only if they are conjugates. Hence, applying conjugacy on the set of Lyndon perfectly clustering words gives the set of perfectly clustering words. Moreover, if a word is perfectly clustering, permutation of letters of that word are necessarily not perfectly clustering word. Hence, we enumerate complete word to ensure that the resulting word is perfectly clustering.

We do not present the proof of Theorem 1, which can be found in [8, Chap. 4], but let us discuss the main ideas of it. In order to prove Theorem 1, we need a result of Ferenczi and Zamboni [5]. A word w is a complete perfectly clustering word if and only if w^2 is a factor in the trajectory of a minimal symmetric discrete interval exchange transformation. Let w be a perfectly clustering word. We use this bijection to prove that the reduced words of $\lambda_\ell(w)$ and $\rho_\ell(w)$ are perfectly clustering words by working on the interval exchange transformation instead of the Burrows-Wheeler transform.

In fact, minimal symmetric discrete interval exchange transformations are defined by a vector, called circular composition, which is the commutative image, i.e., the Parikh vector, of the corresponding perfectly clustering words. We use a tree labelled by all circular compositions introduced in [7] to prove that any Lyndon complete perfectly clustering word can be obtained using a sequence of morphisms as in Theorem 1. This construct along the way a tree labelled by all perfectly clustering words.

3.3   On the Positivity

As perfectly clustering words are defined on the free monoid, not the free group, we want that at each step i in the sequence of morphisms the element $g_i \circ g_k(a)$ is positive. However, the morphisms λ_ℓ and ρ_ℓ are not positive morphisms, but we can determine if $\lambda_\ell(w)$ or $\rho_\ell(w)$ reduces to a positive word only by looking at its factors of length two.

Proposition 1. *Let* $w = w_1 \ldots w_n \in A^*$ *be a word. The reduced word equivalent to* $\lambda_\ell(w)$ *is positive if and only if for all,* $i \in \{1,\ldots,n-1\}$, $w_i < \ell$ *implies that* $w_{i+1} \geq \ell$ *and* $w_n \geq \ell$. *The reduced word equivalent to* $\rho_\ell(w)$ *is positive if and only if, for all* $i \in \{2,\ldots,n\}$, $w_i < \ell$ *implies that* $w_{i-1} \geq \ell$ *and* $w_1 \leq \ell$.

Proof. Suppose that $w_i < \ell$ implies that $w_{i+1} \geq \ell$ for all $i \in \{1,\ldots,n-1\}$ and $w_n \geq \ell$. The only inverse letter in the reduced word of $\lambda_\ell(w)$ is ℓ^{-1}, since w is a positive element of F_A. Moreover, all occurrences of ℓ^{-1} in λ_ℓ are followed by at least one occurrence of the letter ℓ. Hence, in the reduced word equivalent to

$\lambda_\ell(w)$, the letter ℓ^{-1} does not appear. Then, the reduced equivalent to $\lambda_\ell(w)$ is positive.

Suppose that u, the reduced word equivalent to $\lambda_\ell(w)$, is positive. This means that the letter ℓ^{-1} does not appear in u. Then, all letters in the word w are larger than ℓ, or all occurrences of a letter smaller than w is followed by an occurrence of a letter larger than w, i.e., for all $i \in \{1, \ldots, n-1\}$, $w_i < \ell$ implies that $w_{i+1} \geq \ell$ and $w_n \geq \ell$. The proof is similar to the circularly reduced element of $\rho_\ell(w)$. ∎

In the case of a perfectly clustering word, we can say even more as the factor of length two are determined by the Parikh vector of the word.

Proposition 2. *Let w be a Lyndon perfectly clustering word and ℓ a letter. The reduced word equivalent to $\lambda_\ell(w)$ (resp. $\rho_\ell(w)$) is positive if*

$$\sum_{j > \ell} |w|_j > \sum_{j < \ell} |w|_j \qquad \left(resp. \sum_{j > \ell} |w|_j < \sum_{j < \ell} |w|_j \right)$$

with $j \in A$.

For example, the word $acbcacc$ is perfectly clustering and its Parikh vector is $(2,1,4)$. Since $\sum_{j>b} |acbcacc|_j = 4 > \sum_{j<b} |acbcacc|_j = 2$, then $\lambda_b(w) = acbbcacbc$ is positive, but $\rho_b(w) = accacb^{-1}c$ is not positive.

Remark 4. The same condition on the Parikh vector of w means that $\lambda_\ell(w)$ (resp. $\rho_\ell(w)$) is a perfectly clustering word as shown in [8, Theorem 4.12 and Theorem 4.18]

Proof. Let w be a Lyndon perfectly clustering word of length n on the alphabet A_r and its Parikh vector $(d_1, \ldots, d_r)$. Let $w_1 < w_2 < \cdots < w_n$ be the set of conjugates of w in lexicographic order. The word formed by the first letter of each word w_i is $u_1 \ldots u_n = a_1^{d_1} \ldots a_r^{d_r}$ and the word formed by the last letter of each word is $v_1 \ldots v_n = a_r^{d_r} \ldots a_1^{d_1}$, since w is a perfectly clustering word. Then, the factors of length 2 of w are of the form: $v_i u_i \in Fact_2(w)$, for some $i \in \{1, \ldots, n\}$. Since ℓ is a letter in A, there exist a letter a_k for some $k \in \{1, \ldots, r\}$ such that $a_k = \ell$. Suppose that $\sum_{j>\ell} |w|_j > \sum_{j<\ell} |w|_j$, that is if $\sum_{j=k+1}^{m} d_j > \sum_{j=1}^{k-1} d_j$. If $vu \in Fact_2(w)$ and $v < \ell$, then $u > \ell$ as the v_i's are in non-increasing order and the u_i's in non-decreasing order and more letters in w are larger than ℓ. Then, $u_i < \ell$ implies that $u_{i+1} > \ell$. Moreover, $w_n = a_r$ since w is a Lyndon perfectly clustering word. By Lemma 1, the reduced word equivalent to $\lambda_\ell(w)$ is a positive. A similar proof is done for the reduced word equivalent to $\rho_\ell(w)$. ∎

4   Primitive Elements of the Free Group

As for Christoffel words, perfectly clustering words are primitive positive elements of the free group. In fact, we show that λ_ℓ and ρ_ℓ are automorphisms of the free group F_A meaning that they send a basis of F_A to another basis. However, f_ℓ is not an automorphism of the free group.

Proposition 3. *The morphisms λ_ℓ and ρ_ℓ are automorphisms of the free group F_A.*

Proof. Let A_r be a totally ordered alphabet of cardinality r and F_{A_r} the free group with basis A_r. Let $\ell \in A_r$ be a letter. We want to show that the morphism λ_ℓ is the inverse morphism of ρ_ℓ, i.e., $\lambda_\ell \circ \rho_\ell = \rho_\ell \circ \lambda_\ell = Id$. It is enough to check that $\lambda_\ell \circ \rho_\ell(a) = \rho_\ell \circ \lambda_\ell(a) = a$ for any letter $a \in A_r$, since both are group morphisms. There are three cases.

1. If $a < \ell$, then

$$\lambda_\ell \circ \rho_\ell(a) = \lambda_\ell(a\ell) = a\ell^{-1}\ell = a$$
$$\rho_\ell \circ \lambda_\ell(a) = \rho_\ell(a\ell^{-1}) = a\ell\ell^{-1} = a.$$

2. If $a = \ell$, then

$$\lambda_\ell \circ \rho_\ell(a) = \lambda_\ell(a) = a = \rho_\ell(a) = \rho_\ell \circ \lambda_\ell(a).$$

3. If $a > \ell$, then

$$\lambda_\ell \circ \rho_\ell(a) = \lambda_\ell(\ell^{-1}a) = \ell^{-1}\ell a = a$$
$$\rho_\ell \circ \lambda_\ell(a) = \rho_\ell(\ell a) = \ell\ell^{-1}a = a.$$

Therefore λ_ℓ is the inverse of ρ_ℓ and both morphisms are automorphisms of the free group F_{A_r}. ∎

Let us stress the fact that f_ℓ is not an automorphism, but we will show in Lemma 2 that f_ℓ sends a primitive element of A_r to a primitive element of A_{r+1}. Notice that f_ℓ is a restriction to A_r^* of an automorphism of the free group $F_{A_{r+1}}$. Let us check first some properties of primitive elements of F_A.

Lemma 1. *Let x be a primitive element of F_A. If $A \subseteq B$, then x is a primitive element of F_B.*

Proof. Let $A = \{a_1, \ldots, a_r\}$ be an alphabet. We have that $B = \{a_1, \ldots, a_r, b_1, \ldots, b_s\}$ with $a_i \neq b_j, \forall i \in \{1, \ldots, r\}$ and $\forall j \in \{1, \ldots, s\}$, since $A \subseteq B$. Let x be a primitive element of F_A, then there exist elements $h_1, \ldots, h_{r-1} \in F_A$ such that $\{x, h_1, \ldots, h_r\}$ is a basis of F_A. Moreover, x and h_i are elements of F_B where the only letters appearing are in A. Therefore $\{x, h_1, \ldots, h_{r-1}, b_1, \ldots, b_s\}$ is a basis of F_B. Then x is a primitive element of F_B. ∎

Lemma 2. *Let A be a totally ordered alphabet of cardinality r, $b \notin A$ and $B = A \cup \{b\}$, such that for all $a \in A$, $a < b$. If w is a primitive element of F_A, then $\lambda_\ell(w)$, $\rho_\ell(w)$ and $f_\ell(w)$ are primitive elements of F_B for any $\ell \in A$.*

Proof. Let us recall that λ_ℓ and ρ_ℓ are automorphisms by Proposition 3 of the free group F_A. Since λ_ℓ and ρ_ℓ are automorphisms, they send a basis of F_A to another basis, then the reduced words of $\lambda_\ell(w)$ and $\rho_\ell(w)$ are primitive words.

Then, the reduced word of $\lambda_\ell(w)$ and $\rho_\ell(w)$ are primitive elements of F_B by Lemma 1.

Let us show that $f_\ell(w)$ is a primitive element of F_B. Observe that $f_\ell(A) \cup \{\ell\} = A \cup \{b\}$. Thus f_ℓ can be extended to a permutation letting $f_\ell(b) = \ell$. This extension of f_ℓ is an automorphism. Then, for any primitive element w, $f_\ell(w)$ is a primitive element. ∎

Let us state our main theorem.

Theorem 2. *Let w be a perfectly clustering word on the totally ordered alphabet A. Then w is a primitive element of F_A.*

Before doing the demonstration, let us introduce the conjugation on the free group. Two elements $u, v \in F_A$, are *conjugate* if there exists $g \in F_A$ such that $u = gvg^{-1}$. This is not the same definition as the one for conjugate on the free monoid. However, there is a relationship between the notions of (monoid-theoretic) conjugacy in A^* and (group-theoretic) conjugacy in F_A. In the proof of Theorem 2, we will show that if two words are conjugates in A^*, then they are also conjugates in F_A. The opposite statement also holds as proven in [1, Lemma 5.1].

Proof. Let w be a complete perfectly clustering word. If w is a Lyndon word. There exists a sequence of free group morphisms, such that $g(a) = w$ by Theorem 1. Moreover, the word a is a positive primitive element of F_A, if $a \in A$. Moreover, g sends primitive elements to the primitive elements by Lemma 2. Then w is a primitive element of F_A.

If w is not a Lyndon word, then w is conjugate to a Lyndon word since perfectly clustering words are primitive words by definition. Denote by v the conjugate of w which is a Lyndon word. Since w and v are conjugate, there exist two words $x, y \in A^*$ such that $w = xy$ and $v = yx$. Since $ywy^{-1} = y(xy)y^{-1} = yx = v$. Then, w and v are conjugate in F_A. Moreover, v is a primitive element of F_A. There is a basis $\{v, h_1, \ldots, h_r\}$ of F_A with $h_i \in F_A$, namely $\{y^{-1}vy, y^{-1}h_1y, \ldots, y^{-1}h_ry\} = \{w, y^{-1}h_1y, \ldots, y^{-1}h_ry\}$ is a basis of F_A (remark: the conjugation is a group automorphism). Then w is a primitive element of F_A. ∎

5   Conclusion

Theorem 2 extends to a larger alphabet a property of Christoffel words, namely they are positive primitive elements of the free group, and allows to recover the following result.

Corollary 1 ([6,11]). *Any conjugate of a Christoffel word is a primitive positive element of the free group $F_{\{a,b\}}$.*

In fact, the articles [6,11] also show that any primitive element of $F_{\{a,b\}}$ is a conjugate of a Christoffel word. However, on a larger alphabet there is not an equivalence between perfectly clustering words and positive primitive elements of the free group. In F_A with $|A| > 2$, it is possible to find positive primitive elements of F_A which are not perfectly clustering words. Let us construct a basis of the free group whose elements are not perfectly clustering words. Let u be the Tribonacci word, i.e., the infinite word over the alphabet $\{a, b, c\}$ and $a < b < c$ which is the fixed point of the substitution $\phi(a) = ab$, $\phi(b) = ac$ and $\phi(c) = a$. The return words of the factor baa are $\{baabaca, baabacababaca, baabacabaca\}$. Those words are a positive basis of the free group $F_{\{a,b,c\}}$ (see [3]), but none of them is a perfectly clustering word.

To conclude, let us check some known results on Christoffel words and symmetric interval exchange transformations that lead to open questions on perfectly clustering words. In [3], the authors construct bases of the free group with the first return words of dendric words. Since trajectory of minimal interval exchange transformations are dendric words, it means that first return words of those words are also primitive elements of F_A. As perfectly clustering words are also factors of the trajectory of an interval exchange transformation, we are currently investigating the following question supported by experimental data:

1. Are return words of a symmetric interval exchange transformation perfectly clustering words?

It is important to highlight that perfectly clustering words and first return words are proper subsets of the factors of a trajectory of a minimal interval exchange transformation. For example, the word $baab$ is a factor of sturmian words, but it is not a first return word, neither a perfectly clustering word.

As primitive element of $F_{\{a,b\}}$ are Christoffel words or conjugates to a Christoffel word, they can be factorized as a product of two Christoffel words forming a basis of the free group. We would want to factorize similarly perfectly clustering words on a larger alphabet. For example, takes $acbbcacbc$ a perfectly clustering word on the ternary alphabet and the factorization $ac.bbc.acbc$. Each word is a perfectly clustering word and they form a basis of the free group as each letter can be expressed as a product of the words ac, bbc and $acbc$: $(ac)(acbc)^{-1}(ac)(bbc)(acbc)^{-1}(ac) = a$, $(bbc)(acbc)^{-1}(ac) = b$, and $(ac)^{-1}(acbc)$ $(bbc)^{-1}(ac)^{-1}(acbc) = c$. The factorisation of a Christoffel word into two Christoffel words also provide a way to construct a basis from a Christoffel word. Hence, the following questions on perfectly clustering word:

2. Can a complete perfectly clustering word on the alphabet A_r be factorized in a product of r words forming a basis of F_{A_r}?
3. If w is a perfectly clustering word, can we describe a set of words B of cardinality $r - 1$ such that $B \cup \{w\}$ is a positive basis of the free group F_{A_r}?
4. If yes, are the words in B perfectly clustering words?

Acknowledgments. We would like to thank Christophe Reutenauer and Valérie Berthé for their suggestions and their helpful comments. We acknowledge the support of the Natural Sciences and Engineering Research Council of Canada (NSERC), [funding reference number BP–545242–2020], and the Fonds de recherche du Québec – Nature et Technologies (FRQNT).

References

1. Berstel, J., Lauve, A., Reutenauer, C., Saliola, F.V.: Combinatorics on Words. CRM Monograph Series, vol. 27. American Mathematical Society, Providence, RI (2009). Christoffel words and repetitions in words
2. Berstel, J., de Luca, A.: Sturmian words, Lyndon words and trees. Theoret. Comput. Sci. **178**(1–2), 171–203 (1997)
3. Berthé, V., et al.: Acyclic, connected and tree sets. Monatsh. Math. **176**, 521–550 (2015). https://doi.org/10.1007/s00605-014-0721-4
4. Burrows, M., Wheeler, D.J.: A block sorting data compression algorithm, p. 18. Technical report, Digital System Research Center (1994)
5. Ferenczi, S., Zamboni, L.Q.: Structure of K-interval exchange transformations: induction, trajectories, and distance theorems. J. Anal. Math. **112**, 289–328 (2010)
6. Kassel, C., Reutenauer, C.: Sturmian morphisms, the braid group B_4, Christoffel words and bases of F_2. Ann. Mat. Pura Appl. **186**(2), 317–339 (2007)
7. Lapointe, M.: Number of orbits of discrete interval exchanges. Discrete Math. Theor. Comput. Sci. **21**(3), 16 (2019). Paper No. 17
8. Lapointe, M.: Combinatoire des mots: mots parfaitement amassants, triplets de Markoff et graphes chenilles. Ph.D. thesis, Université du Québec à Montréal, Montréal, Québec, Canada (October 2020). https://archipel.uqam.ca/14120/
9. Lyndon, R.C., Schupp, P.E.: Combinatorial Group Theory. CM, vol. 89. Springer, Heidelberg (2001). https://doi.org/10.1007/978-3-642-61896-3
10. Mantaci, S., Restivo, A., Sciortino, M.: Burrows-Wheeler transform and Sturmian words. Inf. Process. Lett. **86**(5), 241–246 (2003)
11. Osborne, R.P., Zieschang, H.: Primitives in the free group on two generators. Invent. Math. **63**(1), 17–24 (1981)
12. Reutenauer, C.: From Christoffel Words to Markoff Numbers. Oxford University Press, Oxford (2019)
13. Simpson, J., Puglisi, S.J.: Words with simple Burrows-Wheeler transforms. Electron. J. Combin. **15**(1), 1–17 (2008). Research Paper 83

On Billaud Words and Their Companions

Szymon Łopaciuk$^{(\boxtimes)}$ and Daniel Reidenbach

Department of Computer Science, Loughborough University,
Loughborough LE11 3TU, UK
{s.p.lopaciuk,d.reidenbach}@lboro.ac.uk

Abstract. The Billaud Conjecture, which has been open since 1993, is a fundamental problem on finite words w and their heirs, i.e., the words obtained by deleting every occurrence of a given letter from w. It posits that every morphically primitive word, i.e. a word which is a fixed point of the identity morphism only, has at least one morphically primitive heir. In this paper, we introduce and investigate the related class of so-called Billaud words, i.e. words whose all heirs are morphically imprimitive. We provide a characterisation of morphically imprimitive Billaud words, using a new concept. We show that there are two phenomena through which words can have morphically imprimitive heirs, and we highlight that only one of those occurs in morphically primitive words. Finally, we examine our concept further, use it to rephrase the Billaud Conjecture and study its difficulty.

Keywords: Billaud conjecture · Morphic primitivity · Fixed point

1 Introduction

In this paper we study the notion of morphic primitivity of words: a word is morphically primitive if the only morphism for which it is a fixed point is the identity morphism. The context of our research is the conjecture posed by Billaud [1] in 1993, which is still open today:

Conjecture 1 (The Billaud Conjecture). There exists at least one letter x in every morphically primitive word w such that the word obtained by deleting all occurrences of x in w is also morphically primitive.

We shall call the word w the *parent*, and the words obtained through a deletion of a letter from w *heirs*. As a simple example, consider the morphically primitive word abcbac: the word resulting from the deletion of the letter c is abba, and it is morphically primitive. The Billaud Conjecture, however, is an implication and not a characterisation. Hence, while there exist words, such as $(abc)^2$, where the deletion of each of the letters leads to a morphically imprimitive heir, there are also words such as abcbca, which are morphically imprimitive, yet have morphically primitive heirs (here abba).

T. Lecroq and S. Puzynina (Eds.): WORDS 2021, LNCS 12847, pp. 129–141, 2021.
https://doi.org/10.1007/978-3-030-85088-3_11

Despite being an open problem, a number of results in relation to the Billaud Conjecture have been established. In the direct response to the question posed by Billaud, the initial insight that we need only to consider idempotent morphisms was due to Geser [3]; an extended version of this statement was proved by Levé and Richomme [9]. Several special cases where the conjecture holds were identified: Zimmermann [15] showed that the conjecture holds for the alphabet size of 3. (We note the case of alphabet size of 2 is trivial.) Levé and Richomme [9], working with the contrapositive of the conjecture, proved that if all morphically imprimitive heirs of a word are fixed points of non-trivial morphisms with exactly one expanding letter each, then the same holds for the parent word; the letter x is expanding if the morphism ϕ for which the word is a fixed point is such that $|\phi(x)| \geq 2$. Walter [14] identified and proved some of the cases necessary for showing that the conjecture holds for alphabet size of 4. Nevisi and Reidenbach [11] proved that the conjecture is correct for all words (with three or more different letters) if they contain each letter exactly twice.

More generally, the study of morphic primitivity of words has been of interest in a variety of contexts. Finite fixed points of morphisms were first characterised by Head [5] in the context of L-systems. An alternative characterisation of the finite fixed points was given by Hamm and Shallit [4], and in the context of erasing pattern languages by Reidenbach [12]. Reidenbach and Schneider [13] introduced the concept of 'morphic imprimitivity' itself, and described a factorisation characteristic to finite fixed points of morphisms. In relation to terminal-free erasing pattern languages, a morphically primitive word is the shortest generator of a pattern language, as shown by Filè [2], and Jiang et al. [7]. Holub [6] gave a polynomial-time algorithm (further refined and analysed by Matocha and Holub [10]) to decide whether or not a word w is morphically imprimitive; the algorithm can yield the imprimitivity factorisation of a morphically imprimitive w. Kociumaka et al. [8] further improved the algorithm to work in linear time.

In this paper we introduce and explore a class of words related to the Billaud Conjecture, the words whose all heirs are morphically imprimitive. We shall call these *Billaud words*, and as an example consider the word abcabc again, whose three heirs abab, acac, and bcbc are all morphically imprimitive. In particular, if the Billaud Conjecture holds, then all Billaud words are morphically imprimitive. After providing some basic definitions, in Sect. 3 we focus on morphically imprimitive Billaud words. We shall give more complex examples of these words and show that there are two fundamental phenomena leading to morphically imprimitive heirs that occur on deletion of a letter. Finally, we provide a characterisation of morphically imprimitive Billaud words. In the subsequent Sect. 4, we shift our focus to examining different aspects surrounding one of the given phenomena. To learn more about the reasons for the difficulty of the Billaud Conjecture, we provide an alternative statement of it and discuss why certain approaches to solving it are unsuitable.

We note that due to space constraints most of the proofs have been omitted.

2 Preliminaries

We shall denote the set of all positive integers with $\mathbb{N}_+$, and non-negative integers with $\mathbb{N}_0$. Moreover, for any two integers $m, n, m \geq n$, we shall denote with $[\![n, m]\!]$ the set of all integers i such that $n \leq i \leq m$. An *alphabet* Σ is an enumerable set of *symbols* or *letters*. Letters typeset in a sans-serif font, i.e. a, b, c, ..., should always be understood as distinct; in other cases, we shall explicitly state such assumptions. As a *word (over an alphabet Σ)* we refer to a finite sequence of elements, symbols, of Σ. The cardinality of a set A shall be denoted by $|A|$, similarly the length of a word w is $|w|$. The *empty word*, denoted by λ, is the special word for which $|\lambda| = 0$. The *concatenation* of two words w and v shall be written as $w \cdot v$ or wv. Given a word w and $n \in \mathbb{N}_0$, w^n denotes the concatenation of n copies of the word w. We say that a word v is a *factor* of w, denoted $v \sqsubseteq w$, if there exist words w_1 and w_2 such that $w = w_1 \cdot v \cdot w_2$. Given words w, v and $i \in \mathbb{N}_0$, we shall call the tuple $\langle i, v \rangle$ an *occurrence (of v) in w (at position i)* if there are words w_1, w_2 such that $w = w_1 \cdot v \cdot w_2$ and $|w_1| = i$. The number of occurrences of v in w is denoted by $|w|_v$. Let $\mathrm{symb}(w)$ denote the set of all letters $x \sqsubseteq w$.

For any alphabets A and B and all words $w, v \in A^*$, a *(homo-)morphism* $\phi : A^* \to B^*$ is a function that satisfies $\phi(w)\phi(v) = \phi(wv)$. A morphism $\phi : A^* \to A^*$ is *idempotent* if $\phi = \phi \circ \phi$. A word w is a *(finite) fixed point* of a morphism ϕ if $\phi(w) = w$. If a morphism ϕ is not idempotent, and there exists a finite fixed point word w of ϕ, then, as shown by Geser [3], there exists an idempotent morphism ϕ' such that $\phi'(w) = w$. There also exists an integer $i \in \mathbb{N}_+$ such that $\phi^i = \phi'$, and we shall call i the *mortality exponent* of ϕ. If x is a letter, then we denote with π_x the morphism *deleting x*, i.e. $\pi_x(y) = y$ for all $y \neq x$, and $\pi_x(x) = \lambda$. The morphism $\iota : A^* \to A^*$ is the *identity* (or *trivial*) if for every $x \in A$, $\iota(x) = x$.

A word w is *morphically primitive* [13] if there is no word w' with $|w'| < |w|$ such that w and w' can be mapped onto each other by morphisms; otherwise w is *morphically imprimitive*. At the end of the present section, we shall explain that this definition is equivalent to the notion of morphic primitivity used at the beginning of Sect. 1.

Let $\phi : \Sigma^* \to \Sigma^*$ be an idempotent morphism; when discussing morphic imprimitivity we shall restrict ourselves only to idempotent morphisms, unless explicitly stated otherwise. We define the following three sets, similarly to Levé and Richomme [9], which form a partition on $\mathrm{symb}(w)$, in the context of ϕ:

- the set of *expanding letters*, $E_\phi = \{x \in \Sigma \mid |\phi(x)| \geq 2\}$,
- the set of *mortal letters*, $M_\phi = \{x \in \Sigma \mid \phi(x) = \lambda\}$,
- the set of *constant letters*, $C_\phi = \{x \in \Sigma \mid \phi(x) = x\}$.

A letter $x \in \Sigma$ is *necessarily expanding* (or *mortal*, or *constant*) if $x \in E_\phi$ (or M_ϕ, or C_ϕ respectively) for all (idempotent) non-identity $\phi : \mathrm{symb}(w)^* \to \mathrm{symb}(w)^*$ with a fixed point w. Let us denote the set of *(all) expanding letters in a word w* with E_w: the letter x is in E_w if there is a morphism ϕ such that $x \in E_\phi$, and

$\phi(w) = w$. Then we say that x is *expanding in (the word)* w. We define M_w and C_w analogously.

An *imprimitivity factorisation* f of w is a tuple $\langle x_1, x_2, \ldots, x_n; v_1, v_2, \ldots, v_n \rangle$, where $v_1, v_2, \ldots, v_n \in \Sigma^+$, $x_1, x_2, \ldots, x_n \in \Sigma$, $n \in \mathbb{N}_+$, such that w can be written as $u_0 v_1 u_1 v_2 u_2 \cdots v_n u_n$, for some $u_0, u_1, \ldots, u_n \in \Sigma^*$, and there is a nontrivial morphism ϕ with $\phi(w) = w$ such that for all $i \in \mathbb{N}_+$, $|v_i|_{x_i} = 1$, $v_i = \phi(x_i)$ and $u_0, u_i \in C_\phi^*$. We call the word v_i an *imprimitivity factor of* f *(corresponding to x_i)*. We shall say that ϕ *determines* f, and define sets E_f, M_f, and C_f to be equal to E_ϕ, M_ϕ, and C_ϕ.

Finally, as we shall operate with the following concepts interchangeably, we restate the following result by Reidenbach and Schneider [13]. For every word w, the following statements are equivalent: w is morphically imprimitive, w is a fixed point of a non-trivial morphism, and w has an imprimitivity factorisation.

3   Morphically Imprimitive Billaud Words

Let us recall that the Billaud Conjecture is stated merely as an implication, and not as a characterisation. In other words, there exist morphically imprimitive words with morphically primitive heirs. In this section, we shall provide a characterisation of morphically imprimitive Billaud words, i.e. words whose all heirs are morphically imprimitive. In the context of the conjecture, this means partitioning the set of all morphically imprimitive words into those relevant to the Billaud Conjecture, hence our name: Billaud words, and the morphically imprimitive words with morphically primitive heirs. Moreover, it is also worth noting that the Billaud Conjecture in this context posits that all Billaud words are morphically imprimitive, i.e. there are no morphically primitive Billaud words in general. We look into morphically primitive words in more detail in Sect. 4. We shall now proceed to formally define the notion of Billaud words:

Definition 1. *The word w is a* Billaud word *if $\pi_x(w)$ is morphically imprimitive for all $x \in \text{symb}(w)$.*

In order to better understand what Billaud words look like, and to motivate our reasoning thereafter, we consider the following example:

Example 1. Let us define the following words w_1 and w_2:

$$w_1 := \mathsf{ab\,c\,ab\,dea\,c\,ab\,dea\,c\,ab}, \quad w_2 := \mathsf{ab\,c\,ab\,de^2a\,c\,ab\,de^2a\,c\,ab}$$

The words w_1, w_2 are fixed points of the following morphisms ϕ_1, ϕ_2 respectively:

$$\phi_1 : \mathsf{b} \mapsto \mathsf{ab}, \mathsf{d} \mapsto \mathsf{dea}, \mathsf{c} \mapsto \mathsf{c}, \mathsf{ae} \mapsto \lambda, \quad \phi_2 : \mathsf{b} \mapsto \mathsf{ab}, \mathsf{d} \mapsto \mathsf{de^2a}, \mathsf{c} \mapsto \mathsf{c}, \mathsf{ae} \mapsto \lambda$$

Neither ϕ_1 nor ϕ_2 is trivial, and the words w_1 and w_2 are morphically imprimitive. The word w_1 is a Billaud word, as all of its heirs are morphically imprimitive. This can be verified by finding a suitable imprimitivity factorisation of each of these heirs or by finding morphisms for which the words are fixed points. On the other hand, the word w_2 is not a Billaud word, as it has an heir $\pi_\mathsf{d}(w_2)$ which is morphically primitive. $\Diamond$

The words in Example 1 are very similar in many respects, and differ superficially by one square. In particular, we can see that the sets of expanding, constant, and mortal letters are equal for ϕ_1 and ϕ_2. Nevertheless, a more detailed consideration of letter roles in a word is centre to our characterisation of Billaud words. In the remainder of this section we shall prove our characterisation of morphically imprimitive Billaud words, which shall culminate with Theorem 1, by systematically considering deletion of letters of different roles.

We shall start by giving a lemma with a basic necessary condition for a word to be a morphically imprimitive Billaud word, namely that a word has to have at least one imprimitivity factorisation with more than one mortal letter in it.

Lemma 1. *Let w be a morphically imprimitive word. If every imprimitivity factorisation of w has exactly one mortal letter, then $\pi_m(w)$ is morphically primitive for every $m \in M_w$.*

We shall note that the class of words which have one mortal letter in each of their imprimitivity factorisations is severely restricted. Generally speaking, if a word has one mortal letter per imprimitivity factorisation, then it is the same mortal letter across all imprimitivity factorisations. There is an exception to this rule, illustrated by the word $(ab)^2$, where the roles of the letters a, b can 'flip' between the two factorisations of the word. The following proposition illustrates this:

Proposition 1. *Let w be a morphically imprimitive word where every imprimitivity factorisation has exactly one mortal letter. Let w have no pair of imprimitivity factorisations f, g such that $E_f \cup M_f = E_g \cup M_g$. Then the word has exactly one mortal letter.*

We shall now commence with the key part of the proof of our characterisation of morphically imprimitive Billaud words. Let w be an arbitrary morphically imprimitive word. We shall investigate what happens when we delete letters from w, and in particular if the resulting heirs are morphically imprimitive, in order to determine whether or not w is a Billaud word. In particular we note that if a letter is neither mortal nor constant in w, it is necessarily expanding.

We introduce Lemmata 2, 3, and 6, each giving a necessary and sufficient condition for the morphic imprimitivity of the heirs obtained by deletions of different classes of letters. The subsequent two lemmata deal with the simpler cases of deletion of constant or mortal letters from w.

Lemma 2 (Any Constant Letter). *Let w be a morphically imprimitive word. For all $x \in C_w$, the word $\pi_x(w)$ is morphically imprimitive.*

Lemma 3 (Any Mortal Letter). *Let w be a morphically imprimitive word. For all $x \in M_w$, the word $\pi_x(w)$ is morphically imprimitive if and only if there is at least one imprimitivity factorisation f of w where $|M_f| \geq 2$.*

So far, we have dealt with situations where the imprimitivity factorisations of the parent and the heir are in a way 'related' to each other. These are the cases when the morphisms determining the imprimitivity of the parent and the heir, let us call them ϕ and χ respectively, are such that either $\chi = \phi$ or $\chi = \pi_x \circ \phi$.

In order to finalise our characterisation of morphically imprimitive Billaud words, the remainder of our reasoning deals with those situations in the word when, as a result of deletion of a letter, we obtain a word which has an imprimitivity factorisation 'unrelated' to the parent word. As we have considered the deletion of constant, and mortal letters, we now address the final class of necessarily expanding letters. We observe that when we delete a necessarily expanding letter in the word, the resulting heir has an alternative parent, in which the deleted letter is not necessarily expanding. The following is an example of such a word, where there is a letter that is necessarily expanding.

Example 2. Let $w := (\mathsf{abacb})^2$; then it can be verified that $\phi : \mathsf{c} \mapsto \mathsf{abacb}, \mathsf{ab} \mapsto \lambda$ is the only non-trivial morphism for which w is a fixed-point. Note that the letter c is thus necessarily expanding in w. The heir obtained by deleting c from w, $\pi_\mathsf{c} = (\mathsf{ab})^2$, is trivially morphically imprimitive, as it is a fixed point of, e.g., the morphism $\chi : \mathsf{a} \mapsto \mathsf{ab}, \mathsf{b} \mapsto \lambda$. Interestingly, χ is not related to ϕ – there is no simple way in which we could uniformly 'add' letters to χ, such that the new morphism determines the morphic imprimitivity of w.

However, let us observe what happens when we apply χ (redefined to map c to itself) to w: $w' := \chi(w) = ((\mathsf{ab})^2\mathsf{c})^2$. The 'new' word w', which shares an heir with w, still has c as an expanding letter, but it is no longer necessarily expanding, as a result of having another imprimitivity factorisation determined by χ. In fact, with Lemma 6, we show that the existence of a morphism where c is a constant letter, and mapping w to w' is characteristic for the word $\pi_\mathsf{c}(w)$ to be morphically imprimitive. $\Diamond$

We shall define this specific relationship between the 'original', and the 'new' parent as follows:

Definition 2. *Let w and w' be words. We call w' a* companion *of w (with respect to x) if $\pi_x(w) = \pi_x(w')$ and there exists a non-identity idempotent morphism ϕ such that $\phi(w) = w'$ and $\phi(x) = x$. We shall refer to such morphism ϕ as a* companion morphism.

Firstly, we present a fundamental lemma that motivates our use of companions, which demonstrates that these words are always morphically imprimitive:

Lemma 4. *Let w' be a companion of some word w, and let ϕ be the corresponding companion morphism. Then the word w' is morphically imprimitive, and ϕ determines an imprimitivity factorisation of w'.*

Having defined this notion of a companion, we now postulate that a word having a companion for a letter is characteristic for it having a morphically imprimitive heir under the deletion of said letter:

Lemma 5. *Let w be a word, and let x be a letter in w. The word $\pi_x(w)$ is morphically imprimitive if and only if the word w has a companion with respect to x.*

However, it can be seen from Lemma 5 that the present definition of companions is quite general and includes the simpler cases otherwise covered in Lemmata 2 and 3. For example, it is worth noting that any morphically imprimitive word which has a constant letter is its own companion with respect to that letter, as trivially the morphism for which the word is a fixed point maps the letter to itself. Moreover, the phenomenon described before, where a new imprimitivity factorisation appears in the word upon deletion of a letter can also happen, even when the letter is constant or mortal. Since we have Lemmata 2 and 3, such an imprimitivity factorisation is in a way redundant to determining the morphic imprimitivity of the heir. Thus, it is advisable to strengthen the definition of a companion to include only those cases when we do not have a trivial imprimitivity factorisation. Definition 3 accomplishes that, but before we shall introduce it, we present an example to better illustrate the discussed phenomena.

Example 3. Let $w := $ abc abxd ebxc ebd. The word has only one imprimitivity factorisation determined by the following morphism:

$$\phi : a \mapsto ab, e \mapsto eb, b \mapsto \lambda, c \mapsto c, d \mapsto d, x \mapsto x$$

In particular the letter x is a constant letter in w, and thus trivially by Lemma 2 the heir $\pi_x(w)$ is morphically imprimitive. It can then be seen that the morphism ϕ satisfies the definition of a companion morphism, and as such the word w is its own companion. However, w has another companion w' with respect to the following morphism:

$$\chi : c \mapsto bc, d \mapsto bd, b \mapsto \lambda, a \mapsto a, e \mapsto e, x \mapsto x$$

Thus $w' := \chi(w) = $ abc axbd exbc ebd, or in other words we have swapped neighbouring b and x. The word w' has an imprimitivity factorisation determined by χ which is 'new': neither the letter c nor d is expanding in w. However, even though this imprimitivity factorisation of w' cannot be derived from any imprimitivity factorisations of w and is as such an interesting case, it is in a way redundant, as already by Lemma 2 we have determined that $\pi_x(w)$ is morphically imprimitive.

$\Diamond$

In other terms, the word in Example 3 possesses a dual nature: it is a witness to both of the previously discussed phenomena simultaneously. The heir of $\pi_x(w)$ of w has an imprimitivity factorisation derived from the imprimitivity factorisation of w, but also has a new unrelated one. In light of this, in order to strengthen the notion of a companion, we present the following definition, which aims to encompass only those companions that are essential for determining the morphic imprimitivity of the heir:

Definition 3. *Let w be a word and $x \in \text{symb}(w)$. A companion of w with respect to x is called* essential *if there do not exist a morphism ϕ and a companion morphism χ of w with respect to x such that $\phi(w) = w$, and $E_\phi = E_\chi$.*

In the context of the previous example it can be seen that the word w is not an essential companion, as it is a companion of itself, and has a companion morphism ϕ equal to the morphism for which it is a fixed point.

We are now ready to state the third main lemma, in which we characterise the morphic imprimitivity of an heir under deletion of a necessarily expanding letter:

Lemma 6 (Necessarily Expanding Letter). *Let w be a morphically imprimitive word, and let x be a necessarily expanding letter in w. The word $\pi_x(w)$ is morphically imprimitive if and only if the word w has an essential companion with respect to x.*

From the above lemmata we can now see that there are two conditions that together are necessary and sufficient for an arbitrary morphically imprimitive word to be a Billaud word, which we formally express in Theorem 1. This is the characterisation for morphically *imprimitive* Billaud words only, particularly due to Lemma 2, and the somewhat unclear nature of the notion of letter roles in morphically primitive words.

Theorem 1. *A morphically imprimitive word w is a Billaud word if and only if it has at least one imprimitivity factorisation with two or more mortal letters, and for every necessarily expanding letter it has an essential companion.*

Proof. If Direction. Let w have at least one imprimitivity factorisation with two or more mortal letters, and let it have an essential companion with respect to each of its necessarily expanding letters.

Since the word w is morphically imprimitive, due to Lemma 2, $\pi_x(w)$ is morphically imprimitive for all letters $x \in C_w$. Given that the word w has at least one imprimitivity factorisation with more than one mortal letter, by Lemma 3, $\pi_x(w)$ is morphically imprimitive for all letters $x \in M_f$. Finally, as w has an essential companion with respect to every necessarily expanding letter, by Lemma 6, $\pi_x(w)$ is morphically imprimitive for all letters $x \in E_w \setminus (C_w \cup M_w)$.

Therefore, as every letter from $\mathrm{symb}(w)$ is constant, mortal, or necessarily expanding, all heirs of w are morphically imprimitive.

Only-if Direction. Let w be a morphically imprimitive Billaud word, and thus let all of its heirs be morphically imprimitive.

In particular, all heirs $\pi_x(w)$, where x is a mortal letter in w, are morphically imprimitive, and thus by Lemma 1 there exists an imprimitivity factorisation f of w, such that f has two or more mortal letters. Moreover, if y is a necessarily expanding letter in w, then by Lemma 6, w has an essential companion with respect to x.

Therefore, w has at least one imprimitivity factorisation with more than one mortal letter, and for every necessarily expanding letter in w (the count of which could be zero), the word w has an essential companion. $\square$

Theorem 1 concludes our characterisation of morphically imprimitive Billaud words. We have shown precisely which morphically imprimitive words have morphically imprimitive heirs only.

4   Some Technical Considerations on Companions

Various questions can be posed about the nature of companion words, and in particular about the essential companions. For instance, after examining our examples, one might be tempted to conclude that necessarily expanding letters are relatively infrequent in words, with many words not having any. The below proposition shows that such a conclusion is not necessarily true:

Proposition 2. *For every $n \in \mathbb{N}_+$ there exists a morphically imprimitive Billaud word with n necessarily expanding letters and $2n + 1$ letters that are not necessarily expanding.*

Proof Idea. We examine the set of words $w_i := (\mathsf{a_1 b_1 c a_1 d_1 b_1})^2 \cdots (\mathsf{a_i b_i c a_i d_i b_i})^2$, and show that for every $i \in \mathbb{N}_+$, the word w_i is a Billaud word and has i necessarily expanding letters, and $2i + 1$ letters that are not necessarily expanding. □

We now present another observation pertaining to the companions and why another approach, or a tool, previously used to prove a special case of the conjecture might not be suitable in the general case. Before we state our proposition, it is necessary for us to define the notion of a letter interrupting an occurrence of an imprimitivity factor:

Definition 4. *Let w' be a companion of some word w with respect to a letter x, and let ϕ be the corresponding companion morphism. We say that x interrupts an occurrence of an imprimitivity factor $\phi(y)$ (at position(s) $i_1, i_2, \ldots$), for some $y \in E_\phi$, if there exists a factor $v \sqsubseteq w$ such that $\pi_x(v) = \phi(y)$, and such that for all $i_j, j \in \mathbb{N}_+, j \leq |w|_x$, v can be factorised as $v_j x v'_j$, for some words v_j, v'_j, such that $i_j = |\pi_x(v_j)|$.*

We point out that every collection of positions mentioned in Definition 4 refers to an *occurrence* of an imprimitivity factor:

Example 4. Let $w := \mathsf{abxc\,axbxc}$ and let $w' := \mathsf{abcx\,abcx}^2$ be a companion of w. The companion morphism ϕ can be defined as follows: $\phi : \mathsf{a} \mapsto \mathsf{abc}, \mathsf{b}, \mathsf{c} \mapsto \lambda, \mathsf{x} \mapsto \mathsf{x}$. We say that x interrupts an occurrence of an imprimitivity factor $\phi(\mathsf{a}) = \mathsf{abc}$ at position 2, as $\mathsf{abxc} = v'xv'' \sqsubseteq w$, for $v' = \mathsf{ab}$, and $v'' = \mathsf{c}$, and $|v'| = 2$. The letter x also interrupts an occurrence of an imprimitivity factor $\phi(\mathsf{a}) = \mathsf{abc}$ at positions 1 and 2, as $\mathsf{axbxc} = v'xv'' \sqsubseteq w$, for $v' = \mathsf{a}$, and $v'' = \mathsf{bxc}$, and for $v' = \mathsf{axb}$, and $v'' = \mathsf{c}$.   ◊

Levé and Richomme [9], in their proof of a special case of the Billaud Conjecture, work with its contrapositive statement. One of the observations, in their special case, is that if a letter interrupts an imprimitivity factor of a word, it does so once and at the same position for every occurrence of the imprimitivity factor of the heir (and thus the companion). We can show that this is not generally true when considering companions:

Proposition 3. *There exists a word w that has an essential companion $w' = \chi(w)$ with respect to some necessarily expanding letter x, where χ is the companion morphism, such that x interrupts two occurrences of an imprimitivity factor of χ at different positions $i, j \in \mathbb{N}_+$.*

Proof Idea. The word $w :=$ abcdbec becdbcf abcdbcf is a fixed point of $\phi :$ a $\mapsto$ abc, e $\mapsto$ bec, f $\mapsto$ bcf, bc $\mapsto$ λ, d $\mapsto$ d. There is also a companion $\chi(w)$ of w with respect to e, where $\chi :$ d $\mapsto$ bcdbc, bc $\mapsto$ λ, a $\mapsto$ a, e $\mapsto$ e, f $\mapsto$ f. The letter e interrupts two occurrences of the imprimitivity factor $\chi(\mathsf{d})$, once at position 4 and once at position 1. $\qquad\square$

Finally, when examining some of the simpler presented examples of morphically imprimitive words, such as the word abaxb, which has a companion ababx with respect to x, one might spot that the necessarily expanding letter in the parent is also a candidate expanding letter in the companion. We can show that such a conclusion does not hold universally:

Proposition 4. *There exists a morphically imprimitive word w that has an essential companion w' with respect to a necessarily expanding letter x, and which has a necessarily expanding letter y (not necessarily different to x) that is not expanding in w'.*

Proof Idea. Consider the example word in the proof of Proposition 3. The letters a, e, f are necessarily expanding in w due to ϕ being the only non-trivial morphism for which w is a fixed point, but they are not in E_χ. $\qquad\square$

We do note, however, that the word used in the above propositions is not a Billaud word. It could be the case that the above propositions are not satisfied by any such word: if that is the case, such observation would serve as a witness to the complexity of Billaud words.

So far we have discussed the morphically imprimitive Billaud words, that is the morphically imprimitive words without morphically primitive heirs. In the introduction to the previous section we mentioned that the Billaud Conjecture is equivalent to stating that there are no morphically primitive Billaud words. We present this alternative statement of the Billaud Conjecture as follows, using our previously introduced concept of essential companions:

Conjecture 2. Let w be a morphically primitive word. Then there is at least one letter $x \in \mathrm{symb}(w)$ such that w does not have an essential companion with respect to x.

Since morphically primitive words do not have any imprimitivity factorisations, the only way an heir of such a word can be morphically imprimitive is if the latter of the phenomena central to Theorem 1 occurs: by default every imprimitivity factorisation of an heir of a morphically primitive word is unrelated to the parent word. We can use this link between the phenomena occurring in morphically imprimitive and in morphically primitive words to demonstrate the inherent complexity of the Billaud Conjecture. We begin by giving the following proposition, which shows that a morphically primitive word can only have essential companions:

Proposition 5. *Let w be a morphically primitive word. If w has a companion w' with respect to some letter x, then w' is essential.*

Knowing the above, and for the sake of completeness, we can show the equivalence of Conjecture 2 and the Billaud Conjecture:

Proposition 6. *Conjecture 2 is true if and only if Conjecture 1 is true.*

An immediate idea for solving Conjecture 2 could be a numerical argument: one could conjecture that a morphically primitive word w has at most $|\text{symb}(w)| - 1$ companions. This could be a viable approach if there was only one and unique companion with respect to every letter. In such a case, showing that a morphically primitive word w has less companions than letters would be sufficient to prove the Billaud Conjecture. However, it is not universally true that companions have to be unique, or that there is only one with respect to a given letter. The following propositions demonstrate these claims, and they therefore show that such an approach would be futile.

In order to show the first claim, we refer to the example word abcacb:

Proposition 7. *There exists a morphically primitive word w which has more than $|\text{symb}(w)|$ essential companions.*

Proof Idea. Let $w :=$ abcacb. There are 4 essential companions of w: $w_1 =$ abcabc and $w_2 =$ abccab, which are companions with respect to c, and $w_3 =$ bacacb and $w_4 =$ acbacb, which are companions with respect to b. $\quad\square$

Moreover, we present a related proposition, where we claim that there can be arbitrarily many essential companions with respect to the same letter:

Proposition 8. *For every alphabet Σ with $|\Sigma| \geq 3$ there exists a morphically primitive word $w \in \Sigma^*$ and a letter $x \sqsubseteq w$ such that w has $|\Sigma| - 1$ essential companions with respect to x.*

Proof Idea. We consider the recursively defined words $w_1, w_2, \ldots$ such that, for any $i \in \mathbb{N}_+$ we have $w_i := ux^2u'$ where $uu' = v_i$, $|u| = 1$, and where $v_0 := a_0$, $v_j := (v_{j-1}a_j)^2$ for $j \in \mathbb{N}_+$. For instance, $w_3 = a_0xa_1a_0a_1a_2 (a_0a_1)^2a_2$. Every word w_i can be shown to have $i + 1 = |\text{symb}(w_i)| + 2$ essential companions with respect to x, corresponding to every a_k, $k \in [\![0, j]\!]$, being the expanding letter in the given companion. $\quad\square$

The following proposition demonstrates our other claim, namely that companions need not be unique:

Proposition 9. *There exists a morphically primitive word w which has the same word as an essential companion with respect to two different letters.*

Proof. Let $w :=$ abcdacbd, and let $w' :=$ abcdabcd. The word w' is an essential companion with respect to the letter c, as there is a morphism $\chi_c : a \mapsto ab, b \mapsto \lambda, c \mapsto c, d \mapsto d$, such that $\chi_c(w) = w$, χ_c is non-trivial and idempotent, and $\pi_c(w) = \pi_c(w')$. The word w' is also an essential companion with respect to the letter b, as there is a morphism $\chi_b : a \mapsto a, b \mapsto b, c \mapsto \lambda, d \mapsto cd$, such that $\chi_b(w) = w$, χ_b is non-trivial and idempotent, and $\pi_c(w) = \pi_c(w')$. $\quad\square$

Finally, to conclude this section, we present a collection of examples demonstrating the varied ways in which roles of letters can differ between the imprimitivity factorisations of the parents and their essential companion morphisms. We consider companions with respect to necessarily expanding letters, as these are of particular interest due to Theorem 1. These examples illustrate vital phenomena that need to be considered when exploring the existence of companions. While our overview naturally refers to morphically imprimitive parents, we anticipate that further progress on Conjecture 2 will be related to the question of whether there exist any words that would fill the gaps in the below table.

We consider words and their companion morphisms with respect to a necessarily expanding letter x. We note the claims of the table can be verified using the definitions of morphic imprimitivity and companion morphisms. For cells marked with a question mark we conjecture the words do not exist, otherwise we provide an example word witnessing a given letter role change; we indicate such letters in square brackets. We shall denote by NE the necessarily expanding letters. In companions this shall be understood as expanding in all companion morphisms; we define NM and NC similarly. The backslash should be read out as 'but not', i.e. E\NE stands for 'expanding, but not necessarily expanding letter'. Some words exhibit multiple changes: $w_1 := (\text{abm}^2\text{axb})^2(\text{ymzb})^2$, $w_2 := (\text{abaxb})^2(\text{ybz})^2$, $w_3 := \text{abc}^2\text{bc}^2\text{m}^2\text{abc}^2\text{bxc}^2$.

Role in Companion Morphism	Role in Parent					
	NE [y]	E\NE	NM	M\NM	NC [c]	C\NC
NE	$(\text{abaxb})^2(\text{yb})^2$	w_1 [z]	ab^2axb^2 [a]	w_1 [z]	$\text{axcax}(\text{yacax})^2$	?
E\NE	?	w_2 [yz]	w_3 [ab]	w_2 [yz]	?	w_2 [yz]
NM	?	?	w_3 [c]	?	?	?
M\NM	?	w_2 [yz]	w_3 [b]	w_2 [yz]	?	w_2 [yz]
NC	$(\text{abm}^2\text{axb})^2(\text{ym})^2$	w_1 [y]	w_3 [m]	w_1 [y]	$(\text{abaxb})^2\text{c}^2$	w_1 [y]
C\NC	?	w_2 [yz]	w_3 [a]	w_2 [yz]	?	w_2 [yz]

Acknowledgements. We thank G. Richomme and A. Geser for providing the relevant correspondence [3,15], and M. Billaud for an interesting background of the problem. We also wish to thank the anonymous reviewers for their thorough and helpful comments.

References

1. Billaud, M.: A problem with words. Newsgroup 'comp.theory' (1993). https://groups.google.com/d/msg/comp.theory/V_xDDtoR9a4/zgcM4We0CisJ
2. Filè, G.: The relation of two patterns with comparable languages patterns. RAIRO - Theor. Inf. Appl. (Informatique Théorique et Applications) **23**(1), 45–57 (1989)

3. Geser, A.: Your 'Problem with Words' (1993). Private communication to M. Billaud
4. Hamm, D., Shallit, J.: Characterization of finite and one-sided infinite fixed points of morphisms on free monoids. Technical report CS-99-17, University of Waterloo, Ontario, Canada (July 1999)
5. Head, T.: Fixed languages and the adult languages of OL schemes. Int. J. Comput. Math. **10**(2), 103–107 (1981)
6. Holub, Š: Polynomial-time algorithm for fixed points of nontrivial morphisms. Discret. Math. **309**(16), 5069–5076 (2009)
7. Jiang, T., Salomaa, A., Salomaa, K., Yu, S.: Decision Problems for Patterns. J. Comput. Syst. Sci. **50**(1), 53–63 (1995)
8. Kociumaka, T., Radoszewski, J., Rytter, W., Waleń, T.: Linear-time version of Holub's algorithm for morphic imprimitivity testing. Theoret. Comput. Sci. **602**, 7–21 (2015)
9. Levé, F., Richomme, G.: On a conjecture about finite fixed points of morphisms. Theoret. Comput. Sci. **339**(1), 103–128 (2005)
10. Matocha, V., Holub, Š: Complexity of testing morphic primitivity. Kybernetika **49**(2), 216–223 (2013)
11. Nevisi, H., Reidenbach, D.: Morphic primitivity and alphabet reductions. In: Yen, H.-C., Ibarra, O.H. (eds.) DLT 2012. LNCS, vol. 7410, pp. 440–451. Springer, Heidelberg (2012). https://doi.org/10.1007/978-3-642-31653-1_39
12. Reidenbach, D.: Discontinuities in pattern inference. Theoret. Comput. Sci. **397**(1), 166–193 (2008)
13. Reidenbach, D., Schneider, J.C.: Morphically primitive words. Theoret. Comput. Sci. **410**(21), 2148–2161 (2009)
14. Walter, T.: Über die Billaudsche Vermutung. Diplomarbeit, Universität Stuttgart, Fakultät Informatik, Elektrotechnik und Informationstechnik, Stuttgart, Germany (August 2011)
15. Zimmermann, P.: A problem with words from Michel Billaud (1993). Private communication to M. Billaud

Counting Ternary Square-Free Words Quickly

Vladislav Makarov[1,2]([⊠])

[1] St. Petersburg State University, 7/9 Universitetskaya nab.,
Saint Petersburg 199034, Russia
[2] Leonhard Euler International Mathematical Institute at St. Petersburg
State University, 14th Line V.O., 29B, Saint Petersburg 199178, Russia

Abstract. An efficient, when compared to exhaustive enumeration, algorithm for computing the number of square-free words of length n over the alphabet $\{a, b, c\}$ is presented.

Keywords: Square-free words · Dynamic programming · Inclusion-exclusion · Aho-Corasick automaton

1 Introduction

A *square* is a string of form ww for some non-empty string w. A string is *square-free*, if it has no square substrings. Over the binary alphabet, there is only a finite number of square-free strings, but the number of square-free strings of length n over the ternary alphabet grows with n slowly, but exponentially [4].

Since the dawn of combinatorics of words, there has been a lot of research on bounding the number of ternary square free words of length n from below and from above (OEIS sequence A006156 [1]). See, for example, a classic review by Berstel [3] and a new review by Shur [9], to see how much the state of art has changed in-between.

Most of the research was focused on estimating the numbers from above and from below, culminating in Kolpakov's [7] and Shur's [9] methods of proving lower and upper bounds on the growth rate of ternary square-free words (and other power-avoiding words as well), that can be made as close as needed, given enough computational resources.

Computing their *exact* number has attracted significantly less attention. Back in 2001, Grimm [5] obtained the desired values up to 110, but mostly in order to prove a new upper bound on their growth rate.

We will go up to $n = 141$ on a completely ordinary laptop. In just a few hours. This paper gives a high-level account of the underlying ideas, for implementation details and possible optimisations refer to the repository with implementation [6].

V. Makarov—Supported by the Ministry of Science and Higher Education of the Russian Federation, agreement №075-15-2019-1619.

T. Lecroq and S. Puzynina (Eds.): WORDS 2021, LNCS 12847, pp. 142–152, 2021.
https://doi.org/10.1007/978-3-030-85088-3_12

Most ideas used here are very classic, especially using the *antidictionaries*, consisting of minimal squares and building an Aho-Corasick automaton for them. The key contributions are Lemmas 5 and 6 and the way they are used in the final algorithm.

As usual, O^* notation suppresses polynomial factors, so $O^*(1)$ is any at most polynomially growing function, $O^*(2^n)$ is $O(2^n \cdot \text{poly}(n))$, et cetera.

2 A Simple (but Mostly Useless) Algorithm

Unless the opposite is explicitly mentioned, all strings in the following text are over the alphabet $\Sigma = \{a, b, c\}$.

It is not necessary to know how Aho-Corasick automaton works in order to understand this paper. The only important part is the following theorem, which is a simple consequence of a more general result by Aho and Corasick:

Theorem A (Aho and Corasick 1975 [2]). *For any finite subset S of Σ^*, there exists a deterministic finite automaton A with at most $1 + \sum_{w \in S} |w|$ states, such that $L(A)$ is exactly the language of all strings that contain at least one string from S as a substring. Moreover, such an automaton can be constructed in $O(\sum_{w \in S} |w|)$ time.*

Definition 1. *A string is a* minimal square *if it is a square, but does not contain any smaller squares as substrings.*

Definition 2. *Denote the set of all square-free strings of length exactly ℓ by L_ℓ.*

Definition 3. *Similarly, denote the set of all minimal squares with half-length at most ℓ by M_ℓ.*

The main problem at hand is computing $|L_n|$. A string of length n is square-free if and only if it does not contain any *minimal* squares of half-length at most $\lfloor n/2 \rfloor$. Indeed, if a string has a non-minimal square substring, it has a smaller square substring by definition of non-minimality.

Let $A = (Q, q_0, \delta, F)$ be a DFA from Theorem A for the set $M_{\lfloor n/2 \rfloor}$. Here, and in the rest of the text, Q is the set of states, q_0 is the starting state, $\delta \colon Q \times \Sigma \to Q$ is the transition function (with the usual extension to function $Q \times \Sigma^* \to Q$) and F is the set of accepting states.

Let $f(\ell, q)$ be the number of strings $w \in \Sigma^\ell$, such that $\delta(q_0, w) = q$. Then, we can compute all values of $f(\ell, q)$ row-by-row. Indeed, we know $f(0, \cdot)$: $f(0, q_0) = 1$ and $f(0, q) = 0$ for $q \neq q_0$. Moreover, we can compute $f(\ell+1, \cdot)$ through $f(\ell, \cdot)$: to compute $f(\ell+1, q)$, sum up $f(\ell, p)$ over all predecessors of q. In other words, $f(\ell+1, q) = \sum_{p \in Q, d \in \Sigma, \delta(p,d)=q} f(\ell, p)$.

In the end, $\sum_{q \in F} f(n, q)$ is the number of strings of length n that are *not* square free. So, to compute $|L_n|$, it is enough to:

1) Find the set $M_{\lfloor n/2 \rfloor}$ of all short minimal squares.
2) Build the automaton A from Theorem A for $M_{\lfloor n/2 \rfloor}$.
3) Compute $f(n, q)$ for all $q \in Q$.

How to do all these things?

1) Iterate over all square-free words of length at most $\lfloor n/2 \rfloor$ in $O^*(|L_{\lfloor n/2 \rfloor}|)$ time and $O^*(1)$ memory, and, for each of them, check whether it is a minimal square when doubled. It is possible to achieve a polynomial in n speed-up by building the automaton from point 2 on the fly [9, Subsection 3.3] or with other similar optimisations. However, the speed-up would still be only polynomial.
2) Just use Theorem A, the resulting automaton will have at most $2\lfloor n/2 \rfloor \cdot |M_{\lfloor n/2 \rfloor}| + 1 = O^*(|M_{\lfloor n/2 \rfloor}|)$ states and can be constructed in $O^*(|M_{\lfloor n/2 \rfloor}|)$ time.
3) Compute the values of $f(\ell, \cdot)$ row-by-row, using above formulas. It is enough to keep only values of $f(\ell, \cdot)$ and $f(\ell + 1, \cdot)$ in the memory. Here, we need $O^*(|M_{\lfloor n/2 \rfloor}|)$ of both time and memory.

In total, we need $O^*(|L_{\lfloor n/2 \rfloor}|)$ time and $O^*(|M_{\lfloor n/2 \rfloor}|)$ memory, as promised.

3   An Improved Algorithm

The main factor that limits the practical usefulness of the above approach is the memory usage (compared to the high running time of a naive algorithm).

So, we want to reduce the memory consumption, possibly at the cost of making running time slightly worse. The key observation that makes this possible is a pretty interesting one and can be seen as an incomplete application of the inclusion-exclusion principle.

Intuitively, the knowledge that some fixed substring of a string is a square, gives a lot of constraints of type "symbols on some positions are equal". Hence, the existence of two long (whatever that means, the exact definition of "long" will come later) square substrings at the same time places too many constraints on the string, meaning that there are a lot of symbols that are "forced" to be equal, hence we can find a smaller square substring.

Example 4. Suppose we have a string s of length 13 and we know that its prefix of length 8 is a square, and so is its suffix of length 12. Hence, we know that $s_i = s_{i+4}$ for $0 \leqslant i < 4$ and, similarly, $s_i = s_{i+6}$ for $1 \leqslant i < 7$. Then, s looks like 1232123232123, where equal digits correspond to symbols that *must* be equal, and different digits correspond to symbols that *can* be different. And, indeed, there is a short square substring 2323 in the middle.

In general, it is not true that all strings have at most one long minimal square substring, there are some counterexamples. Indeed, the string *abcabcab* has three distinct long minimal square substrings: *abcabc*, *bcabca* and *cabcab*. However, all these squares are of the same half-length and start in the consecutive positions of the original string, meaning that a lot of constraints actually coincide. Intuitively, all counterexamples have to look in this, very regular, way.

The following Lemmas 5 and 6 are exact statements that correspond to this intuition.

Lemma 5. *Let s be a string (over any alphabet), such that some proper prefix uu of s is a minimal square and some proper suffix vv of s is a minimal square. Then, either $|s| \geqslant 3 \min(|u|, |v|) + 1$, or $|u| = |v|$ and $s = uup$ for some non-empty prefix p of u.*

Proof (Sketch of an automatic "partial proof"). Suppose that we want to check this lemma for $|u| \leqslant d$ and $|v| \leqslant d$. Let us iterate on the length of s (up to $3d$) and create a graph with $|s|$ vertices, with edges corresponding to "forced" equalities between symbols: edges between i and $i + |u|$ for $0 \leqslant i < |u|$ and edges between i and $i + |v|$ for $|s| - 2|v| \leqslant i < |s| - |v|$. Now, the connected components in this graph tell which symbols have to be equal and which do not. Now, we can just check whether there are any small forced squares that disprove that either uu or vv was a *minimal* square. The whole procedure needs polynomial in d time. Specifically, $O(d^5)$ for the most straightforward implementation. Hence, running the above procedure can actually *prove* the Lemma, but only for small lengths. I verified the Lemma for lengths of $|s|$ up to 200 this way [6, test_overlay.cpp].

Proof (Mathematical proof). The mathematical proof is messier, but works for all lengths. Proof by contradiction. Let $s = s_0 s_1 \ldots s_{|s|-1}$.

Consider the case $|u| = |v|$ first. Then, either $|s| \geqslant 3|u| + 1$ (and we are done), or $|s| \leqslant 3|u|$. In the latter case, we know $s_i = s_{|u|+i}$ for $0 \leqslant i < |u|$, because uu is a prefix of s and $s_i = s_{|u|+i}$ for $|s| - 2|u| \leqslant i \leqslant |s| - |u|$, because vv is a suffix of s. Therefore, $s_i = s_{|u|+i}$ for $0 \leqslant i < |s| - |u|$, because $|s| - 2|u| \leqslant |u|$. Hence, $s_i = s_{i \bmod |u|}$ for $0 \leqslant i < |s|$, meaning that s is a prefix of u^*. Because the length of u is between $2|u| + 1$ and $3|u|$, it has the same exact form as promised by the lemma.

Now, suppose that $|u| \neq |v|$. Without loss of generality, $|u| < |v|$. Then suffix vv overlaps with the first u: otherwise the whole string s has length at least $|u| + 2|v|$, which is at least $|u| + 2(|u| + 1) = 3|u| + 2 \geqslant 3|u| + 1$, contradiction. Hence, $u = fg$ and $v = gh$, where g is the non-empty overlap between the first u and vv. Moreover, f is non-empty, because otherwise uu would be a substring of vv. Finally, because the right square is longer, $|gh| = |v| > |u| = |fg|$, hence $|h| > |f|$.

Now we know almost everything about the relative positions of uu and vv. More specifically, $fgfg = uu$ is a prefix of $fghgh = fvv$. Hence, $u = fg$ is a prefix of hg (here we use that $|h| > |f|$). This, in turn, implies that f is a proper prefix of h: $h = fx$ for some non-empty string x. Therefore, because $fgfg = (fgf)(g)$ is a prefix of $fghgh = fg(fx)gh = (fgf)(xg)h$. Hence, g is a prefix of xg.

Suppose that $|g| \leqslant |x| - 1$. Then, $|s| \geqslant |fghgh| = |f(gfx)(gfx)| = 2|x| + 2|g| + 3|f| \geqslant 4|g| + 2 + 3|f| = 3|fg| + |g| + 2 = 3|u| + |g| + 2 \geqslant 3|u| + 1$. Contradiction.

Now we know that $|g| \geqslant |x|$. This, along with g being a prefix of xg, means that x is a prefix of g: $g = xy$ for some, possibly empty, string y. Hence, $vv = g(h)(g)h = g(fx)(xy)h = gf(xx)yh$ is not a minimal square, because of a square substring xx. Contradiction.

Lemma 6. *Let s be a string that is not square free, but does not have square substrings with half-length strictly less than $|s|/3$ (no rounding here). Then, s has a unique inclusion-maximal substring of form wwp, where ww is a minimal square and p is some, possibly empty, prefix of w. Moreover, s has exactly $|p| + 1$ minimal square substrings—specifically the substrings of wwp with length $2|w|$.*

Proof. The case when s has exactly one minimal square substring corresponds to $p = \varepsilon$. Now, suppose that s has at least two distinct minimal square substrings. Consider any two of them. Because they are minimal squares, neither of them is a substring of another. So, one of them starts and ends earlier than another and we can apply Lemma 5.

Hence, these squares have the same length (otherwise $|s| > 3(|s|/3) + 1 = |s| + 1$). Because the above statement is true for *any* two minimal square substrings of s, *all* minimal square substrings of s have the same length. Consider the leftmost and the rightmost of them. They intersect because s is short enough, and their union has the form wwp with p being a prefix of w by Lemma 5. Then, any substring of their union with length $2|w|$ is a minimal square. Indeed, all substrings of wwp with length $2|w|$ are cyclic shifts of ww. A cyclic shift of a minimal square is also a minimal square; one can prove this either by case analysis or by using Shur's result that a square is minimal if and only if its half is square-free as a *cyclic* string [8, Proposition 1].

Summarising, all substrings of wwp with length $2|w|$ are minimal squares, and s has no other minimal square substrings, because wwp was chosen to be the union of the leftmost one and the rightmost one. $\square$

Remark 7. In Lemmas 5 and 6 slightly better bounds are actually true, but even the best possible bounds lead only to constant factor improvements in the final algorithm.

Let n be the length of square-strings we need to count. Moreover, let A be the automaton from Theorem A for the set $M_{\lfloor n/3 \rfloor}$ and $f(\ell, q)$ be the number of strings with length ℓ that are rejected by A, when the computation starts in the state q. In other words, $f(\ell, q)$ is the number of strings s, such that $|s| = \ell$ and $\delta(q, s) \notin F$.

Definition 8. *A square is said to be* short, *if its half-length is at most $\lfloor n/3 \rfloor$.*

Definition 9. *A string is* promising *if it has no short square substrings.*

Remark 10. A string s is promising if and only if s^R is promising: if s contains square ww, then s^R contains $w^R w^R$ and vice versa.

Definition 11. *For a promising string t with length at most n and an integer number ℓ with $0 \leqslant \ell \leqslant n - |t|$, denote by $g(\ell, t)$ the number of promising strings of length n that have t as a substring, starting with position ℓ.*

Lemma 12. *For any promising string t with $2\lfloor n/3 \rfloor + 1 \leqslant |t| \leqslant n$ and any integer $0 \leqslant \ell \leqslant n - |t|$, $g(\ell, t) = f(\ell, \delta(q_0, t^R)) \cdot f(n - |t| - \ell, \delta(q_0, t))$.*

Proof. Suppose that xty is a string satisfying the conditions of the lemma, with $|x| = \ell$. Then, xty is promising if and only if xt and ty are both promising. Indeed, because $|t| \geqslant 2\lfloor n/3 \rfloor + 1$, any possible short square in xty fully fits in either xt or ty. By definition of function f, ty is promising if and only if A rejects y, starting from $\delta(q_0, t)$. Hence, there is $f(|y|, \delta(q_0, t)) = f(n - |t| - \ell, \delta(q_0, t))$ ways to choose y.

Similarly, by Remark 10, xt is promising if and only if $t^R x^R$ is promising. Hence, there are $f(|x^R|, \delta(q_0, t^R)) = f(\ell, \delta(q_0, t^R))$ ways to choose x. All in all, there are $f(\ell, \delta(q_0, t)) \cdot f(n - |t| - \ell, \delta(q_0, t))$ ways to choose x and y.

Now, everything is in line for the improved algorithm.

Theorem 13. *One can compute $|L_n|$ in $O^*(|L_{\lfloor n/2 \rfloor}|)$ time and $O^*(|M_{\lfloor n/3 \rfloor}|)$ memory.*

Proof. By Lemma 6, there are three types of strings of length n:

1) not promising
2) promising, but not square-free, they have wwp substring as per Lemma 6.
3) square-free

We want to know the number of strings of type 3. By definition of being promising, the total number of strings of types 2 and 3 is $f(n, q_0)$.

Consider any string of type 2. We can try to count them using Lemma 12, by iterating over a minimal square substring and its position. Of course, there is massive overcounting happening here: if wwp is the substring of s that is given by Lemma 6, then we count s exactly $|p| + 1$ times. To deal with this, notice that, for such a string, there are exactly $|p|$ substrings of type xxx_0, where xx is a minimal square: exactly the substrings of wwp with length $2|w| + 1$. Hence, counting them with minus sign fixes the overcounting problem perfectly, because $(|p| + 1) - |p| = 1$.

In the end, there are

$$f(n, q_0) - \left(\sum_{ww} \sum_{i=0}^{n-|ww|} g(i, ww) - \sum_{ww} \sum_{i=0}^{n-|ww|-1} g(i, www_0) \right) \tag{1}$$

strings of type 3, where both summations are over minimal squares with half-length at least $\lfloor n/3 \rfloor + 1$.

Let's trace the steps necessary to complete the algorithm:

1) Find the set $M_{\lfloor n/3 \rfloor}$ and build the automaton A. Takes $O^*(|L_{\lfloor n/3 \rfloor}|)$ time and $O^*(1)$ memory.

2) Compute the values $f(\ell, q)$ for $0 \leqslant \ell \leqslant \lfloor n/3 \rfloor$ and $q \in Q$. Takes $O^*(|M_{\lfloor n/3 \rfloor}|)$ time and memory.

3) Iterate over all minimal squares with half-length at most $\lfloor n/2 \rfloor$, in order to compute the sum (1). This is the slowest part. Iterating over all minimal squares with half-length at most $\lfloor n/2 \rfloor$ takes $O^*(|L_{\lfloor n/2 \rfloor}|)$ time and $O^*(1)$ memory. Notice, that there is no need to actually store them all in memory, knowing only the current one and the values of f is enough. Lemma 12 comes in play here, allowing to express g's through f's.

In the end, total time complexity is still $O^*(|L_{\lfloor n/2 \rfloor}|)$, but the memory complexity is $O^*(|M_{\lfloor n/3 \rfloor}|)$, as promised.

4 Possible Time-Memory Tradeoffs?

I like to think about the algorithm from Sect. 3 as an incomplete application of the inclusion-exclusion principle. Indeed, we take all promising strings of length n and subtract promising strings with at least one minimal square substring (well, up to technical details in form of the wwp substrings). In a normal inclusion-exclusion algorithm, we would need to add back promising strings with at least two different minimal squares, then subtract promising strings with at least three different minimal squares again, et cetera. However, it turns out that, up to some simple counterexamples, there are *no* promising strings with two different minimal squares!

But what will happen if we replace $\lfloor n/3 \rfloor$ with a smaller number, say, $\lfloor n/10 \rfloor$, and do several steps of inclusion-exclusion instead of just one?

As it turns out, this leads to smaller memory consumption at the cost of higher running time. Indeed, let's fix some $k \geqslant 4$.

Definition 14. *A square is said to be k-short, if its half-length is at most $\lfloor n/k \rfloor$.*

Definition 15. *A string is k-promising if it has no k-short square substrings.*

Consider any k-promising string s. Intuitively, Lemma 6 implies that all pairs of minimal square substrings of s either have small intersection or are both a part of a large wwp block. Hence, there ought to be only $O(1)$ such blocks — otherwise some would have large intersection by Dirichlet's principle.

Let us explain the intuition from the previous paragraph formally. Consider *all* minimal square substrings of some k-promising s, sorted by the coordinate of their left end: $s[\ell_1, r_1), s[\ell_2, r_2), \ldots, s[\ell_d, r_d)$, with $\ell_1 < \ell_2 < \ldots < \ell_d$ (all inequalities are strict; otherwise some minimal square would be a prefix of another). Then, $r_1 < r_2 < \ldots < r_d$ (otherwise some minimal square is a substring of another). For each i from 1 to d inclusive, denote the middle position of the i-th minimal square by m_i. In other words, $m_i = (\ell_i + r_i)/2$. It is easy to see that middles are also increasing: if $\ell_i < \ell_{i+1}$, but $m_i \geqslant m_{i+1}$, then $r_i = 2 \cdot m_i - \ell_i > 2 \cdot m_{i+1} - \ell_{i+1} = r_{i+1}$. Finally, denote the square itself by $u_i u_i$. That is, $s[\ell_i, m_i) = s[m_i, r_i) = u_i$.

Indeed, consider some index $1 \leqslant i \leqslant d - 1$. Then, by Lemma 6, there are the following possibilities:

1. Substrings $s[\ell_i, r_i)$ and $s[\ell_{i+1}, r_{i+1})$ do not intersect at all. In other words, $r_i \leqslant \ell_{i+1}$. Then, $m_{i+1} = \ell_{i+1} + |u_{i+1}| > \ell_{i+1} + \lfloor n/k \rfloor \geqslant r_i + \lfloor n/k \rfloor = m_i + |u_i| + \lfloor n/k \rfloor > m_i + 2\lfloor n/k \rfloor$ (the first and the last inequalities corresponds to the fact that all square substrings of s are long enough).

2. Substrings $s[\ell_i, r_i)$ and $s[\ell_{i+1}, r_{i+1})$ intersect, but the length of their union, the substring $s[\ell_i, r_{i+1})$, is at least $3\min(|u_i|, |u_{i+1}|) + 1$. That is,

$$r_{i+1} - \ell_i \geqslant 3\min(|u_i|, |u_{i+1}|) + 1 \tag{2}$$

Because $s[\ell_i, r_{i+1})$ contains both $u_i u_i$ and $u_{i+1} u_{i+1}$ as proper substrings, $(r_{i+1} - \ell_i) \geqslant 2\max(|u_i|, |u_{i+1}|)$. By taking the average with inequality (2),

$$r_{i+1} - \ell_i \geqslant (3 \cdot \min(|u_i|, |u_{i+1}|) + 2\max(|u_i|, |u_{i+1}|))/2 + 1$$
$$= \min(|u_i|, |u_{i+1}|)/2 + (\min(|u_i|, |u_{i+1}|) + \max(|u_i|, |u_{i+1}|)) + 1$$
$$= \min(|u_i|, |u_{i+1}|)/2 + (|u_i| + |u_{i+1}|) + 1 \geqslant \lfloor n/k \rfloor/2 + (|u_i| + |u_{i+1}| + 1).$$

Hence, $m_{i+1} - m_i = (r_{i+1} - \ell_i) - (|u_i| + |u_{i+1}|) \geqslant \lfloor n/k \rfloor/2 + 1$.

3. The string $s[\ell_i, r_{i+1})$ has small length $(r_{i+1} - \ell_i \leqslant 3\min(|u_i|, |u_{i+1}|))$, but $|u_i| = |u_{i+1}|$. Then, by the conclusion of Lemma 6, $s[\ell_i + 1, r_i + 1)$ is a minimal square. Therefore, $\ell_{i+1} = \ell_i + 1$ and $r_{i+1} = r_i + 1$. In this case, the difference between m_{i+1} and m_i is not large, but, like in Sect. 3, we can consider such minimal squares in batches.

Hence, all minimal square substrings of s split into b inclusion-maximal batches for some $b \geqslant 0$, with i-th $(1 \leqslant i \leqslant b)$ of them defined by three parameters L_i, R_i and $T_i \geqslant 1$: the first minimal square in the batch and the size of the batch. Formally speaking, a batch (L_i, R_i, T_i) corresponds to the fact that substrings $s[L_i + j, R_i + j)$ are minimal squares for each $0 \leqslant j < T_i$, but $R_i + T_i > |s| = n$ or $s[L_i + T_i, R_i + T_i)$ is not a minimal square and, similarly, $L_i - 1 < 0$ or $s[L_i - 1, R_i - 1)$ is not a minimal square.

Let $M_i = (L_i + R_i)/2$ be the middle of the first square in each batch. From the above, it follows that M_i's are increasing rather quickly. More specifically, $M_{i+1} - M_i > \lfloor n/k \rfloor/2$ for each $1 \leqslant i \leqslant b - 1$. Hence, $b \leqslant 2k + 1$ — otherwise $M_b > (2k + 1 - 1) \cdot \lfloor n/k \rfloor/2 \geqslant n$.

Hence, for any k-promising string, there are $O(k)$ batches in total. Each batch is uniquely defined by its integer parameters (L_i, R_i, T_i) and a square-free string $s[L_i, M_i)$ of length $M_i - L_i = (R_i - L_i)/2$. Of course, some square-free strings do not correspond to a valid batch, but this it not important right now. From now on, by batch, I mean the tuple (L_i, R_i, T_i, U_i), with U_i being a square-free string of length $(R_i - L_i)/2$. A string s contains a batch (L_i, R_i, T_i, U_i) if $s[L_i, L_i + |U_i|) = U_i$, substrings $s[L_i + j, R_i + j)$ are minimal squares for each $0 \leqslant j < T_i$ and the batch itself is maximal possible by inclusion (in other words, $L_i - 1 < 0$ or $s[L_i - 1, R_i - 1)$ is not a minimal square and $R_i + T_i > n$ or $s[L_i + T_i, R_i + T_i)$ is not a minimal square).

Example 16. For $k = 4$, a string $abcabcabc$ is k-promising and contains exactly one batch: $(0, 6, 4, abc)$. It does not contain batches $(1, 7, 3, bca)$, $(0, 6, 3, abc)$ and $(1, 7, 2, bca)$, because they are not inclusion-maximal.

We want to compute the number of square-free strings, or, in other words, k-promising strings that contain no batches. For a set S of batches let $h(S)$ be the number of k-promising strings of length n that contain all batches from the set S, but *may also contain some other batches*. Then, by inclusion-exclusion, the answer for length n is $\sum\limits_{|S| \leqslant 2k+1} (-1)^{|S|} h(S)$, where the summation is over all possible sets of batches (as we know already, there are no strings that contain $2k + 2$ or more batches). Hence, we are left with the two following subproblems:

1. For a given set S of batches, compute $h(S)$ quickly enough.
2. Iterate over all possible sets of batches efficiently. In particular, prove that there are not too many possible sets.

Let us solve the first subproblem first.

Lemma 17. $h(S)$ *can be computed in* $O^*(2^{O(k)})$ *after precomputation that uses* $O^*(|M_{\lfloor n/k \rfloor}|^3)$ *time and* $O^*(|M_{\lfloor n/k \rfloor}|^2)$ *memory. Moreover, for* $k = 4$, $h(S)$ *can be computed in* $O^*(2^{O(k)}) = O^*(1)$ *time after precomputation that uses* $O^*(|M_{\lfloor n/k \rfloor}|)$ *time and memory.*

Proof. Suppose that some string s contains every batch from S. Then, we already know what some symbols in s are equal to. Moreover, because each batch from S is inclusion-maximal, we know for some symbols what they are *not* equal to. Specifically, for a batch (L, R, T, U) and $M := (L + R)/2$ we know that $s_{L-1} \neq s_{M-1}$ and $s_{R+T-1} \neq s_{M+T-1}$ (if $L - 1 \geqslant 0$ and $R + T - 1 < n$ respectively, of course). For each such symbol (at most $2 \cdot (2k + 1) = O(k)$ of them), iterate over two possibillities. For each of those $2^{O(k)}$ cases, check two things (both can be done in $O^*(1)$ time by simply iterating over all fully-known substrings of s):

– that s does not contain a k-short square consisting only of known symbols,
– that each batch from S indeed is a valid batch contained in s.

Now, we are left with a simpler problem: how many k-promising strings are there, assuming that symbols on some positions are already known? Moreover, positions with known symbols appear in blocks of length at least $2(\lfloor n/k \rfloor + 1)$ each. Hence, any k-short square intersects exactly one block of *unknown* symbols (otherwise it fully contains a block of known symbols and, therefore, cannot be k-short).

Firstly, let us deal with the simpler case of $k = 4$. In this case, there is at most one block of known symbols. Indeed, each such block has length at least $2(\lfloor n/4 \rfloor + 1)$ and there is just not enough space for two of them. Hence, there is an unknown prefix, a fully-known middle and an unknown suffix (each of those three parts may be empty). What we need to know is the number of k-promising strings that conform to this pattern. This situation already appeared before: specifically, see Lemma 12. We can define and compute the functions $f(\cdot, \cdot)$ and $g(\cdot, \cdot)$ in the same way, with only difference being that the automaton we build will corespond to k-short squares and will therefore have size $O^*(|M_{\lfloor n/k \rfloor}|)$.

In the general case, there *may* be blocks of unknown symbols that are surrounded by known symbols from both left and right. However, all blocks of unknown symbols can still be filled independently. Consider a block of unknown symbols of length ℓ that is surrounded by (possibly, empty) blocks w_p and w_q of known symbols. Let $A_k = (Q, q_0, \delta, F)$ be the automaton from Theorem A for k-short squares. Then, we can fill-in unknown symbols with a string $s \in \Sigma^\ell$ if and only if $\delta(q_0, w_p s w_q) \notin F$. In other words, $\delta(\delta(\delta(q_0, w_p), s), w_q) \notin F$.

Hence, let us compute $f_{\text{both}}(\ell, p, q)$: how many strings $s \in \Sigma^\ell$ are there, such that $\delta(s, p) = q$. We can do this in $O^*(|M_{\lfloor n/k \rfloor}|^2)$ by dynamic programming over the states of A_k. To compute the number of ways to fill the block, substitute $p := \delta(q_0, w_p)$ and iterate over all q, such that $\delta(q, w_q) \notin F$.

Unfortunately, this approach takes $O^*(2^{O(k)} \cdot |M_{\lfloor n/k \rfloor}|)$ time to compute $h(S)$ (the second factor comes from iterating over q). To get rid of the second factor, notice the following: for *any* s, whether or not $w_p s w_q$ has any k-short square substrings, depends only on $\delta(q_0, w_p)$ and $\delta(q_0, w_q^R)$, but not on their exact values (this immediately follows from the Theorem A and the fact that $w_p s w_q$ does not have any k-short squares if and only if $(w_p s w_q)^R = w_q^R s^R w_p^R$ also does not.

Hence, the numbers of ways to fill-in the block depends only on its length, $\delta(q_0, w_p)$ and $\delta(q_0, w_q^R)$. We can simply precompute all those $O^*(|M_{\lfloor n/k \rfloor}|^2)$ numbers, each in $O^*(|M_{\lfloor n/k \rfloor}|)$ time.

Finally, we need to iterate over all possible sets of batches somehow. Iterating over the numbers L_i, R_i, T_i takes only $O(n^{O(k)})$ time. To iterate over possible strings U_i, iterate over batches from left to right and fill them in that order. Because each batch consists of consecutive minimal squares, $L_i + T_i \leqslant L_{i+1}$ and $R_i + T_i \leqslant R_{i+1}$ for consecutive batches. Hence, for each batch, some prefix of U_i is already known, and some, possibly empty, suffix is not. The unknown part is a square-free string by itself. Moreover, each symbol in the unknown part corresponds to at least two positions in the string (otherwise we would have figured out this symbol already). Hence, we need to iterate over $O(k)$ strings of total length at most $\lfloor n/2 \rfloor$. It is known that $|L_\ell|$ grows exponentially. In particular, $c_1 \gamma^\ell \leqslant |L_\ell| \leqslant c_2 \gamma^\ell$ for some γ and $c_1, c_2 > 0$. Hence, there are at most $c_2^k \gamma^{\lfloor n/2 \rfloor}$ ways to choose these strings, which is at most $|L_{\lfloor n/2 \rfloor}| \cdot c_2^k / c_1 = O(|L_{\lfloor n/2 \rfloor}| \cdot 2^{O(k)})$. In total, iterating over all possible sets S takes $O(|L_{\lfloor n/2 \rfloor}| \cdot n^{O(k)})$ time.

Hence, we need $O^*(|M_{\lfloor n/k \rfloor}|^2)$ memory and $O^*(|L_{\lfloor n/2 \rfloor}| \cdot n^{O(k)})$ time (precomputation from Lemma 17 is irrelevant for large k). Moreover, for $k = 4$ only $O^*(|M_{\lfloor n/k \rfloor}|)$ memory is needed. Unfortunately, the practical value of this optimisation is questionable. The memory consumption of the algorithm from Sect. 3 is, indeed, quite a problem already for $n = 141$, but adding even *one* extra $O(n)$ factor to the time complexity turns "several hours" into "several *weeks*". Moreover, assuming that $|M_\ell|$ grows exponentially with ℓ, we need to choose either $k = 4$ or $k \geqslant 7$ to get any memory advantage. Because of the above, choosing $k \geqslant 7$ is completely hopeless. Choosing $k = 4$ is an interesting idea that may lead to a better results in the end, but I have not implemented it yet.

The main running time bottleneck of this approach is pretty apparent: even when (L_i, R_i, T_i) are fixed, I do not know any way to avoid iterating over almost a half of the whole string in the worst case. In fact, it seems difficult to compute the number of minimal squares of half-length n in significantly less than $O(|M_n|)$ time. Intuitively, counting only minimal squares corresponds to the first step of inclusion-exclusion and should therefore be easier somehow. However, even such, intuitively simpler, problem seems to be out of reach now.

5   Final Notes

Of course, the same ideas work for larger alphabet sizes.

The algorithm from Sect. 3 is implemented in the linked repository [6], with some constant optimisations and other minor tweaks. There are still several optimisations possible, both in terms of time and memory, but they are more annoying to implement. If you want to suggest some code improvements, contact me via e-mail.

As noted in Sect. 4, any substantial improvement to counting square-free words would likely require a faster way to count minimal squares. I believe that it also works in the opposite direction: any non-trivial algorithm for counting minimal squares will lead to a better algorithm for counting square-free words.

References

1. OEIS sequence A006156. http://oeis.org/A006156
2. Aho, A.V., Corasick, M.J.: Efficient string matching: an aid to bibliographic search. Commun. ACM **18**(6), 333–340 (1975). https://doi.org/10.1145/360825.360855
3. Berstel, Jean: Some recent results on squarefree words. In: Fontet, M., Mehlhorn, K. (eds.) STACS 1984. LNCS, vol. 166, pp. 14–25. Springer, Heidelberg (1984). https://doi.org/10.1007/3-540-12920-0_2
4. Brandenburg, F.J.: Uniformly growing k-th power-free homomorphisms. Theor. Comput. Sci. **23**(1), 69–82 (1988). https://doi.org/10.1016/0304-3975(88)90009-6
5. Grimm, U.: Improved bounds on the number of ternary square-free words. J. Integer Seq. **4** (2001)
6. Kaban-5: square_free_words, a GitHub repository (2019–2021). https://github.com/Kaban-5/square_free_words
7. Kolpakov, R.: Efficient lower bounds on the number of repetition-free words. J. Integer Seq. **10**, 2–3 (2007)
8. Shur, A.M.: On ternary square-free circular words. Electron. J. Comb. **17** (2010). https://doi.org/10.37236/412
9. Shur, A.M.: Growth properties of power-free languages. Comput. Sci. Rev. **6**(5), 187–208 (2012). https://doi.org/10.1016/j.cosrev.2012.09.001

Doubled Patterns with Reversal Are 3-Avoidable

Pascal Ochem[✉]

LIRMM, CNRS, Université de Montpellier, Montpellier, France
ochem@lirmm.fr

Abstract. In combinatorics on words, a word w over an alphabet Σ is said to avoid a pattern p over an alphabet Δ if there is no factor f of w such that $f = h(p)$ where $h : \Delta^* \to \Sigma^*$ is a non-erasing morphism. A pattern p is said to be k-avoidable if there exists an infinite word over a k-letter alphabet that avoids p. A pattern is *doubled* if every variable occurs at least twice. Doubled patterns are known to be 3-avoidable. Currie, Mol, and Rampersad have considered a generalized notion which allows variable occurrences to be reversed. That is, $h(V^R)$ is the mirror image of $h(V)$ for every $V \in \Delta$. We show that doubled patterns with reversal are 3-avoidable.

1 Introduction

The *mirror image* of the word $w = w_1 w_2 \ldots w_n$ is the word $w^R = w_n w_{n-1} \ldots w_1$. A pattern with reversal p is a non-empty word over an alphabet $\Delta = \{A, A^R, B, B^R, C, C^R \ldots\}$ such that $\{A, B, C, \ldots\}$ are the *variables* of p. An *occurrence* of p in a word w is a non-erasing morphism $h : \Delta^* \to \Sigma^*$ satisfying $h(X^R) = (h(X))^R$ for every variable X and such that $h(p)$ is a factor of w. The avoidability index $\lambda(p)$ of a pattern with reversal p is the size of the smallest alphabet Σ such that there exists an infinite word w over Σ containing no occurrence of p. A pattern p such that $\lambda(p) \leq k$ is said to be k-avoidable. To emphasive that a pattern is without reversal (i.e., it contains no X^R), it is said to be *classical*. A pattern is *doubled* if every variable occurs at least twice.

Our result is

Theorem 1. *Every doubled pattern with reversal is 3-avoidable.*

The restriction of Theorem 1 to classical patterns is known to hold.

Theorem 2. *[4–6] Every doubled pattern is 3-avoidable.*

Let $v(p)$ be the number of distinct variables of the pattern p. In the proof of Theorem 2, the set of doubled patterns is partitioned as follows:

1. Patterns with $v(p) \leq 3$: the avoidability index of every ternary pattern has been determined [5].
2. Patterns shown to be 3-avoidable with the so-called power series method:

© Springer Nature Switzerland AG 2021
T. Lecroq and S. Puzynina (Eds.): WORDS 2021, LNCS 12847, pp. 153–159, 2021.
https://doi.org/10.1007/978-3-030-85088-3_13

- Patterns with $v(p) \geq 6$ [4]
- Patterns with $v(p) = 5$ and prefix ABC or length at least 11 [6]
- Patterns with $v(p) = 4$ and prefix $ABCD$ or length at least 9 [6]

3. Ten sporadic patterns with $4 \leq v(p) \leq 5$ whose 3-avoidability cannot be deduced from the previous results: they have been shown to be 2-avoidable [6] using the method in [5].

The proof of Theorem 1 uses the same partition. Each of the last three sections is devoted to one type of doubled pattern with reversal.

2   Preliminaries

A word w is d-*directed* if for every factor f of w of length d, the word f^R is not a factor of w.

Remark 1. If a d-directed word contains an occurrence h of $X.X^R$ for some variable X, then $|h(X)| \leq d - 1$.

A variable that appears only once in a pattern is said to be *isolated*. The *formula* f associated to a pattern p is obtained by replacing every isolated variable in p by a dot. The factors between the dots are called *fragments*. An occurrence of a formula f in a word w is a non-erasing morphism h such that the h-image of every fragment of f is a factor of w. As for patterns, the avoidability index $\lambda(f)$ of a formula f is the size of the smallest alphabet allowing the existence of an infinite word containing no occurrence of f. Recently, the avoidability of formulas with reversal has been considered by Currie, Mol, and Rampersad [2,3] and me [7].

Recall that a formula is *nice* if every variable occurs at least twice in the same fragment. In particular, a doubled pattern is a nice formula with exactly one fragment.

The *avoidability exponent* $AE(f)$ of a formula f is the largest real x such that every x-free word avoids f. Every nice formula f with $v(f) \geq 3$ variables is such that $AE(f) \geq 1 + \frac{1}{2v(f)-3}$ [9].

Let $\simeq$ be the equivalence relation on words defined by $w \simeq w'$ if $w' \in \{w, w^R\}$. Avoiding a pattern up to $\simeq$ has been investigated for every binary formulas [1]. Remark that for a given classical pattern or formula p, avoiding p up to $\simeq$ implies avoiding simultaneously all the variants of p with reversal.

Recall that a word is (β^+, n)-free if it contains no repetition with exponent strictly greater than β and period at least n.

3   Formulas with at Most 3 Variables

For classical doubled patterns with at most 3 variables, all the avoidability indices are known. There are many such patterns, so it would be tedious to consider all their variants with reversal.

However, we are only interested in their 3-avoidability, which follows from the 3-avoidability of nice formulas with at most 3 variables [8].

Thus, to obtain the 3-avoidability of doubled patterns with reversal with at most 3 variables, we show that every minimally nice formula with at most 3 variables is 3-avoidable up to $\simeq$.

The minimally nice formulas with at most 3 variables, up to symmetries, are determined in [8] and listed in the following table. Every such formula f is avoided by the image by a q-uniform morphism of either any infinite $\left(\frac{5}{4}^+\right)$-free word w_5 over Σ_5 or any infinite $\left(\frac{7}{5}^+\right)$-free word w_4 over Σ_4, depending on whether the avoidability exponent of f is smaller than $\frac{7}{5}$.

Formula f	$= f^R$	$AE(f)$	Word	q	d	freeness
$ABA.BAB$	yes	1.5	$g_a(w_4)$	9	9	$\left(\frac{131}{90}^+, 28\right)$
$ABCA.BCAB.CABC$	yes	1.333333333	$g_b(w_5)$	6	8	$\left(\frac{4}{3}^+, 25\right)$
$ABCBA.CBABC$	yes	1.333333333	$g_c(w_5)$	4	9	$\left(\frac{30}{23}^+, 18\right)$
$ABCA.BCAB.CBC$	no	1.381966011	$g_d(w_5)$	9	4	$\left(\frac{62}{45}^+, 37\right)$
$ABA.BCB.CAC$	yes	1.5	$g_e(w_4)^a$	9	4	$\left(\frac{67}{45}^+, 37\right)$
$ABCA.BCAB.CBAC$	yes^b	1.333333333	$g_f(w_5)$	6	6	$\left(\frac{31}{24}^+, 31\right)$
$ABCA.BAB.CAC$	yes	1.414213562	$g_g(w_4)$	6	8	$\left(\frac{89}{63}^+, 61\right)$
$ABCA.BAB.CBC$	no	1.430159709	$g_h(w_4)$	6	7	$\left(\frac{17}{12}^+, 61\right)$
$ABCA.BAB.CBAC$	no	1.381966011	$g_i(w_5)$	8	7	$\left(\frac{127}{96}^+, 41\right)$
$ABCBA.CABC$	no	1.361103081	$g_j(w_5)$	6	8	$\left(\frac{4}{3}^+, 25\right)$
$ABCBA.CAC$	yes	1.396608253	$g_k(w_5)$	6	13	$\left(\frac{4}{3}^+, 25\right)$

^{a} The formula $ABA.BCB.CAC$ seems also avoided up to $\simeq$ by the Hall-Thue word, i.e., the fixed point of $0 \to 012$; $1 \to 02$; $2 \to 1$.

^{b} We mistakenly said in [8] that $ABCA.BCAB.CBAC$ is different from its reverse.

In the table above, the columns indicate respectively, the considered minimally nice formula f, whether is equivalent to its reversed formula, the avoidability exponent of f, the infinite ternary word avoiding f, the value q such that the corresponding morphism is q-uniform, the value such that the avoiding word is d-directed, the suitable property of (β^+, n)-freeness used in the proof that f is avoided. We list below the corresponding morphisms.

g_a	g_b	g_c	g_d	g_e	g_f
002112201	021221	2011	020112122	001220122	012220
001221122	021121	1200	020101112	001220112	012111
001220112	020001	1120	020001222	001120122	012012
001122012	011102	0222	010121222	001120112	011222
	010222	0012	000111222		010002

g_g	g_h	g_i	g_j	g_k
021210	011120	01222112	021121	022110
011220	002211	01112022	012222	021111
002111	002121	01100022	011220	012222
001222	001222	01012220	011112	012021
		01012120	000102	011220

As an example, we show that $ABCBA.CAC$ is avoided by $g_k(w_5)$. First, we check that $g_k(w_5)$ is $\left(\frac{4}{3}^+, 25\right)$-free using the main lemma in [5], that is, we check the $\left(\frac{4}{3}^+, 25\right)$-freeness of the g_k-image of every $\left(\frac{5}{4}^+\right)$-free word of length at most $\frac{2 \times \frac{4}{3}}{\frac{4}{3} - \frac{5}{4}} = 32$. Then we check that $g_k(w_5)$ is 13-directed by inspecting the factors of $g_k(w_5)$ of length 13. For contradiction, suppose that $g_k(w_5)$ contains an occurrence h of $ABCBA.CAC$ up to $\simeq$. Let us write $a = |h(A)|$, $b = |h(B)|$, $c = |h(C)|$.

Suppose that $a \geq 25$. Since $g_k(w_5)$ is 13-directed, all occurrences of $h(A)$ are identical. Then $h(ABCBA)$ is a repetition with period $|h(ABCB)| \geq 25$. So the $\left(\frac{4}{3}^+, 25\right)$-freenes bound $\frac{2a + 2b + c}{a + 2b + c} \leq \frac{4}{3}$, that is, $a \leq b + \frac{1}{2}c$.

In every case, we have

$$a \leq \max\left\{b + \tfrac{1}{2}c, 24\right\}.$$

Similarly, the factors $h(BCB)$ and $h(CAC)$ imply

$$b \leq \max\left\{\tfrac{1}{2}c, 24\right\}$$

and

$$c \leq \max\left\{\tfrac{1}{2}a, 24\right\}.$$

Solving these inequalities gives $a \leq 36$, $b \leq 24$, and $c \leq 24$. Now we can check exhaustively that $g_k(w_5)$ contains no occurrence up to $\simeq$ satisfying these bounds.

Except for $ABCBA.CBABC$, the avoidability index of the nice formulas in the above table is 3. So the results in this section extend their 3-avoidability up to $\simeq$.

4   The Power Series Method

The so-called power series method has been used [4,6] to prove the 3-avoidability of many classical doubled patterns with at least 4 variables and every doubled pattern with at least 6 variables, as mentioned in the introduction.

Let p be such a classical doubled pattern and let p' be a doubled pattern with reversal obtained by adding some $-^R$ to p. Witout loss of generality, the leftmost appearance of every variable X of p remains free of $-^R$ in p'. Then we will see that p' is also 3-avoidable. The power series method is a counting argument that relies on the following observation. If the h-image of the leftmost appearance of the variable X of p is fixed, say $h(X) = w_X$, then there is exactly one possibility for the h-image of the other appearances of X, namely $h(X) = w_X$. This observation can be extended to p', since there is also exactly one possibility for $h(X^R)$, namely $h(X^R) = w_X^R$.

Notice that this straightforward generalization of the power series method from classical doubled patterns to doubled patterns with reversal cannot be extended to avoiding a doubled pattern up to $\simeq$. Indeed, if $h(X) = w_X$ for the leftmost appearance of the variable X and w_X is not a palindrome, then there exist two possibilities for the other appearances of X, namely w_X and w_X^R.

5   Sporadic Patterns

Up to symmetries, there are ten doubled patterns whose 3-avoidability cannot be deduced by the previous results. They have been identified in [6] and are listed in the following table.

Doubled pattern	Avoidability exponent
$ABACBDCD$	1.381966011
$ABACDBDC$	1.333333333
$ABACDCBD$	1.340090632
$ABCADBDC$	1.292893219
$ABCADCBD$	1.295597743
$ABCADCDB$	1.327621756
$ABCBDADC$	1.302775638
$ABACBDCEDE$	1.366025404
$ABACDBCEDE$	1.302775638
$ABACDBDECE$	1.320416579

Let w_5 be any infinite $\left(\frac{5}{4}^+\right)$-free word over Σ_5 and let h be the following 9-uniform morphism.

$$h(0) = 020022221$$
$$h(1) = 011111221$$
$$h(2) = 010202110$$
$$h(3) = 010022112$$
$$h(4) = 000022121$$

First, we check that $h(w_5)$ is 7-directed and $\left(\frac{139}{108}^+, 46\right)$-free. Then, using the same method as in Sect. 3, we show that $h(w_5)$ avoids up to $\simeq$ these ten sporadic patterns simultaneously.

6   Conclusion

Unlike classical formulas, we know that there exist avoidable formulas with reversal of arbitrarily high avoidability index [7]. Maybe doubled patterns and nice formulas are easier to avoid. We propose the following open problems.

– Are there infinitely many doubled patterns up to $\simeq$ that are not 2-avoidable?
– Is there a nice formula up to $\simeq$ that is not 3-avoidable?

A first step would be to improve Theorem 1 by generalizing the 3-avoidability of doubled patterns with reversal to doubled patterns up to $\simeq$. Notice that the results in Sects. 3 and 5 already consider avoidability up to $\simeq$. However, the power series method gives weaker results. Classical doubled patterns with at least 6 variables are 3-avoidable because

$$1 - 3x + \left(\frac{3x^2}{1-3x^2}\right)^v$$

has a positive real root for $v \geq 6$. The (basic) power series for doubled patterns up to $\simeq$ with v variables would be

$$1 - 3x + \left(\frac{6x^2}{1-3x^2} - \frac{3x^2+3x^4}{1-3x^4}\right)^v.$$

The term $\frac{6x^2}{1-3x^2}$ counts for twice the term $\frac{3x^2}{1-3x^2}$ in the classical setting, for $h(V)$ and $h(V)^R$. The term $\frac{3x^2+3x^4}{1-3x^4}$ corrects for the case of palindromic $h(V)$, which should not be counted twice. This power series has a positive real root only for $v \geq 10$. This leaves many doubled patterns up to $\simeq$ whose 3-avoidability must be proved proved with morphisms.

References

1. Currie, J., Mol, L.: The undirected repetition threshold and undirected pattern avoidance. Theor. Comput. Sci. **866**, 56–69 (2021)
2. Currie, J., Mol, L., Rampersad, N.: A family of formulas with reversal of high avoidability index. Int. J. Algebra Comput. **27**(5), 477–493 (2017)
3. Currie, J., Mol, L., Rampersad, N.: Avoidance bases for formulas with reversal. Theor. Comput. Sci. **738**, 25–41 (2018)
4. Bell, J., Goh, T.L.: Exponential lower bounds for the number of words of uniform length avoiding a pattern. Inform. Comput. **205**, 1295–1306 (2007)
5. Ochem, P.: A generator of morphisms for infinite words. RAIRO Theoret. Inform. Appl. **40**, 427–441 (2006)

6. Ochem, P.: Doubled patterns are 3-avoidable. Electron. J. Comb. **23**(1), #P1.19 (2016)
7. Ochem, P.: A family of formulas with reversal of arbitrarily high avoidability index. arXiv:2103.07693
8. Ochem, P.: Rosenfeld, M: On some interesting ternary formulas. Electron. J. Comb. **26**(1), #P1.12 (2019)
9. Ochem, P., Rosenfeld, M.: Avoidability of palindrome patterns. Electron. J. Comb **28**(1), #P1.4 (2021)

A Characterization of Binary Morphisms Generating Lyndon Infinite Words

Gwenaël Richomme[(✉)] and Patrice Séébold

LIRMM, Université Paul-Valéry Montpellier 3, Université de Montpellier, CNRS,
Montpellier, France
{gwenael.richomme,patrice.seebold}@lirmm.fr

Abstract. An infinite word is an infinite Lyndon word if it is smaller, with respect to the lexicographic order, than all its proper suffixes, or equivalently if it has infinitely many finite Lyndon words as prefixes. A characterization of binary endomorphisms generating Lyndon infinite words is provided.

1 Introduction

Finite Lyndon words are the non-empty words which are smaller, w.r.t. (with respect to) the lexicographic order, than all their proper suffixes. They are important tools in many studies (see, *e.g.*, [2,11,12,15]). Infinite Lyndon words are defined similarly. They are also the words that have infinitely many finite Lyndon words as prefixes. They occur in many context (see, *e.g.*, [1,3,5,8,10,13,14]).

The aim of the current paper is to provide a characterization, in the binary case, of endomorphisms that generate infinite Lyndon words. This paper continues the study of links between morphisms and Lyndon words done by the first author. In [16] he studied and characterized the morphisms that preserve Lyndon words, calling them *Lyndon morphisms*: these morphisms are those that map any Lyndon word to another Lyndon word. This study was extended to morphisms that preserve infinite Lyndon words in [17].

Note that being a morphism that preserves finite Lyndon words is a sufficient condition to generate an infinite Lyndon word (if the morphism generates an infinite word). Indeed if f is a morphism that preserves finite Lyndon words and u is a Lyndon word, then, for any $n \geq 0$, $f^n(u)$ is a Lyndon word. Applying this process when $u = a$ with a morphism f that generates from a an infinite word $\mathbf{w}$, we see that $\mathbf{w}$ has infinitely many finite Lyndon words as prefixes: it is an infinite Lyndon word. But the condition is not necessary. For instance, the morphism defined by $f(a) = ababa$ and $f(b) = babbb$ generates an infinite Lyndon word (the proof can be done using Proposition 8) but it does not preserve finite Lyndon words since $f(a)$ is not a Lyndon word.

Our main characterization is Theorem 2: Over $\{a, b\}$ with $a \prec b$, a non-periodic word generated by a morphism f prolongable on a is an infinite Lyndon word if and only if f preserves the lexicographic order on finite words and $f^3(a)$ is a prefix of a Lyndon word. The proof needs to consider separately the

© Springer Nature Switzerland AG 2021
T. Lecroq and S. Puzynina (Eds.): WORDS 2021, LNCS 12847, pp. 160–171, 2021.
https://doi.org/10.1007/978-3-030-85088-3_14

case where aa is a prefix of $f^\omega(a)$ and the case where ab is a prefix of $f^\omega(a)$. After some needed preliminaries in Sect. 2, we prove the following general necessary condition: a binary endomorphism that generates an infinite Lyndon word must preserve the lexicographic order on finite words. In Sect. 4, we characterize morphisms that generate an infinite Lyndon word beginning with aa (Proposition 7). In Sect. 5, we characterize morphisms that generate an infinite Lyndon word beginning with ab (Proposition 8). In Sect. 6, we prove our main result. We conclude with a few words on what happens on larger alphabets.

2   About Lyndon Words and Morphisms

We assume that readers are familiar with combinatorics on words and morphisms (see, e.g., [11,12]). We specify our notation and recall useful results.

An *alphabet* A is a set of symbols called *letters*. Here we consider only finite alphabets. A *word over* A is a sequence of letters from A. The *empty word* ε is the empty sequence. Equipped with the concatenation operation, the set A^* of finite words over A is a free monoid with neutral element ε and set of generators A. We let A^ω denote the set of infinite words over A. As usually, for a finite word u and an integer n, the nth power of u, denoted u^n, is the word ε if $n = 0$ and the word $u^{n-1}u$ otherwise. If u is not the empty word, u^ω denotes the infinite word obtained by infinitely repeating u. Such a word is called *periodic*. A finite word w is said *primitive* if for any word u, the equality $w = u^n$ (with n an integer) implies $n = 1$.

Given a non-empty word $u = a_1 \cdots a_n$ with $a_i \in A$, the *length* $|u|$ of u is the integer n. One has $|\varepsilon| = 0$. If for some words u, v, p, s (possibly empty), $u = pvs$, then v is a *factor* of u, p is a *prefix* of u and s is a *suffix* of u. When $p \neq u$ (resp. $s \neq u$), we say that p is a *proper prefix* (resp. s is a *proper suffix*) of u.

Let us recall two basic results.

Proposition 1 (see, e.g., [11, Prop. 1.3.2]). *For any words u and v, $uv = vu$ if and only if there exist a word w and integers k, ℓ such that $u = w^k$ and $v = w^\ell$.*

Theorem 1 (Fine and Wilf's Theorem, see, e.g., [11, Prop. 1.3.5]). *Let $x, y \in A^*$, $n = |x|$, $m = |y|$, $d = \gcd(n, m)$. Assume there exist integers p and q such that x^p and y^q have a common prefix of length at least equal to $n + m - d$. Then x and y are powers of the same word.*

2.1   Lyndon Words

From now on we consider ordered alphabets. We let $A_n = \{a_1 \prec \ldots \prec a_n\}$ denote the n-letter alphabet $A_n = \{a_1, \ldots, a_n\}$ with order $a_1 \prec \ldots \prec a_n$. Given an ordered alphabet A, we let also $\preceq$ denote the lexicographic order whenever used on A^* or on A^ω. Let us recall that for two different (finite or infinite) words u and v, $u \prec v$ if and only if $u = x\alpha y$, $v = x\beta z$ with $\alpha, \beta \in A$, $\alpha \prec \beta$, $x \in A^*$, $y, z \in A^* \cup A^\omega$, or if (when u is finite) u is a proper prefix of v. For any finite words u, v, w, if $u \prec v$, then $wu \prec wv$. Moreover if u is not a prefix of v and $u \prec v$, then $ux \prec vy$ for any words x and y.

A non-empty finite word w is a *Lyndon word* if for all non-empty words u and v, $w = uv$ implies $w \prec vu$. Equivalently [6,11], a non-empty word w is a Lyndon word if all its non-empty proper suffixes are greater than itself for the lexicographic order. For instance, on the one-letter alphabet $\{a\}$, only a is a Lyndon word. On $\{a \prec b\}$ the Lyndon words of length 6 are $aaaaab$, $aaaabb$, $aaabab$, $aaabbb$, $aababb$, $aabbab$, $aabbbb$, $ababbb$, $abbbbb$. Lyndon words are primitive. Note that Lyndon words have no non-empty border, that is, there is no proper prefix of a Lyndon word u that is also a suffix of u. Also observe that if u is a prefix of a Lyndon word then there cannot exist words v and w such that the following three conditions hold: v is a prefix of u; w is a factor of u which is not a prefix of u; $w \prec v$.

Proposition 2 (see, *e.g.*, *[11,* **Prop. 5.1.3]***). A non-empty word w is a Lyndon word if and only if $|w| = 1$ or $w = uv$ with u and v two Lyndon words such that $u \prec v$.*

Lyndon infinite words were introduced in [19] as the infinite words that have infinitely many prefixes that are Lyndon words. By definition an infinite Lyndon word is not periodic (but it may be ultimately periodic as, for instance, ab^ω is). More generally an infinite word is Lyndon if and only if all its proper suffixes are greater than it w.r.t. the lexicographic order [19, Proposition 2.2].

2.2   Morphisms

Let A and B be two alphabets. A *morphism* f from A^* to B^* is a mapping from A^* to B^* such that for all words u, v over A, $f(uv) = f(u)f(v)$. We say that f is a morphism over A if we don't need to refer to B. When $A = B$, f is an *endomorphism* over A. A morphism is *erasing* if $f(a) = \varepsilon$ for some letter a. For $n \geq 0$ and any finite or infinite word u, $f^n(u)$ is u if $n = 0$ and $f^{n-1}(f(u))$ otherwise.

An endomorphism is said *prolongable* on a if $f(a) = au$ for some word u and if $\lim_{n \to \infty} |f^n(a)| = \infty$. For such a morphism, for all $n \geq 0$, $f^n(a)$ is a prefix of $f^{n+1}(a)$. Then the sequence $(f^n(a))_{n \geq 0}$ defines a unique infinite word, denoted $f^\omega(a)$. This word is a fixed point of f.

A morphism *preserves finite Lyndon words* if and only if the image of any finite Lyndon word is also a Lyndon word. Similarly morphisms that *preserve infinite Lyndon words* can be defined. A morphism *preserves the order on finite words* if, for all words u and v, $u \prec v$ implies $f(u) \prec f(v)$. Such a morphism is injective and so non-erasing. In [16], it is proved that a morphism is a Lyndon morphism if and only if it preserves the lexicographic order on finite words and if the image of each letter is a Lyndon word. We also have the following characterization.

Proposition 3 (*[16,* **Prop. 3.3]***). A morphism f over $\{a \prec b\}$ preserves the lexicographic order on finite words if and only if $f(ab) \prec f(b)$.*

3   A Necessary Condition

In this section we prove the following result that states a necessary condition for a prolongable binary morphism to generate an infinite Lyndon word.

Proposition 4. *Let f be an endomorphism over $\{a \prec b\}$. Assume that f is prolongable on a. If $f^\omega(a)$ is a Lyndon infinite word then f preserves the lexicographic order on finite words.*

We will use the basic fact and the following characterization of prefixes of Lyndon words. As recalled to us by one of the referees, the notion of prefix of a Lyndon word occurs in the literature under various terminologies as pre-necklace, sesquipower of a Lyndon word or preprime word (see for instance respectively [4,9,18] where can be found some results close to Proposition 6). It may be observed that from Proposition 6 (and with c defined as in this result), any power of a Lyndon word is a prefix of a Lyndon word unless it is a power c^k with $k \geq 2$.

Fact 5. *Given any finite Lyndon word x and any proper non-empty prefix p of x, $px \prec x$.*

Proof. Let q be the word such that $x = pq$. Since x is a Lyndon word and since $x \neq q$ and $x \neq \varepsilon$, $x \prec q$. It follows that $px \prec pq = x$. ☐

Proposition 6 ([7, Prop. 1.7]). *Let A be an ordered alphabet with maximal letter c. Let P be the set of prefixes of Lyndon words. The set $P \cup \{c^k \mid k \geq 2\}$ is equal to the set of all words on the form $(uv)^k u$ with $k \geq 1$ an integer and u, v some finite words such that $v \neq \varepsilon$ and uv is a Lyndon word.*

Proof of Proposition 4. Assume by contradiction that f does not preserve the lexicographic order on finite words. By Proposition 3, $f(b) \prec f(ab)$ (the equality cannot hold as $f(a)$ is not empty). Thus, for any integer $n \geq 0$, $f(a^n b) \prec f(a^{n+1}b)$. So, for any integer $n \geq 1$, $f(b) \prec f(a^n b)$.

From now on let i be the integer such that $a^i b$ is a prefix of $f^\omega(a)$. Let also $\mathbf{w}$ be the word such that $f^\omega(a) = f(a^i b)\mathbf{w}$. Note that i exists and $i \geq 1$ since f is prolongable on a and a^ω is not an infinite Lyndon word.

Observe that $f(b)$ is a prefix of $f(a^i b)$. Otherwise, from $f(b) \prec f(a^i b)$, we deduce that $f(b)\mathbf{w} \prec f^\omega(a)$ which contradicts the fact that $f^\omega(a)$ is an infinite Lyndon word since $f(b)\mathbf{w}$ is a proper suffix of $f^\omega(a)$.

As $f^\omega(a)$ is an infinite Lyndon word, it has infinitely many prefixes that are Lyndon words. Thus its prefix $f(a^i b)$ is a prefix of a Lyndon word. Hence by Proposition 6, there exist an integer $k \geq 1$ and words u and v such that $f(a^i b) = (uv)^k u$, $v \neq \varepsilon$ and uv is a Lyndon word. Consequently $f(b) = (uv)^j u'$ for some $j \geq 0$ and some proper prefix u' of uv.

Note that $f^\omega(a) \neq ab^\omega$ since this implies that $i = 1$ and $f(b) \in b^+$, a contradiction with $f(b) \prec f(ab)$. Hence the letter a occurs at least twice in $f^\omega(a)$. Since $f^\omega(a)$ is a Lyndon word, $f^\omega(a) \neq a^\omega$. This implies that $a^i b$ is a prefix of $f(a)$. We deduce that $a^i b$ has a non-prefix occurrence in $f^\omega(a)$. The word $a^{i+1}b$ cannot be a factor of the Lyndon word $f^\omega(a)$ (since a^ib is a prefix of $f^\omega(a)$). Hence ba^ib is a factor of $f^\omega(a)$.

Assume that $u' \neq \varepsilon$. Since ba^ib is a factor of $f^\omega(a)$, the word $u'uv$ is a factor of $f(ba)$ and so of $f^\omega(a)$. By Fact 5, $u'uv \prec uv$: since uv is a prefix of $f^\omega(a)$, this contradicts the fact that $f^\omega(a)$ is an infinite Lyndon word.

Thus $u' = \varepsilon$. This means that $f(b) = (uv)^j$ with $j \geq 0$. If $j = 0$, $f(b) = \varepsilon$ and $f^\omega(a) = f(a)^\omega$ is a periodic word: a contradiction with the fact that it is an infinite Lyndon word. Thus $j \geq 1$. Since $f(b)$ is a suffix of $f(a^ib) = (uv)^ku$, we get $uv = vu$. Remember that $v \neq \varepsilon$. If $u \neq \varepsilon$, by Proposition 1, the word uv is not primitive: a contradiction with the primitivity of the Lyndon word uv. So $u = \varepsilon$.

This implies that both $f(a)$ and $f(b)$ are powers of v. So $f^\omega(a) = v^\omega$. This is a final contradiction with the fact that an infinite Lyndon word cannot be periodic. The morphism f preserves the order on finite words over $\{a \prec b\}$. □

Note that the converse of Proposition 4 does not hold. Consider, for instance, the morphism f defined by $f(a) = abb$ and $f(b) = baa$. This morphism preserves the lexicographic order on infinite words but the word $f^\omega(a)$ is not an infinite Lyndon word.

One could expect a stronger necessary condition as, for instance, a preservation of infinite Lyndon words. The next example shows that this stronger condition is not necessary.

Let f be defined by $f(a) = aab$ and $f(b) = abaabab$. The word $\mathbf{w} = abbabbb^\omega$ is an infinite Lyndon word. Its image by f begins with $ubua$ where $u = aababaaba$. Hence f does not preserve infinite Lyndon words. Nevertheless using Proposition 7, one can verify that f generates an infinite Lyndon word.

4   Generating Infinite Lyndon Words Beginning with aa

We consider here the case of generated words beginning with aa.

Proposition 7. *Let f be an endomorphism over $\{a \prec b\}$ prolongable on a such that $f^\omega(a)$ begins with a^ib for some integer $i \geq 2$.*

The word $f^\omega(a)$ is an infinite Lyndon word if and only if

1. *f preserves the lexicographic order on finite words, and,*
2. *$f(a^ib)$ is a Lyndon word.*

The proof of this proposition is based on the next lemmas.

Lemma 1. *Let f be a morphism that preserves the order on finite words. Let $i \geq 1$. Assume that $f(a^ib)$ is a Lyndon word. For any word v such that a^ibv is a Lyndon word, the word $f(a^ibv)$ is also a Lyndon word.*

Proof. We act by induction on $|v|$.

By hypothesis the result holds when $|v| = 0$. Assume that $|v| \geq 1$. By Proposition 2, there exist Lyndon words ℓ and m such that $a^ibv = \ell m$ and $\ell \prec m$. Let us choose m with the smallest possible length.

Let us prove that $a^i b$ is a prefix of ℓ. Assume that this does not hold. Then $\ell = a$ and $m = a^{i-1}bv$. Consequently, since m is a Lyndon word, a^i is not a factor of $a^{i-1}bv$. Let m' be the word in $a^+ b^+$ such that bm' is a suffix of m. If such a factor does not exist (that is if $m \in a^{i-1}b^+$), let $m' = b$. In all cases, m' is a Lyndon word. Let ℓ' be the word such that $\ell m = \ell' m'$. The word $a^i b$ is a prefix of ℓ' (when $m' = b$, remember that $|v| \geq 1$). Observe that $\ell' \prec m'$. The last letter of ℓ' is b. Indeed, by construction, it could be the letter a only if $m \in a^{i-1}b^+$, that is if $m = a^{i-1}b^k$ for some $k \geq 1$. But then $m' = b$ and $\ell' = a^i b^{k-1}$. As $|v| \geq 1$, we have $k \geq 2$, and so, the last letter of ℓ' is b. Let s be a proper non-empty suffix of ℓ'. Let $j \geq 0$ be the integer such that s begins with $a^j b$. Since $a^i b$ is not a factor $a^{i-1}bv$, we deduce that $j < i$. So $\ell' \prec s$. Hence ℓ' is a Lyndon word: this contradicts the choice made on m and proves that $a^i b$ is a prefix of ℓ.

If $a^i b$ is a prefix of m then, by inductive hypothesis, $f(\ell)$ and $f(m)$ are Lyndon words. Since $\ell \prec m$ and f preserves the order on finite words, $f(\ell) \prec f(m)$. Proposition 2 implies that $f(a^i bv) = f(\ell m)$ is a Lyndon word.

From now on assume that $a^i b$ is not a prefix of m. Observe that this implies that m begins with $a^k b$ for some integer $k < i$. Indeed since $a^i bv = \ell m$ is a Lyndon word, for any factor a^j of ℓm, we have $j \leq i$. Moreover as m is a Lyndon word, for any factor a^j of m, we have $j \leq k < i$. Let s be a proper non-empty suffix of $f(\ell m)$. Assume $|s| \leq |f(m)|$. Let m_0 be the smallest suffix of m such that s is a suffix of $f(m_0)$. Let $j < i$ be the integer and let m' be the word such that $m_0 = a^j bm'$. By choice of m_0, there exists a non-empty suffix s' of $f(a^j b)$ such that $s = s' f(m')$. The word s' is a proper non-empty suffix of the Lyndon word $f(a^i b)$. So $f(a^i b) \prec s'$ and $f(a^i bv) \prec s' \preceq s$. If $|s| > |f(m)|$ then $s = s' f(m)$ with s' a proper non-empty suffix of $f(\ell)$. By inductive hypothesis, $f(\ell)$ is a Lyndon word. Thus $f(\ell) \prec s'$ and consequently $f(a^i bv) = f(\ell m) \prec s' \prec s' f(m) = s$. The word $f(a^i bv)$ is a Lyndon word.   □

Lemma 2. *Let u be a non-empty word. If uu is a prefix of a Lyndon word, then u is a power of a Lyndon word.*

Proof. Since uu is a prefix of a Lyndon word, also u is a prefix of this Lyndon word. By Proposition 6, there exist words x and y such that $y \neq \varepsilon$, xy is a Lyndon word and for some integer $k \geq 1$, $u = (xy)^k x$. If $x \neq \varepsilon$, since xy is a Lyndon word, we have $xy \prec y$ and so $xxy \prec xy$. Then for any word v, the word $x(xy)^k xv$ is a suffix of uuv and $x(xy)^k xv \prec uuv$. This contradicts the fact that uu is a prefix of a Lyndon word. So $x = \varepsilon$. This implies that $u = y^k$ and y is a Lyndon word.   □

Lemma 3. *Assume that f is an endomorphism over $\{a \prec b\}$ prolongable on a such that $f^3(a)$ is a prefix of a Lyndon word, $f^\omega(a)$ begins with the word $a^i b$ for some integer $i \geq 2$ and $f^\omega(a)$ is not periodic. Then $f(a^i b)$ is a Lyndon word.*

Proof. Let us first observe that $f(a)$ begins with $a^i b$. Indeed otherwise $f(a)$ is a power of a contradicting the non-periodicity of $f^\omega(a)$.

Also observe that the word $f(a^i b)$ is a prefix of $f^2(a)$ which itself is a prefix of $f^3(a)$. Hence $f(a^i b)$ is also a prefix of a Lyndon word. By Proposition 6, there exist a Lyndon word v, a proper prefix p of v (p may be empty) and an integer $\ell \geq 1$ such that $f(a^i b) = v^\ell p$. Since $i \geq 2$, by Lemma 2, $f(a)$ is a power of a Lyndon word u.

If $v = u$, from $f(a^i b) = v^\ell p$, we get $f(b) = v^{\ell'} p$ for some integer ℓ'. In particular p is a suffix of $f(b)$. If $p = \varepsilon$, then $f^\omega(a) = v^\omega$ a contradiction with its non-periodicity. Assume now that $p \neq \varepsilon$. Since $i \geq 2$, the word $a^i b$ occurs twice in $f(a^i b)$ which is a prefix of $f^2(a)$. Thus ba is a factor of $f^2(a)$ and $f(ba)$ is a factor of $f^3(a)$. Then the word pu is a factor of $f^3(a)$. Note also that u is a prefix of $f^3(a)$. As p is a proper non-empty prefix of the Lyndon word u, by Fact 5, $pu \prec u$. This contradicts the fact that $f^3(a)$ is a prefix of a Lyndon word. Thus $v \neq u$.

Since $i \geq 2$, u^2 is a prefix of $v^{\ell+1}$. If $|u| \geq |v|$, by Theorem 1, u and v are powers of the same word. This is not possible as $u \neq v$ and both words u and v are primitive (since they are Lyndon words). Thus $|v| > |u|$.

Note that v is not a factor of $f(a^i) = f(a)^i$. Indeed if v is a factor of $f(a)^i$ then it is a prefix of a power of u, and so, a prefix of u is both a prefix and a suffix of v: this is impossible since v is a Lyndon word. It follows that p is a proper suffix of $f(b)$.

Observe that $a^i b$ is a prefix of $f(a)$ and so $f(a)a^i b$ is a prefix of $f(aa)$ and so of $f^2(a)$. Since $f^2(a)$ is a prefix of a Lyndon word, it cannot contain a^{i+1} as a factor and so the last letter of $f(a)$ must be b. Hence $ba^i b$ and $f(ba^i b)$ are factors of $f^3(a)$. This implies that pv is also a factor of $f^3(a)$. By Fact 5, $pv \prec v$ if $p \neq \varepsilon$. This contradicts the fact that $f^3(a)$ is a prefix of a Lyndon word. So $p = \varepsilon$ and $f(a^i b) = v^\ell$. Assume that $\ell \geq 2$.

Since $a^i b$ is a prefix of $f(a)$ and since $f(a^i b) = v^\ell$, $a^i b$ is also a prefix of v. Thus v^ℓ is a prefix of $f(v)$ itself a prefix of $f^2(a)$. Since $\ell \geq 2$, $f(v)f(v)$ and $f(v)v^\ell$ are prefixes of $f^3(a)$.

Let us prove that $f(v)$ is not a prefix of v^ω. Assume by contradiction that $f(v) = v^k p'$ for some proper prefix p' of $f(v)$ and some integer k. Since v^ℓ is a prefix of $f(v)$, we have $k \geq \ell \geq 2$. If $p' \neq \varepsilon$, by Fact 5, $p'v \prec v$. Since $p'v$ is a factor of $f(v)f(v)$, this contradicts the fact that $f^3(a)$ is a prefix of a Lyndon word. So $p' = \varepsilon$ and $f(v) = v^k$. Hence by induction, for all $n \geq 0$, $f^n(v) \in v^+$. Moreover we have $\lim_{n \to \infty} |f^n(v)| = \infty$. So $f^\omega(a) = v^\omega$: a contradiction with the non-periodicity of $f^\omega(a)$.

So $f(v)$ is not a prefix of v^ω. There exist an integer k, a proper prefix π of v and letters α, β such that $v^k \pi \beta$ is a prefix of $f(v)$ and $\pi \alpha$ is a prefix of v. Since $f^3(a)$ is a prefix of a Lyndon word, $\alpha = a$ and $\beta = b$. Note that $v^{k+1} \prec v^k \pi \beta$.

We have already mentioned that v is not a factor of $f(a^i)$. From $f(a^i b) = v^\ell$ and $\ell \geq 2$, we deduce that v is a suffix of $f(b)$. Moreover since v is a Lyndon word beginning with $a^i b$, the last letter of v is b: v is so a suffix of $f(v)$. Since $f(v)f(v)$ is a factor of $f^3(a)$, the word v^{k+1} is a factor of $f^3(a)$. This contradicts the fact that $f^3(a)$ is a prefix of a Lyndon word.

Thus $\ell = 1$: $f(a^i b)$ is a Lyndon word. $\square$

Proof of Proposition 7. We first prove that the two conditions are sufficient. First observe that, for any integer $n \geq 1$, $a^i b$ is a prefix of $f^n(a^i b)$ (this is a direct consequence of the facts that f is prolongable on a and that $f^\omega(a)$ begins with $a^i b$). Thus by induction, using Lemma 1, we get: for any integer $n \geq 0$, $f^n(a^i b)$ is a Lyndon word. As $\lim_{n \to \infty} |f^n(a^i b)| = \infty$, the word $f^\omega(a)$ has infinitely many prefixes that are Lyndon words. By definition, it is an infinite Lyndon word.

From now on assume that $f^\omega(a)$ is an infinite Lyndon word. Proposition 4 shows that f preserves the lexicographic order on finite words. Observe that since it is an infinite Lyndon word, $f^\omega(a)$ is not periodic and $f^3(a)$ is a prefix of a Lyndon word. Lemma 3 states that $f(a^i b)$ is a Lyndon word.   □

5   Generating Infinite Lyndon Words Starting with *ab*

We consider here the case of generated words beginning with ab. The word ab^ω is an infinite Lyndon word. A morphism f generates it if and only if $f(a) = ab^i$ for some integer $i \geq 1$ and if $f(b) \in b^+$. In what follows we only consider the case where $f^\omega(a)$ begins with $ab^i a$ for some $i \geq 1$.

Proposition 8. *Let f be an endomorphism over $\{a \prec b\}$ prolongable on a such that $f^\omega(a)$ begins with $ab^i a$ for some integer $i \geq 1$.*
The word $f^\omega(a)$ is an infinite Lyndon word if and only if

1. *f preserves the lexicographic order on finite words,*
2. *$f(ab^i)$ is a power of a Lyndon word $u \neq ab^i$, and,*
3. *if $i = 1$, $|u| > |f(b^i)|$.*

Here follows an example showing that indeed in item 2, $f(ab^i)$ is not necessarily a Lyndon word.

Example 9. Let f be defined by $f(a) = abababab$ and $f(b) = abb$: $f(ab) = (ababb)^2$ is the square of a Lyndon word (longer than $f(b)$).

We now provide an example showing the necessity of item 3.

Example 10. Let f be defined by $f(a) = aba$ and $f(b) = bbababb$: $f(ab) = u^2$ with $u = ababb$ is the square of a Lyndon word. Condition 3 is not verified and indeed $f^\omega(a)$ is not a Lyndon word. It can be verified that $f^\omega(a)$ begins with $u^4 bbu^5$ and so contains the factor $u^4 a$ which is smaller than the prefix $u^4 b$.

The proof of Proposition 8 is based on the next lemmas.

Lemma 4. *Let f be a morphism that preserves the lexicographic order on finite words over $\{a \prec b\}$. Assume that $i \geq 2$ is an integer and that $f(ab^i)$ is a power of a Lyndon word u. Then $|u| > |f(b)|$.*

Proof. Assume by contradiction that $|u| \leq |f(b)|$. Assume first that $|u| < |f(b)|$. Since $f(b)f(b)$ is a suffix of $f(ab^i)$ so of a power of u, there exist words p and s and an integer $k \geq 1$ such that $u = ps$, $f(b)$ ends with p and $f(b) = su^k$. Since u is a Lyndon word, p cannot be both a prefix and a suffix of u except if $p = \varepsilon$. When $p = \varepsilon$, $f(b)$ is a power of u. If $|u| = |f(b)|$ then $f(b) = u$. In all cases both $f(a)$ and $f(b)$ are powers of u. Hence f is not injective, a contradiction with the fact that f preserves the lexicographic order on finite words.   □

Lemma 5. *Let f be a morphism that preserves the lexicographic order on finite words over $\{a \prec b\}$. Assume that $f(ab^i)$ is a power of a Lyndon word u for some integer $i \geq 1$. Assume also that $|u| > |f(b)|$ if $i = 1$. Then, for any non-empty word v over $\{a \prec b\}$ such that $ab^i v$ is a Lyndon word, the word $f(ab^i v)$ is also a Lyndon word.*

Proof. We act by induction on $|v|$. Let us observe that f is non-erasing and injective since it preserves the lexicographic order on finite words. Let n be the integer such that $f(ab^i) = u^n$. Observe that $|f(b)| < |u|$ (by hypothesis if $i = 1$ and by Lemma 4 if $i \geq 2$).

We first assume that $|v| = 1$. In this case, since $ab^i v$ is a Lyndon word, $v = b$. Any suffix s of $f(ab^{i+1})$ with $|s| \leq |f(b)|$ is also a suffix of the Lyndon word u. Thus $u \prec s$ (and for length reason, u is not a prefix of s). Hence $f(ab^{i+1}) \prec s$. Consider now a suffix s of $f(ab^{i+1})$ such that $|f(b)| < |s| < |f(ab^{i+1})|$. We have $s = s'f(b)$ for some suffix s' of $f(ab^i) = u^n$. If $s' = s''u^k$ for some proper non-empty suffix s'' of u and some integer k then $u \prec s''$ and u is not a prefix of s''. Once again $f(ab^{i+1}) \prec s$. If $s' = u^k$ for some integer k such that $1 \leq k < n$, $s = u^k f(b)$. As $f(b)$ is a proper non-empty suffix of u, $u \prec f(b)$. Hence $u^{k+1} \prec u^k f(b)$. Moreover since $k+1 \leq n$, $f(ab^i) \prec u^k f(b)$. So for any proper non-empty suffix s of $f(ab^{i+1})$, $f(ab^{i+1}) \prec s$: $f(ab^{i+1})$ is a Lyndon word.

From now on assume that $|v| \geq 2$. By Proposition 2, there exist two Lyndon words ℓ and m such that $ab^i v = \ell m$ and $\ell \prec m$. Two cases can hold.

Case $|m| \geq 2$. As m cannot begin with the letter b (as any Lyndon word of length at least 2 over a binary alphabet), ℓ must begin with ab^i. Moreover as $ab^i v = \ell m$ is a Lyndon word, m is on the form ab^k with $k > i$ or begins with a factor $ab^k a$ with $k \geq i$. In both cases, ab^i is a proper prefix of m, and by inductive hypothesis $f(m)$ is a Lyndon word. If $\ell \neq ab^i$, $f(\ell)$ is also a Lyndon word. Moreover, since f preserves the lexicographic order, $f(\ell) \prec f(m)$. By Proposition 2, $f(ab^i v) = f(\ell m)$ is a Lyndon word. If $\ell = ab^i$, $f(\ell) = u^n$. Since f preserves the lexicographical order, $u \preceq f(\ell) \prec f(m)$. Using Proposition 2, one can prove by induction that $u^k f(m)$ is a Lyndon word for any $k \geq m$. Once again, $f(ab^i v) = u^n f(m)$ is a Lyndon word.

Case $|m| = 1$. In this case, $m = b$. Let s be a proper non-empty suffix of $f(ab^i v)$. If $|s| \leq |f(b)|$, then s is a suffix of the Lyndon word u (remember that $|f(b)| < |u|$ and $f(ab^i) = u^n$). This implies that $u \prec s$ and so that $f(ab^i v) \prec s$. If $|f(b)| < |s| < |f(ab^i v)|$, we have $s = s'f(b)$ for some proper non-empty suffix s' of the Lyndon word $f(\ell)$ (since $|v| \geq 2$, $|\ell| > |ab^i|$ and the inductive hypothesis can be applied). Thus $f(\ell) \prec s'$ which implies that $f(ab^i v) \prec s$. Hence $f(ab^i v)$ is a Lyndon word.   □

Lemma 6. *Assume that f is an endomorphism over $\{a \prec b\}$ prolongable on a such that $f^3(a)$ is a prefix of a Lyndon word, $f^\omega(a)$ begins with the word $ab^i a$ for some integer $i \geq 1$ and $f^\omega(a)$ is not periodic. Then $f(ab^i)$ is a power of a Lyndon word $u \neq ab^i$. Moreover if $i = 1$, $|u| > |f(b)|$.*

Proof. The word $ab^i a$ is a prefix of $f^\omega(a)$. Let us prove that the word $f^3(a)$ has a prefix on the form $ab^i ab^k a$. If $f(a)$ has $ab^i a$ as a prefix, then $f^2(a)$ (and so $f^3(a)$) contains at least 4 occurrences of a. Since $f^2(a)$ is a prefix of a Lyndon word, it cannot contain the factor aa. Hence we get the result. Assume now that $f(a) = ab^j$ for some $j < i$. Since f is prolongable on a, $j > 0$. It follows that $f(b)$ begins with $b^{j-i}a$. Then $f^3(a)$ contains at least 3 occurrences of a. And once again $f^3(a)$ has a prefix on the form $ab^i ab^k a$.

Since $f^3(a)$ is a prefix of a Lyndon word, we have $k \geq i$ and so $(ab^i)^2$ is a prefix of $f^3(a)$. Lemma 2 shows that $f(ab^i)$ is a power of a Lyndon word u: $f(ab^i) = u^n$ for an integer $n \geq 1$. If $u = ab^i$, we have $f^\omega(a) = (ab^i)^\omega$ which contradicts the fact that $f^\omega(a)$ is aperiodic. Thus $u \neq ab^i$.

Assume now that $i = 1$ and $|f(b)| \geq |u|$. From $f(ab) = u^n$ and $f(a) \neq \varepsilon$, we get $n \geq 2$. Let s be the proper suffix of u and let $j \geq 1$ be the integer such that $f(b) = su^j$. If $s = \varepsilon$, then both $f(a)$ and (b) are powers of u. This implies that $f^\omega(a) = u^\omega$, a contradiction. Assume now that $s \neq \varepsilon$. Let p be the word such that $u = ps$: $f(a) = u^k p$ for some integer $k \geq 0$ and $p \notin \{\varepsilon, u\}$ since $s \notin \{\varepsilon, u\}$. Since u is a Lyndon word different from ab but beginning with aba, we deduce that u begins with $(ab)^m b$ for some $m \geq 2$.

Since $n \geq 2$, the word $(ab)^m b$ has at least one non prefix occurrence in $f(ab)$ so in $f^3(a)$. This occurrence must be preceded by the letter b since aa cannot occur in $f^3(a)$ which is a prefix of a Lyndon word. Hence the word $f(ab)^m s = u^{nm}s$ is a prefix of $f((ab)^m b)$ itself a prefix of $f^3(a)$, and, the word uu^{nm} which is a suffix of $f(b(ab)^m)$ is a factor of $f^3(a)$. Since $u \prec s$, we have $u^{nm}u \prec u^{nm}s$: this contradicts the fact that $f^3(a)$ is a prefix of a Lyndon word.   □

Proof of Proposition 8. Let us first show that the three conditions imply that $f^\omega(a)$ is an infinite Lyndon word. Since f preserves the order on finite words, f is not erasing. Let u be the word occurring in condition 2 and let k be the integer such that $f(ab^i) = u^k$. Observe that ab^i and the prefix u of $f(ab^i)$ are both prefixes of $f^\omega(a)$. From $|f(ab^i)| \geq |ab^i|$, ab^i is a prefix of $f(ab^i) = u^k$. Hence ab^i is a prefix of u. By hypothesis, we cannot have $ab^i = u$. So ab^i is a proper prefix of u. For any $n \geq 0$, $f^n(u)$ is a prefix of $f^\omega(a)$ and so ab^i is a proper prefix of $f^n(u)$. Due to condition 3, one can apply Lemma 5. Thus it follows by induction that $f^n(u)$ is a Lyndon word for all $n \geq 0$: $f^\omega(a)$ is an infinite Lyndon word.

Let us show that the conditions are necessary. First Proposition 4 shows that f preserves the lexicographic order on finite words. Observe that since it is an infinite Lyndon word, $f^\omega(a)$ is not periodic and $f^3(a)$ is a prefix of a Lyndon word. Lemma 6 states that $f(ab^i)$ is a power of a Lyndon word u and, when $i = 1$, $|u| > |f(b)|$.   □

6   A General Characterization

Let us prove our main characterization.

Theorem 2. *Let f be an endomorphism over $\{a \prec b\}$ prolongable en a. The word $f^\omega(a)$ is an infinite Lyndon word if and only if*

1. *f preserves the lexicographic order on finite words,*
2. *$f^\omega(a)$ is not periodic and*
3. *the word $f^3(a)$ is a prefix of a Lyndon word.*

Proof. Assume first that $f^\omega(a)$ is an infinite Lyndon word. Conditions 2 and 3 are direct consequences of this hypothesis. Proposition 4 states condition 1.

Assume now that the three conditions hold. If $f^\omega(a)$ begins with aa, then it begins with $a^i b$ for some integer $i \geq 2$. Lemma 3 states that $f(a^i b)$ is a Lyndon word. Thus from Proposition 7 $f^\omega(a)$ is an infinite Lyndon word.

If $f^\omega(a) = ab^\omega$, it is an infinite Lyndon word.

If $f^\omega(a)$ begins with $ab^i a$ for some integer $i \geq 1$, Lemma 6 states that $f(ab^i)$ is a power of a Lyndon word $u \neq ab^i$. Moreover if $i = 1$ then $|u| > |f(b)|$. Thus from Proposition 8 $f^\omega(a)$ is an infinite Lyndon word.   □

Example 11. Let f be the morphism defined by $f(a) = aba$ and $f(b) = bab$. This morphism fulfills conditions 1 and 3 of Theorem 2. It generates the periodic word $(ab)^\omega$. This shows the importance of the condition $f^\omega(a)$ *is not periodic* that does not occur in Propositions 7 and 8.

Example 12. Let μ be the Thue-Morse morphism defined by $\mu(a) = ab$ and $\mu(b) = ba$. The word $\mu^2(a)$ is a prefix of the Lyndon word $\mu^2(a)bbb = abbabbb$ but $\mu^3(a)$ begins with $abbaa$ which is not a prefix of a Lyndon word. This example shows the optimality of the exponent 3 in the last condition of Theorem 2.

7   Conclusion

After Theorem 2, a natural problem is to obtain a characterization of morphisms that generate infinite Lyndon words over an alphabet containing at least three letters.

Let us observe that Proposition 4 does not extend to morphisms over alphabets with at least three letters. Indeed consider any endomorphism f such that $f(a) = au$ with u, $f(b)$ and $f(c)$ belonging to $\{b, c\}^*$ (note that one of the two words $f(b)$ and $f(c)$ could be the empty word: we just need that $\lim_{n \to \infty} |f^n(a)|$ is infinite). Then $f^\omega(a)$ is an infinite Lyndon word whatever is f (which may not preserve the lexicographic order). Note that the previous example can include some erasing morphisms. We don't know whether the condition f *preserves the lexicographic order* is necessary if f generates a recurrent word.

Note also that, if an analog of Theorem 2 exists for a larger alphabet A, then the exponent in the last condition would be at least $\#A + 1$ with $\#A$ the cardinality of A. Indeed if $A_n = \{a_1 \prec \ldots \prec a_n\}$, one can extends Example 12 defining the morphism f by $f(a_1) = a_1a_2$, $f(a_i) = a_{i+1}$ for $2 \leq i < n$ and $f(a_n) = a_1$. Then for $1 \leq i \leq n$, $f^i(a_i)$ is a prefix of $a_1a_2 \cdots a_na_1$ and so a prefix of Lyndon word while $f^{n+1}(a)$ is not such a prefix since it begins with $a_1a_2 \cdots a_na_1a_1$.

Acknowledgment. Many thanks to referees for their remarks that slightly improved the paper.

References

1. Allouche, J.-P., Currie, J., Shallit, J.: Extremal infinite overlap-free binary words. Electron. J. Combin. **5**(1), paper R27 (1998)
2. Berstel, J., Lauve, A., Reutenauer, C., Saliola, F.: Combinatorics on Words: Christoffel Words and Repetitions in Words. CRM Monograph Series, vol. 27. American Mathematical Society (2008)
3. Borel, J.P., Laubie, F.: Quelques mots sur la droite projective réelle. J. Théor. Nombres Bordeaux **5**, 23–51 (1993)
4. Champarnaud, J.-M., Hansel, G., Perrin, D.: Unavoidable sets of constant length. Internat. J. Algebra Comput. **14**(2), 241–251 (2004)
5. Charlier, E., Kamae, T., Puzynina, S., Zamboni, L.Q.: Infinite self-shuffling words. J. Combin. Theory Ser. A **128**, 1–40 (2014)
6. Chen, K.T., Fox, R.H., Lyndon, R.C.: Free differential calculus IV - the quotient groups of the lower central series. Ann. Math. **2**(68), 81–95 (1958)
7. Duval, J.-P.: Factorizing words over an ordered alphabet. J. Algorithms **4**(4), 363–381 (1983)
8. Glen, A., Levé, F., Richomme, G.: Quasiperiodic and Lyndon episturmian words. Theoret. Comput. Sci. **409**(3), 578–600 (2008)
9. Knuth, D.E.: Generating All Tuples and Permutations. The Art of Computer Programming, vol. 4, Fascicle 2. Addison-Wesley, Reading (2005)
10. Levé, F., Richomme, G.: Quasiperiodic Sturmian words and morphisms. Theoret. Comput. Sci. **372**(1), 15–25 (2007)
11. Lothaire, M.: Combinatorics on Words. Encyclopedia of Mathematics and Its Applications, vol. 17. Addison-Wesley, Boston (1983).Reprinted in the Cambridge Mathematical Library, Cambridge University Press, UK (1997)
12. Lothaire, M.: Algebraic Combinatorics on Words. Encyclopedia of Mathematics and Its Applications, vol. 90. Cambridge University Press, Cambridge (2002)
13. Paquin, G.: A characterization of infinite smooth Lyndon words. Discrete Math. Theor. Comput. Sci. **12**(5), 25–62 (2010)
14. Postic, M., Zamboni, L.Q.: Reprint of: ω-Lyndon words. Theoret. Comput. Sci. **834**, 60–65 (2020)
15. Reutenauer, C.: From Christoffel Words to Markoff Numbers. Oxford University Press, Oxford (2019)
16. Richomme, G.: Lyndon morphisms. Bull. Belg. Math. Soc. Simon Stevin **10**(5), 761–786 (2003)
17. Richomme, G.: On morphisms preserving infinite Lyndon words. Discret. Math. Theor. Comput. Sci. **9**(2), 89–108 (2007)
18. Ruskey, F., Savage, C., Wang, T.M.Y.: Generating necklaces. J. Algorithms **13**(3), 414–430 (1992)
19. Siromoney, R., Mathew, L., Dare, V.R., Subramanian, K.G.: Infinite Lyndon words. Inf. Process. Lett. **50**, 101–104 (1994)

Inside the Binary Reflected Gray Code: Flip-Swap Languages in 2-Gray Code Order

Joe Sawada[1(✉)], Aaron Williams[2(✉)], and Dennis Wong[3(✉)]

[1] Computing and Information Science, University of Guelph, Guelph, Canada
jsawada@uoguelph.ca
[2] Computer Science, Williams College, Williamstown, USA
aaron@cs.williams.edu
[3] School of Applied Science, Macao Polytechnic Institute, Macao, China
cwong@uoguelph.ca

Abstract. A *flip-swap language* is a set S of binary strings of length n such that $S \cup \{0^n\}$ is closed under two operations (when applicable): (1) Flip the leftmost 1; and (2) Swap the leftmost 1 with the bit to its right. Flip-swap languages model many combinatorial objects including necklaces, Lyndon words, prefix normal words, left factors of k-ary Dyck words, and feasible solutions to 0-1 knapsack problems. We prove that any flip-swap language forms a cyclic 2-Gray code when listed in binary reflected Gray code (BRGC) order. Furthermore, a generic successor rule computes the next string when provided with a membership tester. The rule generates each string in the aforementioned flip-swap languages in $O(n)$-amortized per string, except for prefix normal words of length n which require $O(n^{1.864})$-amortized per string. Our work generalizes results on necklaces and Lyndon words by Vajnovski [Inf. Process. Lett. 106(3):96−99, 2008].

1 Introduction

Combinatorial generation studies the efficient generation of each instance of a combinatorial object, such as the $n!$ permutations of $\{1, 2, \ldots, n\}$ or the $\frac{1}{n+1}\binom{2n}{n}$ binary trees with n nodes. The research area is fundamental to computer science and it has been covered by textbooks such as *Combinatorial Algorithms for Computers and Calculators* by Nijenhuis and Wilf [26], *Concrete Mathematics: A Foundation for Computer Science* by Graham, Knuth, and Patashnik [9], and *The Art of Computer Programming, Volume 4A, Combinatorial Algorithms* by Knuth [12]. The subject is important to every day programmers, and Arndt's *Matters Computational: Ideas, Algorithms, Source Code* is an excellent practical resource [1]. A primary consideration is listing the instances of a combinatorial object so that consecutive instances differ by a specified *closeness condition*. Lists of this type are called *Gray codes*. This terminology is due to the eponymous *binary reflected Gray code* (*BRGC*) by Frank Gray, which orders the 2^n binary strings of length n so that consecutive strings differ in one bit. The BRGC was patented for a pulse code communication system in 1953 [10]. For example, the order for $n = 4$ is

$$0000, 1000, 1100, 0100, 0110, 1110, 1010, 0010,$$
$$0011, 1011, 1111, 0111, 0101, 1101, 1001, 0001. \tag{1}$$

D. Wong—Most of this work was done while the author was at the State University of New York, Korea.

© Springer Nature Switzerland AG 2021
T. Lecroq and S. Puzynina (Eds.): WORDS 2021, LNCS 12847, pp. 172–184, 2021.
https://doi.org/10.1007/978-3-030-85088-3_15

Variations that reverse the entire order or the individual strings are also commonly used in practice and in the literature. We note that the order in (1) is *cyclic* because the last and first strings also differ by the closeness condition, and this property holds for all n.

One challenge facing combinatorial generation is its relative surplus of breadth and lack of depth[1]. For example, [1,12], and [26] have separate subsections for different combinatorial objects, and the majority of the Gray codes are developed from first principles. Thus, it is important to encourage simple frameworks that can be applied to a variety of combinatorial objects. Previous work in this direction includes the following:

1. the ECO framework developed by Bacchelli, Barcucci, Grazzini, and Pergola [2] that generates Gray codes for a variety of combinatorial objects such as Dyck words in constant amortized time per instance;
2. the twisted lexico computation tree by Takaoka [21] that generates Gray codes for multiple combinatorial objects in constant amortized time per instance;
3. loopless algorithms developed by Walsh [24] to generate Gray codes for multiple combinatorial objects, which extend algorithms initially given by Ehrlich in [8];
4. greedy algorithms observed by Williams [27] that provide a uniform understanding for many previous published results;
5. the reflectable language framework by Li and Sawada [13] for generating Gray codes of k-ary strings, restricted growth strings, and k-ary trees with n nodes;
6. the bubble language framework developed by Ruskey, Sawada and Williams [17] that provides algorithms to generate shift Gray codes for fixed-weight necklaces and Lyndon words, k-ary Dyck words, and representations of interval graphs;
7. the permutation language framework developed by Hartung, Hoang, Mütze and Williams [11] that provides algorithms to generate Gray codes for a variety of combinatorial objects based on encoding them as permutations.

We focus on an approach that is arguably simpler than all of the above: Start with a known Gray code and then *filter* or *induce* the list based on a subset of interest. In other words, the subset is listed in the relative order given by a larger Gray code, and the resulting order is a *sublist (Gray code)* with respect to it. Historically, the first sublist Gray code appears to be the *revolving door* Gray code for combinations [25]. A *combination* is a length n binary string with *weight* (i.e. number of ones) k. The Gray code is created by filtering the BRGC, as shown below for $n = 4$ and $k = 2$ (cf. (1))

$$0000, 1000, 1100, 0100, 0110, 1110, 1010, 0010,$$
$$0011, 1011, 1111, 0111, 0101, 1101, 1001, 0001. \tag{2}$$

This order is a *transposition Gray code* as consecutive strings differ by transposing two bits (i.e. swapping the positions of two bits)[2]. It can be generated *directly* (i.e. without filtering) by an efficient algorithm [25]. Vajnovszki [22] proved that necklaces and Lyndon words form a cyclic 2-Gray code in BRGC order, and efficient algorithms

---

[1] This is not to say that combinatorial generation is always easy. For example, the 'middle levels' conjecture was confirmed by Mütze [14] after 30 years and effort by hundreds of researchers.
[2] When each string is viewed as the incidence vector of a k-subset of $\{1, 2, \ldots, n\}$, then consecutive k-subsets change via a "revolving door" (i.e. one value enters and one value exits).

can generate these sublist Gray codes directly [20]. Our goal is to expand upon the known languages that are 2-Gray codes in BRGC order, and which can be efficiently generated. To do this, we introduce a new class of languages.

A *flip-swap language* (with respect to 1) is a set **S** of length n binary strings such that **S** $\cup \{0^n\}$ is closed under two operations (when applicable): (1) Flip the leftmost 1 (flip-first); and (2) Swap the leftmost 1 with the bit to its right (swap-first). A flip-swap language with respect to 0 is defined similarly. Flip-swap languages encode a wide variety of combinatorial objects. The formal definitions of these languages are given in Sect. 3.

Theorem 1. *The following sets of length n binary strings are flip-swap languages:*

Flip-Swap languages (with respect to 1)

i. all strings

ii. strings with weight $\leq k$

iii. strings $\leq \gamma$

iv. strings with $\leq k$ inversions re: 0^*1^*

v. strings with $\leq k$ transpositions re: 0^*1^*

vi. strings $<$ their reversal

vii. strings $\leq$ their reversal (neckties)

viii. strings $<$ their complemented reversal

ix. strings $\leq$ their complemented reversal

x. strings with forbidden 10^t

xi. strings with forbidden prefix 1γ

xii. 0-prefix normal words

xiii. necklaces (smallest rotation)

xiv. Lyndon words

xv. prenecklaces (smallest rotation)

xvi. pseudo-necklaces with respect to 0^*1^*

xvii. left factors of k-ary Dyck words

xviii. feasible solutions to 0–1 knapsack problems

Flip-Swap languages (with respect to 0)

i. all strings

ii. strings with weight $\geq k$

iii. strings $\geq \gamma$

iv. strings with $\leq k$ inversions re: 1^*0^*

v. strings with $\leq k$ transpositions re: 1^*0^*

vi. strings $>$ their reversal

vii. strings $\geq$ their reversal

viii. strings $>$ their complemented reversal

ix. strings $\geq$ their complemented reversal

x. strings with forbidden 01^t

xi. strings with forbidden prefix 0γ

xii. 1-prefix normal words

xiii. necklaces (largest rotation)

xiv. aperiodic necklaces (largest rotation)

xv. prenecklaces (largest rotation)

xvi. pseudo-necklaces with respect to 1^*0^*

Our second result is that every flip-swap language forms a cyclic 2-Gray code when listed in BRGC order. This generalizes the previous sublist BRGC results [20,22].

Theorem 2. *When a flip-swap language* **S** *is listed in BRGC order the resulting listing is a 2-Gray code. If* **S** *includes* 0^n *then the listing is cyclic.*

Our third result is a generic successor rule, which efficiently computes the next string in the 2-Gray code of a flip-swap language, as long as a fast membership test is given.

Theorem 3. *The languages in Theorem 1 can be generated in $O(n)$-amortized time per string, with the exception of prefix normal words which require $O(n^{1.864})$-time.*

Table 1. Flip-swap languages ordered as sublists of the binary reflected Gray code. Theorem 1 covers each language, so the resulting orders are 2-Gray codes.

$n = 4$ BRGC	all i.	necklaces xiii.	0-PNW xii.	≤ 1001 iii.	$k \leq 2$ ii.	neckties vii.
0000	✓	✓	✓	✓	✓	✓
1000	✓			✓	✓	
1100	✓	-			✓	
0100	✓			✓	✓	
0110	✓		✓	✓	✓	✓
1110	✓					
1010	✓				✓	
0010	✓		✓	✓	✓	✓
0011	✓	✓	✓	✓	✓	✓
1011	✓					✓
1111	✓	✓				✓
0111	✓	✓	✓	✓		✓
0101	✓	✓	✓	✓	✓	✓
1101	✓					
1001	✓			✓	✓	✓
0001	✓	✓	✓	✓	✓	✓

(a) String membership in 6 flip-swap languages.   (b) Visualizing the 2-Gray codes in (a).

In Sect. 2, we formally define our version of the BRGC. In Sect. 3, we prove Theorem 1, and define the flip-swap partially ordered set. In Sect. 4, we give our generic successor rule and prove Theorem 2. In Sect. 5, we present a generic generation algorithm that list out each string of a flip-swap language, and we prove Theorem 3.

2 The Binary Reflected Gray Code

Let $\mathbf{B}(n)$ denote the set of length n binary strings. Let $BRGC(n)$ denote the listing of $\mathbf{B}(n)$ in BRGC order. Let $\overline{BRGC}(n)$ denote the listing $BRGC(n)$ in reverse order. Then $BRGC(n)$ can be defined recursively as follows, where $\mathcal{L} \cdot x$ denotes the listing $\mathcal{L}$ with the character x appended to the end of each string:

$$BRGC(n) = \begin{cases} 0, 1 & \text{if } n = 1; \\ BRGC(n-1) \cdot 0, \overline{BRGC}(n-1) \cdot 1 & \text{if } n > 1. \end{cases}$$

For example, $BRGC(2) = 00, 10, 11, 01$ and $\overline{BRGC}(2) = 01, 11, 10, 00$, thus

$$BRGC(3) = 000, 100, 110, 010, 011, 111, 101, 001.$$

This definition of BRGC order is the same as the one used by Vajnovzski [22]. When the strings are read from right-to-left, we obtain the classic definition of BRGC order [10]. For flip-swap languages with respect to 0, we interchange the roles of the 0s and 1s; however, for our discussions we will focus on flip-swap languages with respect to 1. Table 1 illustrates $BRGC(4)$ and six flip-swap languages listed in Theorem 1.

3 Flip-Swap Languages

In this section, we formalize some of the non-obvious flip-swap languages stated in Theorem 1. We also prove Theorem 1 for a subset of the listed languages (with respect to 1) including necklaces, Lyndon words, prefix normal words, and feasible solutions to the 0–1 knapsack problems.

Consider a binary string $\alpha = b_1 b_2 \cdots b_n$. Let $flip_\alpha(i)$ be the string obtained by complementing b_i. Let $swap_\alpha(i, j)$ be the string obtained by swapping b_i and b_j. When the context is clear we use $flip(i)$ and $swap(i, j)$ instead of $flip_\alpha(i)$ and $swap_\alpha(i, j)$. Also, let $\ell_0(\alpha)$ denote the position of the leftmost 0 of α or $n + 1$ if no such position exists. Similarly, let $\ell_1(\alpha)$ denote the position of the leftmost 1 of α or $n + 1$ if no such position exists. To simplify the notation, we define $\ell_\alpha = \ell_1(\alpha)$.

Binary strings with weight $\leq k$: The *weight* of a binary string is the number of 1s it contains.

Binary strings with $\leq k$ inversions: An *inversion* with respect to 0^*1^* in a binary string $\alpha = b_1 b_2 \cdots b_n$ is any $b_i = 1$ and $b_j = 0$ such that $i < j$. For example $\alpha = 100101$ has 4 inversions: $(b_1, b_2), (b_1, b_3), (b_1, b_5), (b_4, b_5)$.

Binary strings with $\leq k$ transpositions: The number of *transpositions* of a binary string $\alpha = b_1 b_1 \cdots b_n$ with respect to 0^*1^* is the minimum number of $swap(i, j)$ operations required to change α into the form 0^*1^*. For example, the number of transpositions of the string 100101 is 1.

Necklaces: A *necklace* is the lexicographically smallest (largest) string in an equivalence class under rotation. Let $N(n)$ be the set of necklaces of length n and $\alpha = 0^j 1 b_{j+2} b_{j+3} \cdots b_n$ be a necklace in $N(n)$. By the definition of necklace, it is easy to see that $flip_\alpha(\ell_\alpha) = 0^{j+1} b_{j+2} b_{j+3} \cdots b_n \in N(n)$ and thus $N(n)$ satisfies the flip-first property. For the swap-first operation, observe that if $\alpha \neq 0^{n-1}1$ and $b_{j+2} = 1$, then the swap-first operation produces the same necklace. Otherwise if $\alpha \neq 0^{n-1}1$ and $b_{j+2} = 0$, then the swap-first operation produces the string $0^{j+1} 1 b_{j+3} b_{j+4} \cdots b_n$ which is clearly a necklace. Thus, the set of necklaces is a flip-swap language.

Lyndon words: An *aperiodic necklace* is a necklace that cannot be written in the form β^j for some $j < n$. A *Lyndon word* is an aperiodic necklace when using the lexicographically smallest string as the representative. Let $L(n)$ denote the set of Lyndon words of length n. Since $N(n)$ is a flip-swap language and $L(n) \cup \{0^n\} \subseteq N(n)$, it suffices to show that applying the flip-first or the swap-first operation on a Lyndon word either yields an aperiodic string or the string 0^n. Clearly $L(n) \cup \{0^n\}$ satisfies the two closure properties when $\alpha \in \{0^n, 0^{n-1}1\}$. Thus in the remaining of the proof, $\alpha \notin \{0^n, 0^{n-1}1\}$. We first prove by contradiction that $L(n) \cup \{0^n\}$ satisfies the flip-first property. Let $\alpha = 0^j 1 b_{j+2} b_{j+3} \cdots b_n$ be a string in $L(n) \cup \{0^n\}$. Suppose that $L(n) \cup \{0^n\}$ does not satisfy the flip-first property and $flip_\alpha(\ell_\alpha)$ is periodic. Thus $flip_\alpha(\ell_\alpha) = (0^{j+1}\beta)^t$ for some string β and $t \geq 2$. Observe that $\alpha = 0^j 1 \beta (0^{j+1}\beta)^{t-1}$ which is clearly not a Lyndon word, a contradiction. Therefore $L(n) \cup \{0^n\}$ satisfies the flip-first property. Then similarly we prove by contradiction that $L(n) \cup \{0^n\}$ satisfies the swap-first property. If $b_{j+2} = 1$, then applying the swap-first operation on α produces the same Lyndon word. Thus in the remaining of the proof, $b_{j+2} = 0$. Suppose that $L(n) \cup \{0^n\}$ does not satisfy the swap-first property such that $\alpha \in L(n) \cup \{0^n\}$ but $swap_\alpha(\ell_\alpha, \ell_\alpha + 1)$ is periodic. Thus $swap_\alpha(\ell_\alpha, \ell_\alpha + 1) = (0^{j+1} 1 \beta)^t$ for some string β and $t \geq 2$. Thus α contains the prefix $0^j 1$ but also the substring $0^{j+1} 1$ in its suffix which is clearly not a Lyndon word, a contradiction. Thus, $\mathbf{L}(n)$ is a flip-swap language.

Prenecklaces: A *prenecklace* is a prefix of a necklace.

Pseudo-necklaces: A *block* with respect to 0^*1^* is a maximal substring of the form 0^*1^*. Each block B_i with respect to 0^*1^* can be represented by two integers (s_i, t_i) corresponding to the number of 0s and 1s respectively. For example, the string $\alpha = 000110100011001$ can be represented by $B_4 B_3 B_2 B_1 = (3,2)(1,1)(3,2)(2,1)$. A block $B_i = (s_i, t_i)$ is said to be *lexicographically smaller* than a block $B_j = (s_j, t_j)$ (denoted by $B_i < B_j$) if $s_i < s_j$ or $s_i = s_j$ with $t_i < t_j$. A string $\alpha = b_1 b_2 \cdots b_n = B_b B_{b-1} \cdots B_1$ is a *pseudo-necklace* with respect to 0^*1^* if $B_b \leq B_i$ for all $1 \leq i < b$.

Prefix normal words: A binary string α is *prefix normal* with respect to 0 (also known as 0-prefix normal word) if no substring of α has more 0s than its prefix of the same length. For example, the string 001010010111011 is a 0-prefix normal word but the string 001010010011011 is not because it has a substring of length 5 with four 0s while the prefix of length 5 has only three 0s. Observe that the set of 0-prefix normal words of length n satisfies the two closure properties of a flip-swap language as the flip-first and swap-first operations either increases or maintain the number of 0s in its prefix. Thus, the set of 0-prefix normal words of length n is a flip-swap language.

Feasible solutions to 0–1 knapsack problems: The input to a 0–1 knapsack problem is a knapsack capacity W, and a set of n items each of which has a non-negative weight $w_i \geq 0$ and a value v_i. A subset of items is *feasible* if the total weight of the items in the subset is less than or equal to the capacity W. Typically, the goal of the problem is to find a feasible subset with the maximum value, or to decide if a feasible subset exists with value $\geq c$. Given the input to a 0–1 knapsack problem, we reorder the items by non-increasing weight. That is, $w_i \geq w_{i+1}$ for $1 \leq i \leq n-1$. Notice that the incidence vectors of feasible subsets are now a flip-swap language. More specifically, flipping any 1 to 0 causes the subset sum to decrease, and so does swapping any 1 with the bit to its right. Hence, the language satisfies the flip-first and the swap-first closure properties and is a flip-swap language.

Left factors of k-ary Dyck words: A k-ary Dyck word is a binary string of length $n = tk$ with t copies of 1 and $t(k-1)$ copies of 0 such that every prefix has at least $k-1$ copies of 0 for every 1. A string is said to be a *left factor of a k-ary Dyck word* if it is the prefix of some k-ary Dyck word. It is well-known that k-ary Dyck words are in one-to-one correspondence with k-ary trees with t internal nodes. When $k = 2$, Dyck words are counted by the Catalan numbers and are equivalent to *balanced parentheses*. As an example, 001011 is a 2-ary Dyck word while 011001 is not. k-ary Dyck words and balanced parentheses strings are well studied and have lots of applications including trees and stack-sortable permutations [4, 16, 18, 23].

3.1 Flip-Swap Poset

In this section we introduce a poset whose ideals correspond to a flip-swap language which includes the string 0^n.

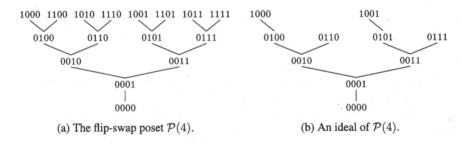

(a) The flip-swap poset $\mathcal{P}(4)$.   (b) An ideal of $\mathcal{P}(4)$.

Fig. 1. Flip-swap languages are the ideals of the flip-swap poset. The ideal in (b) contains the 4-bit binary strings that are ≤ 1001 with respect to lexicographic order.

Let $\alpha = b_1 b_2 \cdots b_n$ be a length n binary string. We define $\tau(\alpha)$ as follows:

$$\tau(\alpha) = \begin{cases} \alpha & \text{if } \alpha = 0^n, \\ flip_\alpha(\ell_\alpha) & \text{if } \alpha \neq 0^n \text{ and } (\ell_\alpha = n \text{ or } b_{\ell_\alpha+1} = 1) \\ swap_\alpha(\ell_\alpha, \ell_\alpha + 1) & \text{otherwise.} \end{cases}$$

Let $\tau^t(\alpha)$ denote the string that results from applying the τ operation t times to α. We define the binary relation $<_R$ on $\mathbf{B}(n)$ to be the transitive closure of the cover relation τ, that is $\beta <_R \alpha$ if $\beta \neq \alpha$ and $\beta = \tau^t(\alpha)$ for some $t > 0$. It is easy to see that the binary relation $<_R$ is irreflexive, anti-symmetric and transitive. Thus $<_R$ is a strict partial order. The relation $<_R$ on binary strings defines our flip-swap poset.

Definition 1. *The* flip-swap poset $\mathcal{P}(n)$ *is a strict poset with* $\mathbf{B}(n)$ *as the ground set and* $<_R$ *as the strict partial order.*

Figure 1 shows the Hasse diagram of $\mathcal{P}(4)$ with the ideal for binary strings of length 4 that are lexicographically smaller or equal to 1001 in bold. Observe that $\mathcal{P}(n)$ is always a tree with 0^n as the unique minimum element, and that its ideals are the subtrees that contain this minimum.

Lemma 1. *A set* $\mathbf{S}$ *over* $\mathbf{B}(n)$ *that includes* 0^n *is a flip-swap language if and only if* $\mathbf{S}$ *is an ideal of* $\mathcal{P}(n)$.

Proof. Let $\mathbf{S}$ be a flip-swap language and α be a string in $\mathbf{S}$. Since $\mathbf{S}$ is a flip-swap language, $\mathbf{S}$ satisfies the flip-first and swap-first properties and thus $\tau(\alpha)$ is a string in $\mathbf{S}$. Therefore every string $\gamma <_R \alpha$ is in $\mathbf{S}$ and hence $\mathbf{S}$ is an ideal of $\mathcal{P}(n)$. The other direction is similar. □

If $\mathbf{S}$ is a set of binary strings and γ is a binary string, then the *quotient* of $\mathbf{S}$ and γ is $\mathbf{S}/\gamma = \{\alpha \mid \alpha\gamma \in \mathbf{S}\}$.

Lemma 2. *If* $\mathbf{S}_1$ *and* $\mathbf{S}_2$ *are flip-swap languages and* γ *is a binary string, then* $\mathbf{S}_1 \cap \mathbf{S}_2$, $\mathbf{S}_1 \cup \mathbf{S}_2$ *and* $\mathbf{S}_1/\gamma$ *are flip-swap languages.*

Proof. Let $\mathbf{S}_1$ and $\mathbf{S}_2$ be two flip-swap languages and let γ be a binary string. The intersection and union of ideals of any poset are also ideals of that poset, so $\mathbf{S}_1 \cap \mathbf{S}_2$ and $\mathbf{S}_1 \cup \mathbf{S}_2$ are flip-swap languages. Now consider $\alpha \in \mathbf{S}_1/\gamma$.

Suppose $\alpha \in \mathbf{S}_1/\gamma$ for some non-empty γ where $j = |\alpha|$. This means that $\alpha\gamma \in \mathbf{S}_1$. Consider three cases depending $\ell_{\alpha\gamma}$. If $\ell_{\alpha\gamma} < j$, then clearly $\tau(\alpha\gamma) = \tau(\alpha)\gamma$. From Lemma 1, $\tau(\alpha)\gamma \in \mathbf{S}_1$ and thus $\tau(\alpha) \in \mathbf{S}_1/\gamma$. If $\ell_{\alpha\gamma} = j$, then $\alpha = 0^{j-1}1$ and $\tau(\alpha) = 0^j$. Since $\mathbf{S}_1$ is a flip-swap language $0^j\gamma \in \mathbf{S}_1$. Again this implies that $\tau(\alpha) \in \mathbf{S}_1/\gamma$. If $\ell_{\alpha\gamma} > j$ then $\alpha = 0^j$ and $\tau(\alpha) = \alpha$ in this case. For each case we have shown that $\tau(\alpha) \in \mathbf{S}_1/\gamma$ and thus $\mathbf{S}_1/\gamma$ is a flip-swap language by Lemma 1. □

Corollary 1. *Flip-swap languages are closed under union, intersection, and quotient.*

Proof. Let $\mathbf{S}_A$ and $\mathbf{S}_B$ be flip-swap languages and γ be a binary string. Since $\mathbf{S}_A$ and $\mathbf{S}_B$ can be represented by ideals of the flip-swap poset, possibly excluding 0^n, by Lemma 2 the sets $\mathbf{S}_A \cap \mathbf{S}_B$, $\mathbf{S}_A \cup \mathbf{S}_B$ and $\mathbf{S}_A/\gamma$ are flip-swap languages. □

Lemma 3. *If $\alpha\gamma$ is a binary string in a flip-swap language $\mathbf{S}$, then $0^{|\alpha|}\gamma \in \mathbf{S}$.*

Proof. This result follows from the flip-first property of flip-swap languages. □

4   A Generic Successor Rule for Flip-Swap Languages

Consider any flip-swap language $\mathbf{S}$ that includes the string 0^n. Let $\mathcal{BRGC}(\mathbf{S})$ denote the listing of $\mathbf{S}$ in BRGC order. Given a string $\alpha \in \mathbf{S}$, we define a generic *successor rule* that computes the string following α in the cyclic listing $\mathcal{BRGC}(\mathbf{S})$.

Let $\alpha = b_1 b_2 \cdots b_n$ be a string in $\mathbf{S}$. Let t_α be the leftmost position of α such that $flip_\alpha(t_\alpha) \in \mathbf{S}$ when $|\mathbf{S}| > 1$. Such a t_α exists since $\mathbf{S}$ satisfies the flip-first property and $|\mathbf{S}| > 1$. Recall that ℓ_α is the position of the leftmost 1 of α (or $|\alpha| + 1$ if no such position exists). Notice that $t_\alpha \leq \ell_\alpha$ when $|\mathbf{S}| > 1$ since $\mathbf{S}$ is a flip-swap language.

Let $flip2_\alpha(i, j)$ be the string obtained by complementing both b_i and b_j. When the context is clear we use $flip2(i, j)$ instead of $flip2_\alpha(i, j)$. Also, let $w(\alpha)$ denote the number of 1s of α. We claim that the following function f computes the next string in the cyclic ordering $\mathcal{BRGC}(\mathbf{S})$:

$$f(\alpha) = \begin{cases} 0^n & \text{if } \alpha = 0^{n-1}1; & (4a) \\ flip_\alpha(t_\alpha) & \text{if } w(\alpha) \text{ is even and } (t_\alpha = 1 \text{ or } flip2_\alpha(t_\alpha - 1, t_\alpha) \notin \mathbf{S}); & (4b) \\ flip2_\alpha(t_\alpha - 1, t_\alpha) & \text{if } w(\alpha) \text{ is even and } flip2_\alpha(t_\alpha - 1, t_\alpha) \in \mathbf{S}; & (4c) \\ flip2_\alpha(\ell_\alpha, \ell_\alpha + 1) & \text{if } w(\alpha) \text{ is odd and } flip_\alpha(\ell_\alpha + 1) \notin \mathbf{S}; & (4d) \\ flip_\alpha(\ell_\alpha + 1) & \text{if } w(\alpha) \text{ is odd and } flip_\alpha(\ell_\alpha + 1) \in \mathbf{S}. & (4e) \end{cases}$$

Thus, successive applications of the function f on a flip-swap language $\mathbf{S}$, starting with the string 0^n, list out each string in $\mathbf{S}$ in BRGC order. As an illustration of the function f, successive applications of this rule for the set of necklaces of length 6 starting with the necklace 000000 produce the listing in Table 2.

Table 2. The necklaces of length 6 induced by successive applications the function f starting from 000000. The sixth column of the table lists out the corresponding rules in f that apply to each necklace to obtain the next necklace.

Necklaces	Parity of $w(\alpha)$	t_α	ℓ_α	Successor	Case
000000	Even	6		$flip2(5,6)$	(4c)
000011	Even	3		$flip2(2,3)$	(4c)
011011	Even	2		$flip(2)$	(4b)
001011	Odd		3	$flip(4)$	(4e)
001111	Even	2		$flip2(1,2)$	(4c)
111111	Even	1		$flip(1)$	(4b)
011111	Odd		2	$flip(3)$	(4e)
010111	Even	3		$flip(2)$	(4b)
000111	Odd		4	$flip(5)$	(4e)
000101	Even	2		$flip(2)$	(4b)
010101	Odd		2	$flip2(2,3)$	(4d)
001101	Odd		3	$flip(4)$	(4e)
001001	Even	3		$flip(3)$	(4b)
000001	Odd			$flip(6)$	(4a)

Theorem 4. *If* **S** *is a flip-swap language including the string* 0^n *and* $|\mathbf{S}| > 1$, *then* $f(\alpha)$ *is the string immediately following the string* α *in* **S** *in the cyclic ordering* $\mathcal{BRGC}(\mathbf{S})$.

We will provide a detailed proof of this theorem in the next subsection. Observe that each rule in f complements at most two bits and thus successive strings in **S** differ by at most two bit positions. Observe that when 0^n is excluded from **S**, then $\mathcal{BRGC}(\mathbf{S})$ is still a 2-Gray code (although not necessarily cyclic). This proves Theorem 2.

4.1 Proof of Theorem 4

This section proves Theorem 4. We begin with a lemma by Vajnovszki [22], and a remark that is due to the fact that $0^{n-1}1$ is in a flip-swap language **S** when $|\mathbf{S}| > 1$.

Lemma 4. *Let* $\alpha = b_1 b_2 \cdots b_n$ *and* β *be length* n *binary strings such that* $\alpha \neq \beta$. *Let* r *be the rightmost position in which* α *and* β *differ. Then* α *comes before* β *in BRGC order (denoted by* $\alpha \prec \beta$) *if and only if* $w(b_r b_{r+1} \cdots b_n)$ *is even.*

Remark 1. A flip-swap language **S** in BRGC order ends with $0^{n-1}1$ when $|\mathbf{S}| > 1$.

Let $succ(\mathbf{S}, \alpha)$ be the *successor* of α in **S** in BRGC order (i.e. the string after α in the cyclic ordering $\mathcal{BRGC}(\mathbf{S})$). Next we provide two lemmas, and then prove Theorem 4.

Lemma 5. *Let* **S** *be a flip-swap language with* $|\mathbf{S}| > 1$ *and* α *be a string in* **S**. *Let* t_α *be the leftmost position such that* $flip_\alpha(t_\alpha) \in \mathbf{S}$. *If* $w(\alpha)$ *is even, then* t_α *is the rightmost position in which* α *and* $succ(\mathbf{S}, \alpha)$ *differ.*

Proof. By contradiction. Let $\alpha = b_1 b_2 \cdots b_n$ and $\beta = succ(\mathbf{S}, \alpha)$. Let r be the rightmost position in which α and β differ with $r \neq t_\alpha$. If $t_\alpha > r$, then β has the suffix $1 b_{r+1} b_{r+2} \cdots b_n$ since $b_r = 0$ because $r < t_\alpha \leq \ell_\alpha$. Thus by the flip-first property, $0^{r-1} 1 b_{r+1} b_{r+2} = flip_\alpha(r) \in \mathbf{S}$ and $r < t_\alpha$, a contradiction.

Otherwise if $t_\alpha < r$, then let $\gamma = flip_\alpha(t_\alpha)$. Clearly $\gamma \neq \alpha$. Now observe that $w(b_t b_{t+1} \cdots b_n)$ is even because $t_\alpha \leq \ell_\alpha$ and $w(\alpha)$ is even, and thus by Lemma 4, $\alpha \prec \gamma$. Also, γ has the suffix $b_r b_{r+1} \cdots b_n$ and $w(b_r b_{r+1} \cdots b_n)$ is even because $\alpha \prec \beta$ and r is the rightmost position α and β differ, and thus also by Lemma 4, $\gamma \prec \beta$. Thus $\alpha \prec \gamma \prec \beta$, a contradiction. Therefore $r = t_\alpha$. □

Lemma 6. *Let* $\mathbf{S}$ *be a flip-swap language with* $|\mathbf{S}| > 1$ *and* $\alpha \neq 0^{n-1}1$ *be a string in* $\mathbf{S}$. *If* $w(\alpha)$ *is odd, then* $\ell_\alpha + 1$ *is the rightmost position in which* α *and* $succ(\mathbf{S}, \alpha)$ *differ.*

Proof. Since $\alpha \neq 0^{n-1}1$ and $w(\alpha)$ is odd, $\ell_\alpha < n - 1$. We now prove the lemma by contradiction. Let $\alpha = b_1 b_2 \cdots b_n$ and $\beta = succ(\mathbf{S}, \alpha)$. Let $r \neq \ell_\alpha + 1$ be the rightmost position in which α and β differ. If $r < \ell_\alpha + 1$, then $w(b_r b_{r+1} \cdots b_n)$ is odd but $\alpha \prec \beta$, a contradiction by Lemma 4. Otherwise if $r > \ell_\alpha + 1$, then let $\gamma = flip2_\alpha(\ell_\alpha, \ell_\alpha + 1)$. Clearly $\gamma \neq \alpha$, and by the flip-first and swap-first properties, $\gamma \in \mathbf{S}$. Also, observe that $w(b_{\ell_\alpha+1} b_{\ell_\alpha+2} \cdots b_n)$ is even because $w(\alpha)$ is odd, and thus by Lemma 4, $\alpha \prec \gamma$. Further, γ has the suffix $b_r b_{r+1} \cdots b_n$ and $w(b_r b_{r+1} \cdots b_n)$ is even because $\alpha \prec \beta$ and r is the rightmost position α and β differ, and thus also by Lemma 4, $\gamma \prec \beta$. Thus $\alpha \prec \gamma \prec \beta$, a contradiction. Therefore $r = \ell_\alpha + 1$. □

Proof of Theorem 4. Let $\alpha = a_1 a_2 \cdots a_n$ and $\beta = succ(\mathbf{S}, \alpha) = b_1 b_2 \cdots b_n$. Let t_α be the leftmost position such that $flip_\alpha(t_\alpha) \in \mathbf{S}$. First we consider the case when $\alpha = 0^{n-1}1$. Recall that the first string in $\mathbf{B}(n)$ in BRGC order is 0^n [15] and 0^n is a string in $\mathbf{S}$ by Lemma 3. Also, the last string in $\mathbf{S}$ in BRGC order is $0^{n-1}1$ by Remark 1 when $|\mathbf{S}| > 1$. Thus the string that appears immediately after α in the cyclic ordering $\mathcal{BRGC}(\mathbf{S})$ is $f(\alpha)$ when $\alpha = 0^{n-1}1$. In the remainder of the proof, $\alpha \neq 0^{n-1}1$ and we consider the following two cases.

Case 1: $w(\alpha)$ is even: If $t_\alpha = 1$, then clearly $\beta = flip_\alpha(t_\alpha) = f(\alpha)$. For the remainder of the proof, $t_\alpha > 1$.
Since $t_\alpha \leq \ell_\alpha$, $flip2_\alpha(t_\alpha - 1, t_\alpha)$ has the prefix $0^{t_\alpha - 2}1$. We now consider the following two cases. If $flip2_\alpha(t_\alpha - 1, t_\alpha) \notin \mathbf{S}$, then $flip_\alpha(t_\alpha)$ is the only string in $\mathbf{S}$ that has t_α as the rightmost position that differ with α and has the prefix 0^{t-2}. Therefore, $\beta = flip_\alpha(t_\alpha) = f(\alpha)$. Otherwise, $flip2_\alpha(t_\alpha - 1, t_\alpha)$ and $flip_\alpha(t_\alpha)$ are the only strings in $\mathbf{S}$ that have t_α as the rightmost position that differ with α and have the prefix $0^{t_\alpha - 2}$. By Lemma 4, $flip2_\alpha(t_\alpha - 1, t_\alpha) \prec flip_\alpha(t_\alpha)$ since $w(1\bar{a}_{t_\alpha} a_{t_\alpha+1} a_{t_\alpha+2} \cdots a_n)$ is even. Thus, $\beta = flip2_\alpha(t_\alpha - 1, t_\alpha) = f(\alpha)$.

Case 2: $w(\alpha)$ is odd: By Lemma 6, β has the suffix $\bar{a}_{\ell_\alpha+1} a_{\ell_\alpha+2} a_{\ell_\alpha+3} \cdots a_n$. Observe that if $flip_\alpha(\ell_\alpha + 1) \notin \mathbf{S}$, then by the flip-first and swap-first properties, $flip2_\alpha(\ell_\alpha, \ell_\alpha + 1)$ is the only string in $\mathbf{S}$ that has $\ell_\alpha + 1$ as the rightmost position that differ with β. Thus, $\beta = flip2_\alpha(\ell_\alpha, \ell_\alpha + 1) = f(\alpha)$. Otherwise by Lemma 4, any string $\gamma \in \mathbf{S}$ with the suffix $\bar{a}_{\ell_\alpha+1} a_{\ell_\alpha+2} a_{\ell_\alpha+3} \cdots a_n$ and $\gamma \neq flip_\alpha(\ell_\alpha + 1)$ has $flip_\alpha(\ell_\alpha + 1) \prec \gamma$ because $w(1\bar{a}_{\ell_\alpha+1} a_{\ell_\alpha+2} a_{\ell_\alpha+3} \cdots a_n)$ is even. Thus, $\beta = flip_\alpha(\ell_\alpha + 1) = f(\alpha)$.

Therefore, the string immediately after α in the cyclic ordering $\mathcal{BRGC}(\mathbf{S})$ is $f(\alpha)$. □

Algorithm 1. Pseudocode of the implementation of the function f.

1: **function** $f(\alpha)$
2:    **if** $\alpha = 0^{n-1}1$ **then** $flip_\alpha(n)$
3:    **else if** $w(\alpha)$ is even **then**
4:        $t_\alpha \leftarrow \ell_\alpha$
5:        **while** $t_\alpha > 1$ **and** $flip_\alpha(t_\alpha - 1) \in \mathbf{S}$ **do** $t_\alpha \leftarrow t_\alpha - 1$
6:        **if** $t_\alpha \neq 1$ **and** $flip2_\alpha(t_\alpha - 1, t_\alpha) \in \mathbf{S}$ **then** $\alpha \leftarrow flip2_\alpha(t_\alpha - 1, t_\alpha)$
7:        **else** $\alpha \leftarrow flip_\alpha(t_\alpha)$
8:    **else**
9:        **if** $flip_\alpha(\ell_\alpha + 1) \notin \mathbf{S}$ **then** $\alpha \leftarrow flip2_\alpha(\ell_\alpha, \ell_\alpha + 1)$
10:       **else** $\alpha \leftarrow flip_\alpha(\ell_\alpha + 1)$

Algorithm 2. Algorithm to list out each string of a flip-swap language $\mathbf{S}$ in BRGC order.

1: **procedure** *BRGC*
2:    $\alpha = b_1 b_2 \cdots b_n \leftarrow 0^n$
3:    **do**
4:        **if** $\alpha \neq 0^n$ **or** $0^n \in \mathbf{S}$ **then** Print(α)
5:        $f(\alpha)$
6:        $w(\alpha) \leftarrow 0$
7:        **for** i **from** n **down to** 1 **do**
8:            **if** $b_i = 1$ **then** $w(\alpha) \leftarrow w(\alpha) + 1$
9:            **if** $b_i = 1$ **then** $\ell_\alpha \leftarrow i$
10:       **while** $\alpha \neq 0^n$

---

5   Generation Algorithm for Flip-Swap Languages

In this section we present a generic algorithm to generate 2-Gray codes for flip-swap languages via the function f.

A naïve approach to implement f is to find t_α by test flipping each bit in α to see if the result is also in the set when $w(\alpha)$ is even; or test flipping the $(\ell_\alpha + 1)$-th bit of α to see if the result is also in the set when $w(\alpha)$ is odd. Since $t_\alpha \leq \ell_\alpha$, we only need to examine the length $\ell_\alpha - 1$ prefix of α to find t_α. Such a test can be done in $O(nm)$ time, where $O(m)$ is the time required to complete the membership test of the set under consideration. Pseudocode of the function f is given in Algorithm 1.

To list out each string of a flip-swap language $\mathbf{S}$ in BRGC order, we can repeatedly apply the function f until it reaches the starting string. We also maintain $w(\alpha)$ and ℓ_α which can be easily maintained in $O(n)$ time for each string generated. We also add a condition to avoid printing the string 0^n if 0^n is not a string in $\mathbf{S}$. Pseudocode for this algorithm, starting with the string 0^n, is given in Algorithm 2. The algorithm can easily be modified to generate the corresponding counterpart of $\mathbf{S}$ with respect to 0.

A simple analysis shows that the algorithm generates $\mathbf{S}$ in $O(nm)$-time per string. A more thorough analysis improves this to $O(n + m)$-amortized time per string.

Theorem 5. *If* **S** *is a flip-swap language, then the algorithm BRGC produces* $\mathcal{BRGC}(S)$ *in* $O(n + m)$-*amortized time per string, where* $O(m)$ *is the time required to complete the membership tester for* **S**.

Proof. Let $\alpha = a_1 a_2 \cdots a_n$ be a string in **S**. Clearly f can be computed in $O(n)$ time when $w(\alpha)$ is odd. Otherwise when $w(\alpha)$ is even, the **while** loop in line 5 of Algorithm 1 performs a membership tester on each string $\beta = b_1 b_2 \cdots b_n$ in **S** with $b_{\ell_\alpha} b_{\ell_\alpha+1} \cdots b_n = a_{\ell_\alpha} a_{\ell_\alpha+1} \cdots a_n$ and $w(b_1 b_2 \cdots b_{\ell_\alpha-1}) = 1$. Observe that each of these strings can only be examined by the membership tester once, or otherwise the **while** loop in line 5 of Algorithm 1 produces the same t_α which results in a duplicated string, a contradiction. Thus, the total number of membership testers performed by the algorithm is bound by $|S|$, and therefore f runs in $O(m)$-amortized time per string. Finally, since the other part of the algorithm runs in $O(n)$ time per string, the algorithm *BRGC* runs in $O(n + m)$-amortized time per string.   □

The membership tests in this paper can be implemented in $O(n)$ time and $O(n)$ space; see [3,7,19] for necklaces, Lyndon words, prenecklaces and pseudo-necklaces of length n. One exception is the test for prefix normal words of length n requires $O(n^{1.864})$ time and $O(n)$ space [5]. Together with the above theorem, this proves Theorem 3.

Visit the Combinatorial Object Server [6] for a C implementation of our algorithms.

Acknowledgements. The research of Joe Sawada is supported by the *Natural Sciences and Engineering Research Council of Canada* (NSERC) grant RGPIN-2018-04211. The research of Dennis Wong is supported by the MSIT (Ministry of Science and ICT), Korea, under the ICT Consilience Creative Program (IITP-2019-H8601-15-1011) supervised by the IITP (Institute for Information & communications Technology Planning & Evaluation).

A part of this work was done while the third author was visiting IBS discrete mathematics group in Korea.

References

1. Arndt, J.: Matters Computational: Ideas, Algorithms, Source Code. Springer, Heidelberg (2011). https://doi.org/10.1007/978-3-642-14764-7
2. Bacchelli, S., Barcucci, E., Grazzini, E., Pergola, E.: Exhaustive generation of combinatorial objects by ECO. Acta Informatica **40**(8), 585–602 (2004)
3. Booth, K.S.: Lexicographically least circular substrings. Inf. Process. Lett. **10**(4/5), 240–242 (1980). https://doi.org/10.1007/s00236-004-0139-x
4. Bultena, B., Ruskey, F.: An Eades-McKay algorithm for well-formed parenthesis strings. Inf. Process. Lett. **68**(5), 255–259 (1998)
5. Chan, T.M., Lewenstein, M.: Clustered integer 3SUM via additive combinatorics. In: Proceedings of the Forty-seventh Annual ACM Symposium on Theory of Computing, STOC 2015, pp. 31–40, New York, NY, USA (2015)
6. COS++. The Combinatorial Object Server. http://combos.org/brgc
7. Duval, J.P.: Factorizing words over an ordered alphabet. J. Algorithms **4**(4), 363–381 (1983)
8. Ehrlich, G.: Loopless algorithms for generating permutations, combinations, and other combinatorial configurations. J. ACM **20**(3), 500–513 (1973)

9. Graham, R., Knuth, D., Patashnik, O.: Concrete Mathematics: A Foundation for Computer Science. Addison-Wesley Longman Publishing Co., Inc., Boston (1994)
10. Gray, T.: Pulse code communication (1953). US Patent 2,632,058
11. Hartung, E., Hoang, H.P., Mütze, T., Williams, A.: Combinatorial generation via permutation languages. In: Proceedings of the Thirty-First Annual ACM-SIAM Symposium on Discrete Algorithms, SODA 2020, pp. 1214–1225, USA (2020)
12. Knuth, D.: The Art of Computer Programming. Volume 4, fascicule 2, Generating all tuples and permutations. The Art of Computer Programming. Addison-Wesley, Upper Saddle River (N.J.), 2005. Autre tirage (2010)
13. Li, Y., Sawada, J.: Gray codes for reflectable languages. Inf. Process. Lett. **109**(5), 296–300 (2009)
14. Mütze, T.: Proof of the middle levels conjecture. In: Proceedings of the London Mathematical Society, vol. 112, pp. 677–713 (2016)
15. Ruskey, F.: Combinatorial Generation. Working version (1j-CSC 425/520) edition (2003)
16. Ruskey, F., Proskurowski, A.: Generating binary trees by transpositions. J. Algorithms **11**(1), 68–84 (1990)
17. Ruskey, F., Sawada, J., Williams, A.: Binary bubble languages and cool-lex order. J. Comb. Theory Ser. A **119**(1), 155–169 (2012)
18. Ruskey, F., Williams, A.: Generating balanced parentheses and binary trees by prefix shifts. In: Proceedings of the Fourteenth Symposium on Computing: The Australasian Theory, CATS 2008, vol.77 pp. 107–115, Darlinghurst, Australia (2008)
19. Sawada, J., Williams, A.: A Gray code for fixed-density necklaces and Lyndon words in constant amortized time. Theor. Comput. Sci. **502**, 46–54 (2013)
20. Sawada, J., Williams, A., Wong, D.: Necklaces and Lyndon words in colexicographic and reflected Gray code order. J. Discrete Algorithms **46–47**, 25–35 (2017)
21. Takaoka, T.: An $O(1)$ time algorithm for generating multiset permutations. In: ISAAC 1999. LNCS, vol. 1741, pp. 237–246. Springer, Heidelberg (1999). https://doi.org/10.1007/3-540-46632-0_25
22. Vajnovszki, V.: More restrictive Gray codes for necklaces and Lyndon words. Inf. Process. Lett. **106**(3), 96–99 (2008)
23. Vajnovszki, V., Walsh, T.: A loop-free two-close Gray-code algorithm for listing k-ary Dyck words. J. Discrete Algorithms **4**(4), 633–648 (2006)
24. Walsh, T.: Generating Gray codes in $O(1)$ worst-case time per word. In: Calude, C.S., Dinneen, M.J., Vajnovszki, V. (eds.) DMTCS 2003. LNCS, vol. 2731, pp. 73–88. Springer, Heidelberg (2003). https://doi.org/10.1007/3-540-45066-1_5
25. Wilf, H.S.: A unified setting for sequencing, ranking, and selection algorithms for combinatorial objects. Adv. Math. **24**, 281–291 (1977)
26. Wilf, H.S., Nijenhuis, A.: Combinatorial Algorithms: for Computers and Calculators. Academic Press, New York (1978)
27. Williams, A.: The greedy Gray code algorithm. In: 13th International Symposium, WADS 2013, London, ON, Canada, 12–14 August 2013, Proceedings, pp. 525–536 (2013)

Equations over the k-Binomial Monoids

Markus A. Whiteland$^{(\boxtimes)}$ (ID)

Max Planck Institute for Software Systems, Saarland Informatics Campus,
Saarbrücken, Germany
mawhit@mpi-sws.org

Abstract. Two finite words u and v are k-binomially equivalent if each
word of length at most k appears equally many times in u and v as
a subword, or scattered factor. We consider equations in the so-called
k-binomial monoid defined by the k-binomial equivalence relation on
words. We remark that the k-binomial monoid possesses the compactness property, namely, any system of equations has a finite equivalent
subsystem. We further show an upper bound, depending on k and the
size of the underlying alphabet, on the number of equations in such a
finite subsystem. We further consider commutativity and conjugacy in
the k-binomial monoids. We characterise 2-binomial conjugacy and 2-
binomial commutativity. We also obtain partial results on k-binomial
commutativity for $k > 2$.

1 Introduction

Word equations (equations over free monoids), or string equations, are of fundamental importance in mathematics and theoretical computer science, for example in program verification. The area is actively studied both in theoretical and
practical areas (see the recent papers [2,3,6,12] and the references therein).

The notion of compactness is a foundational in numerous areas of mathematics. In semigroup theory, the compactness property takes the following form: a
semigroup is said to possess the compactness property if every system of equations with finitely many variables has an equivalent finite subsystem of equations.
That is to say, any solution to the finite subsystem satisfies all the equations of
the original system.

Famously, the free monoid Σ^* possesses the so-called *compactness property*,
independently proved in [1] and [4]. The latter work also shows that free groups
possess the compactness property. In [5] it was shown that all commutative
semigroups possess the compactness property. Nevertheless, not all semigroups
possess the compactness property. Some such examples are the monoid of finite
languages [9], the so-called *bicyclic monoid*, and the *Baumslag–Solitar group* [5].
An interesting non-example is given in [20]: Shevlyakov gives a semigroup over
which each consistent system of equations (i.e., has a solution) has an equivalent
finite subsystem, yet the semigroup does not possess the compactness property.
Namely, there is an inconsistent system of equations such that each of its finite
subsystems is consistent.

© Springer Nature Switzerland AG 2021
T. Lecroq and S. Puzynina (Eds.): WORDS 2021, LNCS 12847, pp. 185–197, 2021.
https://doi.org/10.1007/978-3-030-85088-3_16

A system of equations is called *independent* if it is not equivalent to any of its subsystems. Now if a semigroup possesses the compactness property, then any independent system of equations is finite. The aspect of considering sizes of independent systems of equations in semigroups has been previously treated, e.g., in the paper [7]. See also [16], and references therein, concerning the free semigroup.

In this note we consider equations over the *k-binomial monoids*. Two words $u, v \in \Sigma^*$ are called k-binomially equivalent, in symbols $u \equiv_k v$, if each word e of length at most k occurs as a subword, or scattered factor, equally many times in both u and v. The notion was introduced in [18], and has attracted a lot interest in contemporary research areas in combinatorics on words [10,11,17]. The k-binomial equivalence actually defines a congruence on Σ^*, the free semigroup [18]. Hence $\Sigma^*/\equiv_k$ defines a monoid. Our aim is to study basic equations over this monoid. Our main motivation is to discover algebraic properties of these k-binomial monoids. In particular, we consider when two words commute and when they are conjugate in the k-binomial monoid. These are well understood equations over Σ^* [14]: for two word $u, v \in \Sigma^*$ we have $uv = vu$ if and only if there exists $r \in \Sigma^*$ such that $u, v \in r^*$. Thus the set $\mathrm{Sol}(xy = yx)$ of solutions to the equation $xy = yx$ in Σ^* equals $\{\alpha : x \mapsto r^i, y \mapsto r^j : r \in \Sigma^*, i, j \geqslant 0\}$. Similarly, for words $x, y, z \in \Sigma^*$ we have $xz = zy$ if and only if there exist $p, q \in \Sigma^*$ such that $x = pq$, $y = qp$, and $z \in (pq)^*p$ (or $x = y = \varepsilon$ and z is arbitrary). In the free monoid, we thus have $\mathrm{Sol}(xz = zy) = \{(x, y, z) \mapsto (pq, qp, (pq)^r p,) : p, q \in \Sigma^*, r \in \mathbb{N}\}$. The results we obtain in this paper differ quite a bit to these characterisations.

The paper is organised as follows. In Sect. 2 we introduce some basic properties of the k-binomial equivalence. In Sect. 3 we study the equation $xy \equiv_k yx$, i.e., when do two words commute in the k-binomial monoid. We characterise the solutions in case $k = 2$, and give partial results for $k \geqslant 3$. In Sect. 4 we study the equation $xz \equiv_k zy$, i.e., when are two words conjugate in the k-binomial monoid. In Sect. 5 we show that the k-binomial monoids possess the compactness property. In fact, we observe that this already follows from results in the literature. We give another proof of this fact and, furthermore, give a bound on the number of equations in an independent system of equations when k and the number of variables is fixed. We conclude the paper with some open problems in Sect. 6.

This paper is based on results appearing in the author's PhD thesis [21].

2   Preliminaries and Notation

We recall some notation and basic terminology from the literature of combinatorics on words. We refer the reader to [13,14] for more on the subject.

For a finite alphabet Σ we let Σ^* denote the set of finite words of Σ. We use ε to denote the empty word. We let Σ^+ denote the set of non-empty words. For a word $w \in \Sigma^*$, $|w|$ denotes the length of w. By convention we set $|\varepsilon| = 0$. For a letter $a \in \Sigma$, we use $|w|_a$ to denote the number of occurrences of a in w.

The Parikh vector of w is defined by $\Psi(w) := (|s|_a)_{a \in \Sigma}$. A word $u = u_0 \cdots u_t$, $u_i \in \Sigma$ is called a subword of $w = a_0 \cdots a_n$, if $u = a_{i_1} \cdots a_{i_t}$ for some indices $0 \leqslant i_1 < \ldots < i_t \leqslant n$. We let $\binom{w}{u}$ denote the number of occurrences of u in w as a subword. Basic properties of binomial coefficients $\binom{u}{v}$ are presented in [13, Chapter 6]. We repeat the main properties here. Define, for $a, b \in \Sigma$, $\delta_{a,b} = 1$ if $a = b$, otherwise $\delta_{a,b} = 0$. For all $p, q \in \mathbb{N}$, $u, v \in \Sigma^*$, and $a, b \in \Sigma$ we have

$$\binom{a^p}{a^q} = \binom{p}{q}; \quad \binom{u}{\varepsilon} = 1; \quad |u| < |v| \text{ implies } \binom{u}{v} = 0; \quad \binom{ua}{vb} = \binom{u}{vb} + \delta_{a,b}\binom{u}{v}.$$

The last three relations completely determine the binomial coefficient $\binom{u}{v}$ for all $u, v \in \Sigma^*$.

A mapping $\varphi : \Delta^* \to \Sigma^*$ from the language Δ^* to the language Σ^* is called a *morphism* if $\varphi(uv) = \varphi(u)\varphi(v)$ for all $u, v \in \Delta^*$. Let Ξ be a finite non-empty set of *variables* and S a semigroup. An element $(u, v) \in (\Xi \cup \Sigma)^+ \times (\Xi \cup \Sigma)^+$ is called an *equation* over S with variables Ξ. A *solution* to an equation (u, v) over S with variables Ξ is a morphism $\alpha : \Xi \to S$ such that $\alpha(u) = \alpha(v)$. Here we extend α to $\Xi \cup S$ so that α acts as the identity morphism on S. An equation $e = (u, v)$ is often denoted by $e : u = v$. The set of solutions to the equation e is denoted by $\mathrm{Sol}(e)$.

A set $E \subseteq \Xi^+ \times \Xi^+$ is called a system of equations. The solutions to E are defined as

$$\mathrm{Sol}(E) = \bigcap_{e \in E} \mathrm{Sol}(e).$$

We say that two systems E_1 and E_2 of equations are *equivalent* if $\mathrm{Sol}(E_1) = \mathrm{Sol}(E_2)$. Further, we say that a system of equations E is *independent* if E is not equivalent to any of its *finite* proper subsystems $E' \subseteq E$.

Let us turn to the main notion of the paper. Two words $u, v \in \Sigma^*$ are k-binomially equivalent if $\binom{u}{e} = \binom{v}{e}$ for all $e \in \Sigma^*$ with $|e| \leqslant k$. As noted in the introduction, the k-binomial monoid is defined as the quotient monoid $\Sigma^*/\equiv_k$. We recall a basic result on k-binomial equivalence from [18].

Proposition 2.1. *Let $u, v, e \in \Sigma^*$ and $a \in \Sigma$.*

- *We have $\binom{uv}{e} = \sum_{e_1 e_2 = e} \binom{u}{e_1}\binom{v}{e_2}$.*
- *Let $\ell \geqslant 0$. We have $\binom{u}{a^\ell} = \binom{|u|_a}{\ell}$ and $\sum_{|v|=\ell} \binom{u}{v} = \binom{|u|}{\ell}$.*

The second point can be refined further:

Lemma 2.2. *Let $u, v \in \Sigma^*$. Then $\sum_{v' \equiv_1 v} \binom{u}{v'} = \prod_{a \in \Sigma} \binom{|u|_a}{|v|_a}$, where the summation runs through all words v' for which $\Psi(v') = \Psi(v)$.*

Proof. We count the number of choices of subwords v' of u having $|v'|_a = |v|_a$ for each $a \in \Sigma$. For each $a \in \Sigma$, we may choose the occurrences of a in $\binom{|u|_a}{|v|_a}$ ways. Since the choices of distinct letters are independent, the total number of choices equals $\prod_{a \in \Sigma} \binom{|u|_a}{|v|_a}$. Each of these choices corresponds to an occurrence of a subword $v' \equiv_1 v$ of u. $\square$

Let $\lhd$ be a lexicographic order on Σ.

Corollary 2.3. *Given two words $x, y \in \Sigma^*$, we have that $x \equiv_2 y$ if and only if $x \equiv_1 y$ and $\binom{x}{ab} = \binom{y}{ab}$ for all pairs of letters $a, b \in \Sigma$ with $a \lhd b$.*

Proof. Clearly $x \equiv_2 y$ implies the weaker condition. Now $x \equiv_1 y$ implies that $\binom{x}{aa} = \binom{|x|_a}{2} = \binom{|y|_a}{2} = \binom{y}{aa}$, and Lemma 2.2 implies that, for $a \lhd b$,

$$\binom{x}{ba} = |x|_a |x|_b - \binom{x}{ab} = |y|_a |y|_b - \binom{y}{ab} = \binom{y}{ba}.$$
$\square$

3   On Commutativity in the k-Binomial Monoids

We first study when two words commute in the k-binomial monoid. Let us begin with a straightforward characterisation of commutativity in the 2-binomial monoids.

Proposition 3.1. *For all $x, y \in \Sigma^*$, $xy \equiv_2 yx$ if and only if $\Psi(x)$ and $\Psi(y)$ are collinear.*

Proof. Notice first that $xy \equiv_2 yx$ is equivalent to

$$\binom{x}{ab} + \binom{x}{a}\binom{y}{b} + \binom{y}{ab} = \binom{xy}{ab} = \binom{yx}{ab} = \binom{x}{ab} + \binom{y}{a}\binom{x}{b} + \binom{y}{ab}$$

for all $a, b \in \Sigma$. This, in turn is equivalent to

$$|x|_a |y|_b = |y|_a |x|_b, \quad a, b \in \Sigma. \tag{1}$$

Assume now that $xy \equiv_2 yx$, i.e., (1) holds. Summing both sides over $b \in \Sigma$ yields $|x|_a |y| = |y|_a |x|$ for all $a \in \Sigma$, which is equivalent to $|y|\Psi(x) = |x|\Psi(y)$, and so the vectors are collinear.

For the converse assume that $\Psi(x) = \alpha\Psi(y)$ for some $\alpha \in \mathbb{Q}$. If $\alpha = 0$, then $x = \varepsilon$ and there is nothing to prove. Otherwise, we observe that the property $|x|_a = \alpha|y|_a$ for all $a \in \Sigma$ implies $|x|_a|y|_b = \alpha|y|_a\frac{1}{\alpha}|x|_b = |y|_a|x|_b$ for all $a, b \in \Sigma$, so (1) holds. Therefore $xy \equiv_2 yx$.
$\square$

It is immediate that if $x \equiv_k r^m$ and $y \equiv_k r^n$ for some $r \in \Sigma^*$, $m, n \in \mathbb{N}$, then x and y commute in the k-binomial monoid. It is straightforward to see that the above proposition can be stated as follows: in case $k = 2$, the elements x and y commute in $\Sigma^*/\equiv_k$ if and only if there exist a word $r \in \Sigma^*$ and non-negative integers m and n such that $x \equiv_{k-1} r^m$ and $y \equiv_{k-1} r^n$ (we call such r a *common $(k-1)$-binomial root*). It is natural to consider whether such characterisation holds for larger k. Unfortunately, only one direction generalises.

Let us first give a counterexample of 3-binomially commuting words with no common 2-binomial root.

Example 3.2. Let $x = aaabbb$ and $y = aaabbabb$. As their Parikh vectors are collinear, we have $xy \equiv_2 yx$ by Proposition 3.1. One can further check that $\binom{xy}{e} = \binom{yx}{e}$ for all $e \in \Sigma^3$:

$$aaa; 35, \quad aab; 81, \quad aba; 48, \quad abb; 82, \quad baa; 18, \quad bab; 46, \quad bba; 19, \quad bbb; 35.$$

Now $\gcd(|x|, |y|) = 2$, which implies that a possible common 2-binomial root r must have length at most 2. Clearly it cannot be a single letter, so has length 2 and contains both a and b. Hence $r = ab$ or $r = ba$. Now $\binom{x}{ba} = 0$, while $\binom{r^3}{ba} > 0$. Therefore, x and y do not have a common 2-binomial root.

The other implication of Proposition 3.1 does generalise to arbitrary $k \geqslant 2$:

Proposition 3.3. *Let $k \geqslant 2$ be an integer, $r \in \Sigma^*$, and $m, n \geqslant 0$. For any $x \equiv_{k-1} r^m$ and $y \equiv_{k-1} r^n$ we have $xy \equiv_k yx$.*

Proof. For all $a \in \Sigma$, we clearly have $|xy|_a = |yx|_a$. Further, for each word $e \in \Sigma^{\leqslant k}$ of length at least two,

$$\binom{xy}{e} - \binom{x}{e} - \binom{y}{e} = \sum_{\substack{e_1 e_2 = e \\ e_1, e_2 \in \Sigma^+}} \binom{x}{e_1}\binom{y}{e_2} = \sum_{\substack{e_1 e_2 = e \\ e_1, e_2 \in \Sigma^+}} \binom{r^m}{e_1}\binom{r^n}{e_2}$$

$$= \binom{r^{m+n}}{e} - \binom{r^m}{e} - \binom{r^n}{e} = \sum_{\substack{e_1 e_2 = e \\ e_1, e_2 \in \Sigma^+}} \binom{r^n}{e_1}\binom{r^m}{e_2}$$

$$= \sum_{\substack{e_1 e_2 = e \\ e_1, e_2 \in \Sigma^+}} \binom{y}{e_1}\binom{x}{e_2} = \binom{yx}{e} - \binom{x}{e} - \binom{y}{e},$$

where the second and fifth equalities above follow from $x \equiv_{k-1} r^m$ and $y \equiv_{k-1} r^n$ and the observation that $e_1, e_2 \in \Sigma^{\leqslant k-1}$ in the summations. $\square$

Example 3.4. Let $x = aba$ and $y = baaaab$. Now $y \equiv_2 x^2$, by simply counting the occurrences of subwords of length at most two:

$$a; 4, \quad b; 2, \quad aa; 6, \quad ab; 4, \quad ba; 4, \quad bb; 1$$

By the above proposition we have $xy \equiv_3 yx$.

Let us end this section on a positive note: we characterise k-binomial commutation among words of equal length.

Theorem 3.5. *Let $x, y \in \Sigma^*$ with $|x| = |y|$. Then $xy \equiv_k yx$ if and only if $x \equiv_{k-1} y$.*

Proof. Note that $x \equiv_{k-1} y$ implies $xy \equiv_k yx$ by Proposition 3.3. We shall prove the converse by induction on k. Note that the case of $k = 2$ follows from applying Proposition 3.1 with $|x| = |y|$. Assume that the claim holds for some $k \geqslant 2$ and .

suppose $xy \equiv_{k+1} yx$. It follows that $xy \equiv_k yx$ so that $x \equiv_{k-1} y$ by induction. Let then $a, b \in \Sigma$ and $e \in \Sigma^{k-1}$. We have

$$\binom{xy}{aeb} = \binom{x}{aeb} + \binom{y}{aeb} + \binom{x}{a}\binom{y}{eb} + \binom{x}{ae}\binom{y}{b} + \sum_{\substack{e_1 e_2 = e \\ e_1, e_2 \in \Sigma^+}} \binom{x}{ae_1}\binom{y}{e_2 b} \text{ and}$$

$$\binom{yx}{aeb} = \binom{y}{aeb} + \binom{x}{aeb} + \binom{y}{a}\binom{x}{eb} + \binom{y}{ae}\binom{x}{b} + \sum_{\substack{e_1 e_2 = e \\ e_1, e_2 \in \Sigma^+}} \binom{y}{ae_1}\binom{x}{e_2 b}.$$

Putting $\binom{xy}{aeb} = \binom{yx}{aeb}$ and noting that $\binom{y}{ae_1}\binom{x}{e_2 b} = \binom{x}{ae_1}\binom{y}{e_2 b}$ for all terms in the summation (as $x \equiv_{k-1} y$), we obtain, after rearranging,

$$|x|_a \left(\binom{y}{eb} - \binom{x}{eb} \right) = |x|_b \left(\binom{y}{ae} - \binom{x}{ae} \right).$$

Note that the above equation holds for all $a, b \in \Sigma$ and $e \in \Sigma^{k-1}$. Assume without loss of generality that $|x|_a \neq 0$. Letting $e = e_1 \cdots e_{k-1}$ and repeatedly applying the above (to possibly different letters a, b and words $e \in \Sigma^{k-1}$), we obtain

$$\binom{y}{eb} - \binom{x}{eb} = \left(\binom{y}{ae_1 \cdots e_{k-1}} - \binom{x}{ae_1 \cdots e_{k-1}} \right) \frac{|x|_b}{|x|_a}$$

$$= \left(\binom{y}{aae_1 \cdots e_{k-2}} - \binom{x}{aae_1 \cdots e_{k-2}} \right) \frac{|x|_{e_{k-1}}}{|x|_a} \frac{|x|_b}{|x|_a}$$

$$= \cdots$$

$$= \left(\binom{y}{a^k} - \binom{x}{a^k} \right) \frac{|x|_{e_1} \cdots |x|_{e_{k-1}} |x|_b}{|x|_a^k} = 0,$$

since $\binom{y}{a^k} = \binom{x}{a^k}$ follows from $x \equiv_1 y$. It thus follows that $\binom{y}{eb} = \binom{x}{eb}$ for all $b \in \Sigma$ and $e \in \Sigma^{k-1}$, and consequently $x \equiv_k y$. This concludes the proof. $\square$

Corollary 3.6. Let $k \geqslant 2$ and $x, y \in \Sigma^*$. If $xy \equiv_k yx$, then there exist $m, n \in \mathbb{N}$ such that $x^m \equiv_{k-1} y^n$.

Proof. Since $xy \equiv_k yx$ it follows that $x^m y^n \equiv_k y^n x^m$ for all $m, n \in \mathbb{N}$. We may choose $m = \text{lcm}(|x|, |y|)/|x|$ and $n = \text{lcm}(|x|, |y|)/|y|$, whence $|x^m| = |y^n|$. By the above proposition we have that $x^m \equiv_{k-1} y^n$ as was claimed. $\square$

4   Conjugacy in the 2-Binomial Monoids

Here we consider conjugacy in the 2-binomial monoids. Two words $x, y \in \Sigma^*$ are k-binomially conjugate if there exists $z \in \Sigma^*$ such that $xz \equiv_k zy$. Notice that for such a z to exist, we must have $x \equiv_1 y$. Furthermore, for $k \geqslant 2$, z cannot contain any letters not occurring in x and y. Indeed, if $|x|_b = |y|_b = 0$, $|z|_b \geqslant 1$, and $|x|_a \geqslant 1$, then $\binom{xz}{ab} = \binom{x}{a}\binom{z}{b} + \binom{z}{ab} > \binom{z}{ab} = \binom{zy}{ab}$.

Let us consider first the case when $\Sigma = \{a, b\}$.

Proposition 4.1. *Let $x, y \in \{a, b\}^*$. Then there exists $z \in \{a, b\}^*$ such that $xz \equiv_2 zy$ if and only if $x \equiv_1 y$ and $\gcd(|x|_a, |x|_b)$ divides $\binom{x}{ab} - \binom{y}{ab}$.*

Proof. Assume first there exists z such that $xz \equiv_2 zy$. It immediately follows that $x \equiv_1 y$. We also have

$$\binom{x}{ab} + \binom{x}{a}\binom{z}{b} + \binom{z}{ab} = \binom{xz}{ab} = \binom{zy}{ab} = \binom{y}{ab} + \binom{z}{a}\binom{y}{b} + \binom{z}{ab}, \qquad (2)$$

which implies that $\binom{x}{ab} - \binom{y}{ab} = |z|_a|y|_b - |z|_b|x|_a = |z|_a|x|_b - |z|_b|x|_a$. It now follows that $\gcd(|x|_a, |x|_b)$ divides $\binom{x}{ab} - \binom{y}{ab}$.

Let $d = \gcd(|x|_a, |x|_b)$ and assume that $x \equiv_1 y$ and $\binom{x}{ab} - \binom{y}{ab} = kd$ for some $k \in \mathbb{Z}$. By Bezout's identity there exist $i, j \in \mathbb{Z}$, such that $kd = i|x|_b - j|x|_a$. Here we may assume that $i, j \geqslant 0$ since otherwise we may replace (i, j) with $(h|x|_a + i, h|x|_b + j)$ for some suitably large h. We claim that $z = a^i b^j$ satisfies $\binom{xz}{ab} = \binom{zy}{ab}$. Indeed, $\binom{x}{ab} - \binom{y}{ab} = i|x|_b - j|x|_a$ which is equivalent to

$$\binom{x}{ab} + |z|_b|x|_a + \binom{z}{ab} = \binom{y}{ab} + |z|_a|x|_b + \binom{z}{ab}.$$

The latter is equivalent to $\binom{xz}{ab} = \binom{zy}{ab}$ as seen above. By Lemma 2.2, we further have $\binom{xz}{ba} = \binom{zy}{ba}$ and, since $y \equiv_1 x$, we have $xz \equiv_2 zy$ as claimed. $\qquad \square$

Example 4.2. Let $x = aabaaaabbbab$ and $y = bbaababaaaba$. As $y \equiv_1 x$ and $\gcd(|x|_a, |x|_b) = 1$, there exists $z \in \Sigma^*$ such that $xz \equiv_2 zy$. Now $\binom{x}{ab} - \binom{y}{ab} = 16$ and $3|x|_b - 2|x|_a = 1$; therefore, the proof above gives us, for example, $z = a^{48}b^{32}$. Note that $z' = a^6 b^2$ satisfies $xz' \equiv_2 z'y$ as well.

On the other hand, for $x = aabb$ and $y = abab$, we have $x \equiv_1 y$ and $\gcd(|x|_a, |x|_b) = 2$, which does not divide $\binom{x}{ab} - \binom{y}{ab} = 1$. Thus x and y are not 2-binomial conjugate, in other words, $xz \not\equiv_2 zy$ for all $z \in \Sigma^*$.

We now discuss the generalisation of the above characterisation for larger alphabets. Notice that if $xz \equiv_2 zy$, then (2) holds for all $a, b \in \Sigma$. Taking into account Corollary 2.3, we have

Lemma 4.3. *For $x, y, z \in \Sigma^*$, we have $xz \equiv_2 zy$ if and only if $x \equiv_1 y$ and $\binom{x}{ab} - \binom{y}{ab} = |z|_a|x|_b - |z|_b|x|_a$ for all pairs of letters $a, b \in \Sigma$ with $a \lhd b$.*

Hence, deciding whether x and y are 2-binomially conjugate reduces to solving a system of linear equations with integer coefficients. Let us formalise this observation. Let $x, y \in \Sigma^*$ and assume that $x \equiv_1 y$. Assume further that each letter of Σ occurs in x, otherwise we consider a sub-alphabet instead. Fix an ordering on Σ and define the vector $\mathbf{D}_{x,y}$ indexed by pairs of letters $a, b \in \Sigma$, $a \lhd b$, defined as follows: $\mathbf{D}_{x,y}[(a, b)] = \binom{x}{ab} - \binom{y}{ab}$. Let then M_x be the $\binom{|\Sigma|}{2} \times |\Sigma|$-matrix (rows indexed by pairs $a, b \in \Sigma$ with $a \lhd b$, columns by letters $a \in \Sigma$) defined as $M_x[(a, b), a] = |x|_b$, $M_x[(a, b), b] = -|x|_a$, and $M_x[(a, b), c] = 0$ for $c \neq a, b$. Let $\mathbf{X}$ be a vector of $|\Sigma|$ unknowns indexed by the letters $a \in \Sigma$. We consider solutions to the equation

$$M_x \mathbf{X} = \mathbf{D}_{x,y}. \qquad (3)$$

Let us give a brief example of the entities defined above.

Example 4.4. Let $\Sigma = \{0, 1, 2\}$ and let $x, y \in \Sigma^*$ such that $x \equiv_1 y$ and $|x|_a \geqslant 1$ for each $a \in \Sigma$. Then Eq. (3) is defined as

$$\begin{pmatrix} |x|_1 - |x|_0 & 0 \\ |x|_2 & 0 & -|x|_0 \\ 0 & |x|_2 & -|x|_1 \end{pmatrix} \begin{pmatrix} X[0] \\ X[1] \\ X[2] \end{pmatrix} = \begin{pmatrix} \binom{x}{01} - \binom{y}{01} \\ \binom{x}{02} - \binom{y}{02} \\ \binom{x}{12} - \binom{y}{12} \end{pmatrix}.$$

In general, observe that for any word $z \in \Sigma^*$ we have

$$M_x \Psi(z)^\top = \sum_{c \in \Sigma} M_x[(a, b), c] \cdot |z|_c = (|x|_b |z|_a - |x|_a |z|_b)_{(a,b), a \vartriangleleft b}. \tag{4}$$

Now, for x and y as defined above, if there exists $z \in \Sigma^*$ such that $xz \equiv_k zy$, then $X = \Psi(z)^\top$ is a solution to Eq. (3). Indeed, recall that

$$|x|_b |z|_a - |x|_a |z|_b = \binom{x}{ab} - \binom{y}{ab} = D_{x,y}[(a, b)].$$

On the other hand, if X is a solution to Eq. (3) having non-negative entries, then the word $z = \prod_{a \in \Sigma} a^{X[a]}$ is a solution to $xz \equiv_2 zy$.

We are in the position to characterise 2-binomial conjugacy over arbitrary alphabets.

Theorem 4.5. *Let $x, y \in \Sigma^*$ and assume that each letter of Σ occurs in x. Then there exists $z \in \Sigma^*$ such that $xz \equiv_2 zy$ if and only if $x \equiv_1 y$ and Eq. (3) has solution X having integer entries.*

Proof. If there exists z such that $xz \equiv_2 zy$, then $\Psi(z)^\top$ is an integer solution to the equation, as was asserted previously.

Conversely, assume that X is an integer solution to Eq. (3). Notice that some entries of X could be negative. However, plugging $z = x$ in Eq. (4), we have $M_x \Psi(x)^\top = \mathbf{0}$.[1] Thus, for each $n \geqslant 0$, $X + n\Psi(x)^\top$ is also an integer solution to the equation. Moreover, taking n large enough, each entry is a non-negative integer, since all entries of $\Psi(x)$ are assumed to be positive. Now the word $z = \prod_{a \in \Sigma} a^{X[a] + n|x|_a}$ satisfies $xz \equiv_2 zy$ (and is well-defined). $\qquad\square$

Remark 4.6. One can compute an (translated) integer basis for the set of solutions to Eq. (4) in polynomial time (see, e.g., [19, Cor. 5.3c]).

5   Bounds on Sizes of Independent Systems of Equations

In this section we show that the k-binomial monoids possess the compactness property. We further give an upper bound on the size of an independent system of equations. The main results of this section are the following.

Theorem 5.1. *The k-binomial monoids possess the compactness property.*

---

[1] It is not hard to verify that $\operatorname{Ker}(M_x) = \operatorname{Span}(\Psi(x))$ (compare to Proposition 3.1).

Theorem 5.2. *The number of equations in an independent system of equations (without constants) over the semigroup $\Sigma^+/\equiv_k$ with variables Ξ is at most $|\Xi^{\leqslant k}|$.*

As a consequence of the latter theorem, for k fixed, the size of an independent system of equations has a polynomial upper bound with respect to the number of unknowns. On the other hand, the upper bound is exponential when the number $|\Xi|$ of unknowns Ξ is fixed and k is allowed to vary. We remark that these bounds do not depend on the size of the alphabet Σ, when the equations have no constants, that is, the system of equations is a subset of $\Xi^+ \times \Xi^+$.

Let us quickly explain why Theorem 5.2 implies Theorem 5.1.

When considering the compactness property, we remark that there is no loss of generality assuming that h in the above is non-erasing (which implies that we are considering solutions to equations in which each variable is assigned a non-empty word). Indeed, it is not hard to see that a semigroup (possibly without a unit element) possesses the compactness property if and only if the monoid obtained from S by adding a unit element possesses it (see, e.g., [14, Problem 13.5.2]).

Note also that there is no loss in generality assuming that the equations have no constants, as we are dealing with finitely generated monoids: any system E of equations (with or without constants) over S may be modified into a system without constants by identifying each generator $g \in G$ with a new variable X_g. The set of solutions of the original system are obtained from the solutions to the modified system by choosing the solutions where $X_g \mapsto g$ for each generator. Further, if the number of equations in an independent system of equations without constants using n variables is at most $f(n)$ for each n, then the number of equations in an independent system of equations is at most $f(n + \#G)$.

Let us still begin with a short proof of the first main result.

Proof of Theorem 5.1. It is known that if a semigroup S can be embedded in the ring of integer matrices, then S possesses the compactness property [14, Chapter 13]. In [18] such an embedding is explicitly constructed.   □

The rest of this section is devoted to proving Theorem 5.2.

Our approach for upper bounding the size of an independent system of equations over $\Sigma^*/\equiv_k$ is identical to the approach taken for showing a similar result for the so-called k-*abelian monoids* [8]. We interpret the solutions to a system as a subset of a finite dimensional subspace. Basic results from linear algebra are then utilised to show that actually only finitely many equations are required to define all solutions.

Let us fix some notation. Let $k \geqslant 1$ be fixed. Consider a word $u \in \Xi^+$ and define the $\frac{|\Xi|^{k+1}-|\Xi|}{|\Xi|-1}$-dimensional vector $\boldsymbol{u}$ as

$$\boldsymbol{u} = \left(\binom{u}{Y}\right)_{Y \in \Xi^{\leqslant k} \setminus \{\varepsilon\}}.$$

For any non-erasing morphism $h : \Xi \to \Sigma^*/\equiv_k$ we define, for each word $w \in \Sigma^{\leqslant k}$, the $\frac{|\Xi|^{k+1}-|\Xi|}{|\Xi|-1}$-dimensional vector $\boldsymbol{h}_w$ (components indexed by non-empty words in $\Xi^{\leqslant k}$) as

$$h_w[Y] = \sum_{\substack{w = w_1 \cdots w_\ell \\ w_j \in \Sigma^+}} \binom{h(Y_1)}{w_1} \cdots \binom{h(Y_\ell)}{w_\ell},$$

for each $Y = Y_1 \cdots Y_\ell \in \Xi^{\leqslant k}$ with $Y_i \in \Xi$ for all $i = 1, \ldots, \ell$. Note that $h_e[Y] = 0$ for all Y for which $|Y| > |e|$, as e does not have a factorisation into $|Y|$ non-empty words.

The following lemma is crucial in the endeavours that follow. Here (x, y) denotes the inner product of vectors x, y.

Lemma 5.3. Let $h : \Xi \to \Sigma^* / \equiv_k$ be a non-erasing morphism, $u \in \Xi^+$, and $w \in \Sigma^{\leqslant k}$. Then $\binom{h(u)}{w} = (h_w, u)$.

Proof. To avoid cluttering the text, we set $\widehat{X} := h(X)$ for each $X \in \Xi$. Let $u = X_1 \cdots X_n$, where $X_i \in \Xi$ for each $i = 1, \ldots, n$. For any subset S of $\{1, \ldots, n\}$, by the sequence $S_1, \ldots, S_{|S|}$ we mean the sequence of elements of S arranged in increasing order. Now, for each $w \in \Sigma^{\leqslant k}$, we observe that

$$\binom{h(u)}{w} = \sum_{\substack{S \subseteq [1,n] \\ |S| \leqslant |w|}} \sum_{\substack{w = w_1 \cdots w_{|S|} \\ w_j \in \Sigma^+}} \binom{\widehat{X}_{S_1}}{w_1} \cdots \binom{\widehat{X}_{S_{|S|}}}{w_{|S|}}.$$

Indeed, for each occurrence of w as a subword, there exists a subset $S \subseteq [1, n]$ of length at most k such that $w = w_1 \cdots w_{|S|}$, where $w_i \in \Sigma^+$ and the indices of w_i in u are a subset of the indices of $\widehat{X}_{S_i}$ in $h(u)$. For each subset S of $[1, n]$ having $|S| \geqslant |e|$, there exists no such factorisation, and thus the corresponding sum contributes nothing to the total sum. Now for two subsets $S, S' \subseteq [1, n]$ having $Y_{S_1} \cdots Y_{S_{|S|}} = Y_{S'_1} \cdots Y_{S'_{|S'|}} = Y$, the corresponding sums contribute the same value. The number of distinct such sets equals $\binom{u}{Y}$. We may thus rewrite the above equation as

$$\sum_{Y \in \Xi^{\leqslant k}} \binom{u}{Y} \sum_{\substack{w = w_1 \cdots w_{|Y|} \\ w_j \in \Sigma^+}} \binom{\widehat{Y}_1}{w_1} \cdots \binom{\widehat{Y}_{|Y|}}{w_{|Y|}} = \sum_{Y \in \Xi^{\leqslant k}} h_w[Y] \cdot u[Y] = (h_w, u),$$

as claimed. $\qquad\square$

For a vector x, we let $x^\perp$ denote the orthogonal complement of x.

Lemma 5.4. Let $e : u = v$ be an equation and let $h : \Xi \to \Sigma^* / \equiv_k$ be a non-erasing morphism. Then h is a solution to e over $\Sigma^* / \equiv_k$ if and only if $h_w \in e^\perp$ for all $w \in \Sigma^{\leqslant k}$, where $e = u - v$.

Proof. We have $h(u) \equiv_k h(v)$ if and only if $\binom{h(u)}{w} - \binom{h(v)}{w} = 0$ for all non-empty $w \in \Sigma^{\leqslant k}$ if and only if $(h_w, u - v) = (h_w, e) = 0$ for each $w \in \Sigma^{\leqslant k}$, by the lemma above. $\qquad\square$

We may now bound the number of equations in an independent system.

Proof of Theorem 5.2. Let $E = \{e_i : u_i = v_i\}_{i \in I}$ be an independent system of equations over Ξ. Assume again that $\mathrm{Sol}(E)$ is not empty. The case of $\mathrm{Sol}(E)$

having no solutions is analogous to the k-abelian case. Now h is a solution to E if and only if $\boldsymbol{h}_w \in \bigcap_{e \in E} \boldsymbol{e}^\perp = U$ for all $w \in \Sigma^{\leqslant k}$. Since U is a finite dimensional vector space, there exist equations $e_1, \ldots, e_f \in E$ such that $U = \cap_{i=1}^f \mathbf{e}_\mathbf{i}^\perp$, where $f \leqslant |\Xi^{\leqslant k}| - 1$. We claim that $E' = \{e_1, \ldots, e_f\}$ is an equivalent subsystem of E.

Let $e \in E$. Let then h be a solution to E'. It follows that $\boldsymbol{h}_w \in \boldsymbol{e}_i^\perp$ for all $i = 1, \ldots, f$, so that $\boldsymbol{h}_w \in U$ for all $w \in \Sigma^*$. Furthermore $\boldsymbol{h}_w \in \boldsymbol{e}^\perp$ which is equivalent to h being a solution to e by the above lemma. as claimed. $\qquad\square$

6 Conclusions and Future Work

We have considered basic equations over the k-binomial monoids. For commutativity, we obtain a characterisation only in the case $k = 2$. The problem is open for $k > 2$, though we obtain some partial results here. We plan to attack the problem in the future:

Problem 6.1. Characterise when $xy \equiv_k yx$ for $k > 2$.

The mixture of positive and negative results obtained relating to this problem seem to suggest that the problem is quite intricate.

As seen in Theorem 4.5, characterising k-binomial conjugacy of two words is already quite involved even for $k = 2$. It is not immediate how to translate the result into a word combinatorial statement. Furthermore, we suspect that the methods used in the case $k = 2$ do not extend to cases with $k > 2$ without substantial new insights. The following problem is thus left open.

Problem 6.2. Characterise when, for words $x, y, z \in \Sigma^*$ and $k > 2$, we have $xz \equiv_k zy$.

Finally for independent systems of equations, it will be interesting to answer the following question.

Question 6.3. What is the maximal number of equations in an independent system of equations in the k-binomial monoid?

The analogous problem over the free semigroups is notoriously open. There is a constant upper bound given in the case of equations with no constants when the alphabet has size 3 (see [15]).

Acknowledgements. The author would like to thank Juhani Karhumäki and Svetlana Puzynina for interesting discussions on the topic. I also thank the anonymous reviewers for helpful comments.

References

1. Albert, M.H., Lawrence, J.: A proof of Ehrenfeucht's conjecture. Theor. Comput. Sci. **41**, 121–123 (1985). https://doi.org/10.1016/0304-3975(85)90066-0

2. Amadini, R.: A survey on string constraint solving (2021) https://arxiv.org/abs/2002.02376v4

3. Day, J.D., Ganesh, V., He, P., Manea, F., Nowotka, D.: The satisfiability of word equations: decidable and undecidable theories. In: Potapov, I., Reynier, P.-A. (eds.) RP 2018. LNCS, vol. 11123, pp. 15–29. Springer, Cham (2018). https://doi.org/10.1007/978-3-030-00250-3_2

4. Guba, V.S.: Equivalence of infinite systems of equations in free groups and semigroups to finite subsystems. Matematicheskie Zametki **40**(3), 321–324, 428 (1986). in Russian

5. Harju, T., Karhumäki, J., Plandowski, W.: Compactness of systems of equations in semigroups. Int. J. Algebra Comput. **7**(4), 457–470 (1997). https://doi.org/10.1142/S0218196797000204

6. Jeż, A.: Solving word equations (and other unification problems) by recompression (invited talk). In: Fernández, M., Muscholl, A. (eds.) 28th EACSL Annual Conference on Computer Science Logic (CSL 2020). Leibniz International Proceedings in Informatics (LIPIcs), vol. 152, pp. 3:1–3:17. Schloss Dagstuhl-Leibniz-Zentrum fuer Informatik, Dagstuhl (2020). https://doi.org/10.4230/LIPIcs.CSL.2020.3

7. Karhumäki, J., Plandowski, W.: On the size of independent systems of equations in semigroups. Theor. Comput. Sci. **168**(1), 105–119 (1996). https://doi.org/10.1016/S0304-3975(96)00064-3

8. Karhumäki, J., Whiteland, M.A.: A compactness property of the k-abelian monoids. Theor. Comput. Sci. **834**, 3–13 (2020). https://doi.org/10.1016/j.tcs.2020.01.023

9. Lawrence, J.: The non-existence of finite test sets for set-equivalence of finite substitutions. Bull. EATCS **28**, 34–36 (1986)

10. Lejeune, M., Rigo, M., Rosenfeld, M.: The binomial equivalence classes of finite words. Int. J. Algebra Comput. **30**(07), 1375–1397 (2020). https://doi.org/10.1142/s0218196720500459

11. Lejeune, M., Rigo, M., Rosenfeld, M.: Templates for the k-binomial complexity of the tribonacci word. Adv. Appl. Math. **112** (2020). https://doi.org/10.1016/j.aam.2019.101947

12. Lin, A.W., Majumdar, R.: Quadratic word equations with length constraints, counter systems, and presburger arithmetic with divisibility. In: Lahiri, S.K., Wang, C. (eds.) ATVA 2018. LNCS, vol. 11138, pp. 352–369. Springer, Cham (2018). https://doi.org/10.1007/978-3-030-01090-4_21

13. Lothaire, M.: Combinatorics on Words, Encyclopedia of Mathematics and its Applications. Advanced Book Program, World Science Division, vol. 17. Addison-Wesley, Boston (1983)

14. Lothaire, M.: Algebraic Combinatorics on Words, Encyclopedia of Mathematics and its Applications, vol. 90. Cambridge University Press, Cambridge (2002). https://doi.org/10.1017/CBO9781107326019

15. Nowotka, D., Saarela, A.: An optimal bound on the solution sets of one-variable word equations and its consequences. In: Chatzigiannakis, I., Kaklamanis, C., Marx, D., Sannella, D. (eds.) 45th International Colloquium on Automata, Languages, and Programming (ICALP 2018). Leibniz International Proceedings in Informatics (LIPIcs), vol. 107, pp. 136:1–136:13. Schloss Dagstuhl-Leibniz-Zentrum fuer Informatik, Dagstuhl (2018). https://doi.org/10.4230/LIPIcs.ICALP.2018.136

16. Nowotka, D., Saarela, A.: One-variable word equations and three-variable constant-free word equations. Int. J. Found. Comput. Sci. **29**(5) (2018). https://doi.org/10.1142/S0129054118420121

17. Rao, M., Rigo, M., Salimov, P.: Avoiding 2-binomial squares and cubes. Theor. Comput. Sci. **572**, 83–91 (2015). https://doi.org/10.1016/j.tcs.2015.01.029

18. Rigo, M., Salimov, P.: Another generalization of abelian equivalence: binomial complexity of infinite words. Theor. Comput. Sci. **601**, 47–57 (2015). https://doi.org/10.1016/j.tcs.2015.07.025

19. Schrijver, A.: Theory of Linear and Integer Programming. Wiley-Interscience Series in Discrete Mathematics and Optimization. Wiley, Hoboken (1999)

20. Shevlyakov, A.: Elements of algebraic geometry over a free semilattice. Algebra Logic **54**(3), 258–271 (2015). https://doi.org/10.1007/s10469-015-9345-6

21. Whiteland, M.A.: On the k-abelian Equivalence Relation of Finite Words. TUCS Dissertations, vol. 241. Turku Centre for Computer Science (2019). https://urn.fi/URN:ISBN:978-952-12-3837-6, PhD Dissertation (University of Turku)

Author Index

Printed in the United States
by Baker & Taylor Publisher Services